Disaster Management

Disaster Management

Editor

C.A. Brebbia
Wessex Institute, UK

WIT*PRESS* **Southampton, Boston**

Editor:

C.A. Brebbia
Wessex Institute, UK

Published by

WIT Press
Ashurst Lodge, Ashurst, Southampton, SO40 7AA, UK
Tel: 44 (0) 238 029 3223; Fax: 44 (0) 238 029 2853
E-Mail: witpress@witpress.com
http://www.witpress.com

For USA, Canada and Mexico

Computational Mechanics International Inc
25 Bridge Street, Billerica, MA 01821, USA
Tel: 978 667 5841; Fax: 978 667 7582
E-Mail: infousa@witpress.com
http://www.witpress.com

British Library Cataloguing-in-Publication Data

A Catalogue record for this book is available
from the British Library

Library of Congress Catalog Card Number: 2017958177

ISBN: 978-1-78466-279-0
eISBN: 978-1-78466-280-6

The texts of the papers in this volume were set individually by the authors or under their supervision. Only minor corrections to the text may have been carried out by the publisher.

Printed in Great Britain by Lightning Source, UK.

The material contained herein is reprinted from special editions of the International Journal of Safety and Security Engineering, Volume 13, Numbers 2 and 3, published by WIT Press.

Preface

This series originated with the need for academia and practitioners to exchange knowledge and experience on the way to handle the increasing risk of natural and human-made disasters. Recent major earthquakes, tsunamis, hurricanes, floods and other natural phenomena have resulted in huge losses in terms of human life and property destruction. A new range of human-made disasters have afflicted humanity in modern times; terrorist activities have been added to more classical disasters such as those due to the failure of industrial installations for instance.

It is important to understand the nature of these global risks to be able to develop strategies to prepare for these events and plan effective responses in terms of disaster management and the associated human health impacts.

The papers contained in this volume have been written by academics and professionals and represent some of the latest developments in the field.

The Editors

Contents

STATISTICAL ANALYSIS OF FAILURE CONSEQUENCES FOR OIL AND GAS PIPELINES

CHIARA BELVEDERESI & MARKUS R. DANN
Department of Civil Engineering, University of Calgary, Calgary, Canada.

ABSTRACT

Pipelines are among the safest methods to transport oil and gas, but when an incident occurs, it can lead to disaster. Pipeline failures often cause injuries, fatalities, explosions and fires due to product ignition, property damage, and spills that can lead to environmental impact. The likelihood and consequence analyses of pipeline failures from past events are necessary for the development of realistic risk models. For this reason, a statistical analysis of failure consequences between 2010 and 2015 based on the Pipeline and Hazardous Materials Safety Administration (PHMSA) database is provided in this paper. Relationships between the pipeline failure consequences and the basic pipeline design variables are investigated. They provide a valuable contribution to pipeline risk modeling. Results show that recently installed hazardous liquid pipelines of large diameters and high operating pressure are more likely to cause ignitions. In contrast, older installed hazardous liquid pipelines of small diameters cause larger release volumes and more expensive property damages. The portion of fatalities and injuries that is caused by distribution pipeline accidents is higher for the public than workers compared to other pipeline types.

Keywords: consequences, pipeline, pipeline failure, risk, statistical analysis.

1 INTRODUCTION

The International Energy Outlook 2016 [1] estimates that the worldwide consumption of fossil fuels will increase from 90 million barrels per day in 2012 to 121 million barrels per day in 2040. The growth in demand for these fuels increases the importance of identifying, analyzing and evaluating the risks associated with the transportation of combustible materials.

Oil and gas pipelines play a significant role in our society; they are among the safest methods for transporting oil and gas. However, accidents occur and they can lead to significant consequences.

Data related to pipeline accidents in the United States are collected by the Pipeline and Hazardous Materials Safety Administration (PHMSA) of the US Department of Transportation (DOT). PHMSA provides accident details, annual summaries, multi-year trends of safety performance metrics and pipeline mileages. Pipelines operators are required to submit annual reports by the Code of Federal Regulations (49 CFR Parts 191, 195) to PHMSA. These reports, which contain pipeline information, accident preventive actions and post-accident response, are useful for researchers, governmental agencies and industry professionals, for inspection planning and risk assessment.

This paper presents a statistical analysis of the oil and gas pipeline failures in the United States from 2010 to 2015 based on the PHMSA database. The analysis focuses on the relationships between pipeline design variables and common consequences of pipeline accidents, such as ignition, injuries and fatalities, property damage and spilled volume. Risk models can be developed based on such relationships.

The paper is organized as follows. Section 2 presents the PHMSA accident reporting criteria and an overview of the basic pipeline design variables. Section 3 shows the results of

DOI: 10.2495/SAFE-V7-N2-103-112

the analysis of accidents consequences with a focus on ignition, injuries, fatalities, property damage and spilled volume. Conclusions and future developments are presentd in Section 4.

2 PHMSA DATA COLLECTION

Approximately 3.7×10^6 km of gas transmission and gathering, gas distribution and hazardous liquid pipelines cross the US. PHMSA regulates approximately 76% of the total length of pipelines in the US [2]. The pipeline network is divided into three main types: gathering, transmission and distribution pipelines. Gathering pipelines transport materials form production wells to processing facilities; afterwards, transmission pipelines carry oil and gas for long distances from processing facilities to many locations in the US and, eventually, fuels reach customers through distribution pipelines.

PHMSA maintains an oil and gas pipeline accident database since the 1980s [3]. Reporting criteria have changed over time. Incidents between 2010 and 2015 are considered in this study for all pipeline types because events contained in this time window are consistent within the same reporting criteria described in the next section.

2.1 Accident reporting criteria

Operators under the PHMSA jurisdiction must submit a form no later than 30 days after the accident is detected. Title 49 of the Code of Federal Regulations [4] contains reporting criteria for all types of pipelines.

According to Title 49 CFR §191.3 for gas transmission, gathering and distribution pipeline, an accident must be reported if it involves a release of gas from a pipeline, or a liquefied natural gas, liquefied petroleum gas, refrigerant gas, or gas from a liquefied natural gas facility. An accident must result in one or more of the following outcomes to be reported:

(i) a death, or personal injury necessitating in-patient hospitalization;
(ii) estimated property damage of $50,000 or more, including loss to the operator and others, or both, but excluding cost of gas lost;
(iii) unintentional estimated gas loss of three million cubic feet or more;

Similarly, an event that results in an emergency shutdown of a liquefied natural gas facility must be reported as well as an event that is significant in the judgment of the operator, even though it does not meet the aforementioned criteria.

Title 49 CFR §195.50 defines accidents for hazardous liquid pipelines. An accident report is required if a release of the hazardous liquid or carbon dioxide transported results in:

(a) explosion or fire not intentionally set by the operator.
(b) release of 5 gallons (19 litres) or more of hazardous liquid or carbon dioxide, except that no report is required for a release of less than 5 barrels (0.8 cubic meters) resulting from a pipeline maintenance activity if the release is:
    (1) not otherwise reportable under this section;
    (2) not one described in §195.52(a)(4);
    (3) confined to company property or pipeline right-of-way; and
    (4) cleaned up promptly;
(c) death of any person;
(d) personal injury necessitating hospitalization;

(e) estimated property damage, including cost of clean-up and recovery, value of lost product, and damage to the property of the operator or others, or both, exceeding \$50,000.

Table 1 summarizes the reporting criteria for gas gathering, transmission, distribution lines and hazardous liquid pipelines.

2.2 Pipeline design variables

Pipeline operators are required to provide details about the pipeline design characteristics when they complete the incident form. In this paper, the class location, installation year, pipe diameter, and maximum allowable operating pressure are considered because they are the basic variables to characterize the design of a pipeline.

2.2.1 Class location

Section §192.5 CFR defines a class location unit as an onshore area that extends 200 m on either side of the centreline of any continuous 1.6 km length of pipeline. In total, there are four class locations organised as follows [4]:

1. Class 1 location
   a. An offshore area; or
   b. Any class location unit that has 10 or fewer buildings intended for human capacity.
2. Class 2 location includes any location unit that has more than 10 but less than 46 buildings intended for human occupancy
3. Class 3 location
   a. Any class location unit that has 46 or more buildings intended for human occupancy: or
   b. An area where the pipeline lies within 91 m of either a building or a small, well-defined outside area (such as a playground, recreation area, outdoor theatre, or other place of public assembly) that is occupied by 20 or more persons on at least 5 days a week for 10 weeks in any 12-month period. (The days and weeks need not be consecutive.)

Table 1: Oil and gas pipelines failures reporting criteria.

	Gas gathering, transmission and distribution pipelines	Hazardous liquid pipelines
Title 49 CFR	§191.3	§195.50
Death	Yes	Yes
Injury	Yes	Yes
Property damage	≥ \$50000 (not considering cost of gas lost)	≥ \$50000 (considering cost of product lost)
Release	Gas loss ≥ 3 million ft^3	Product loss ≥ 5 gallons
Explosion	No	Yes
Fire	No	Yes
Other	Liquefied natural gas facility shutdown	No

4. Class 4 location is any class location unit where buildings with four or more stories above ground are prevalent.

The length of Class locations 2, 3 and 4 may be adjusted as follows:

1. A Class 4 location ends 220 yards (200 m) from the nearest building with four or more stories above ground.
2. When a cluster of buildings intended for human occupancy requires a Class 2 or 3 location, the class location ends 220 yards (200 m) from the nearest building in the cluster.

2.2.2 Installation year

The installation year reported in the database indicates the year when part of the system involved in the accident was installed. In this paper, installation years were divided into decades for the only purpose of a better comprehension of results. It must be noted that information on installation year in the accident database is often missing and this may lead to bias in the results.

2.2.3 Pipeline diameter

The size of pipelines is identified by the nominal pipe size (NPS) [5]; NPS refers to the North American set of standard pipeline sizes. Pipeline diameters were divided into three groups for gas transmission and gathering, gas distribution, and hazardous liquid pipelines. Table 2 shows for each pipeline type the range selected and the number of accidents under each range. The criterion to select each range is based on the number of occurrences so that each group contains a comparable number of cases. As for the installation year, missing information on pipeline diameter may cause bias in the results.

2.2.4 Maximum allowable operating pressure (MAOP)

According to the Code of Federal Regulations (CFR) Title 49 §192.619 [4], no person may operate a segment of steel or plastic pipeline at a pressure that exceeds a maximum allowable operating pressure (MAOP).

In order to facilitate the comparison among failure pressures, MAOP were divided into three groups, similarly to the pipe diameter. Table 3 shows pressure ranges and number of cases for each range per pipeline type.

Table 2: Pipe diameter ranges per pipeline type.

		Gas transmission pipeline	Gas distribution pipeline	Gas gathering pipeline	Hazardous liquid pipeline
D1	range [in]	[0, 12]	[0, 2]	[0, 10]	[0, 8]
	number of cases	127	172	12	161
D2	range [in]	(12, 21]	(2, 5]	(10, 14]	(8, 12]
	number of cases	130	77	14	188
D3	range [in]	(21, 42]	(5, 30]	(14, 26]	(12, 48]
	number of cases	97	130	12	204

Table 3: Maximum allowable operating pressure (MAOP) ranges per pipeline type.

		Gas transmission pipeline	Gas distribution pipeline	Gas gathering pipeline	Hazardous liquid pipeline
P1	range [psig]	[0, 800]	[0, 40]	[0, 1100]	[0, 280]
	number of cases	278	202	15	497
P2	range [psig]	(800, 1000]	(40, 60]	(1100, 1400]	(280, 980]
	number of cases	190	318	30	517
P3	range [psig]	(1000, 2600]	(60, 880]	(1400, 5000]	(980, 3600]
	number of cases	216	129	21	578

3 ANALYSIS OF INCIDENT EVENTS

3.1 Annual failure rate

The annual pipeline failure rate indicates the frequency with which the pipeline system fails per year. It is calculated as the annual number of accidents over the mileage. Figure 1 shows the annual failure rate per pipeline category; distribution pipelines have the lowest failure with nearly constant rates between 2.9×10^{-5} and 3.6×10^{-5} accidents per km per year. In contrast, hazardous liquid pipelines have the highest failure rates around 1.2×10^{-3} to 1.4×10^{-3} accidents per km per year in the recent past. Incidents often led to major consequences, such as ignitions, injuries, fatalities, property damage and release of product. The next sections investigate the relationship between pipeline incident consequences and pipe design variables, such as class location, installation year, pipeline diameter and maximum allowable operating pressure.

3.2 Annual ignition rate

The occurrence of product ignition is a possible failure outcome. Figure 2 shows the annual ignition rate per pipeline type. Hazardous liquid pipelines have the highest ignition rates. The ignition rate increases compared to the failure rate with the exception for 2015. For this

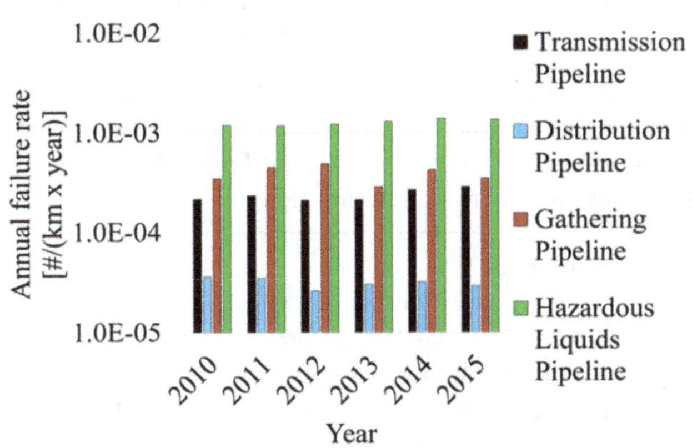

Figure 1: Annual failure rate per pipeline type.

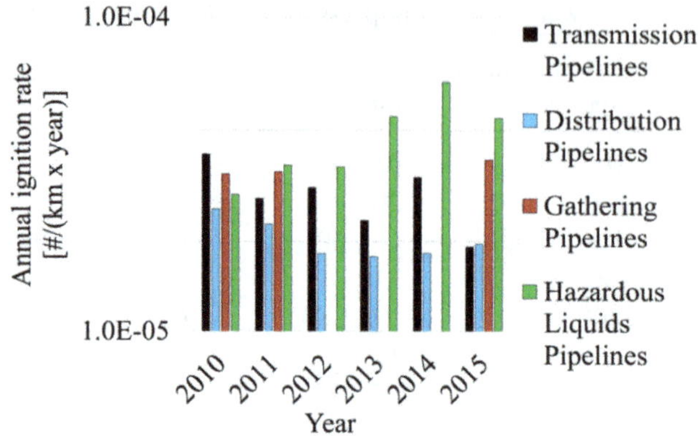

Figure 2: Annual ignition rate per pipeline type.

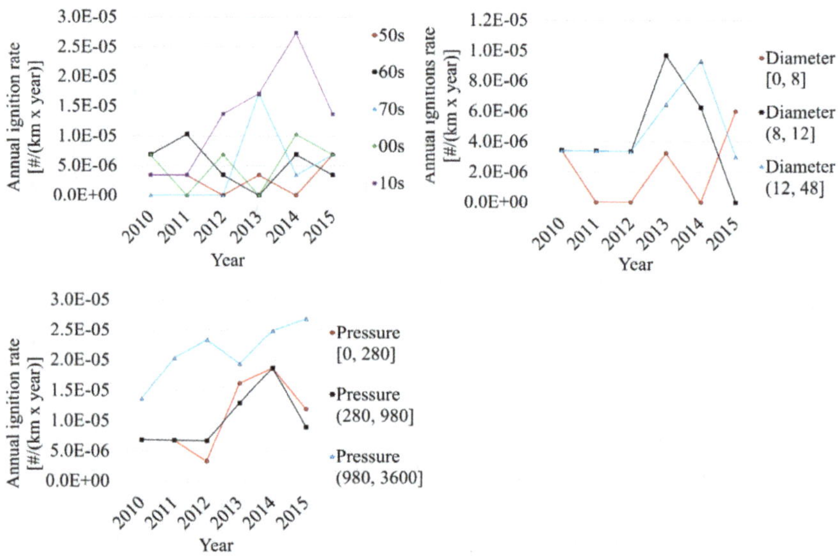

Figure 3: Annual number of ignitions for design variables per hazardous liquid pipelines.

reason, the analysis of relationships between pipeline design variables and ignition is focused on hazardous liquid pipelines in this section. Note that hazardous liquids pipelines are not divided into population classes. Figure 3 shows that larger diameters, higher MAOP, and recently installed pipelines have incresed ignition rates.

3.3 Annual fatality and injury rate

The major consequence of oil and gas pipeline failures are fatalities and injuries. Distribution pipelines have steady annual injury and fatality rates, as shown in Figure. 4. In fact, distribution lines are often located in areas with medium and high population densities, where the general public, such as residents, may be involved in accidents. Figure 5 shows the ratio of workers and general public involved in accidents. Similarly to Lam *et al.* [6], the general

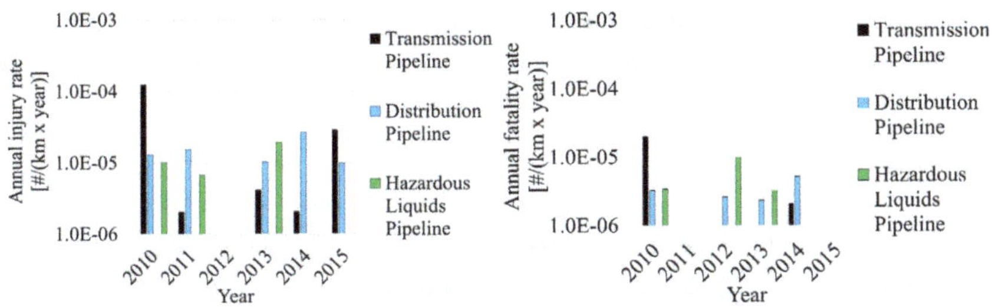

Figure 4: Annual injury and fatality rates per pipeline type.

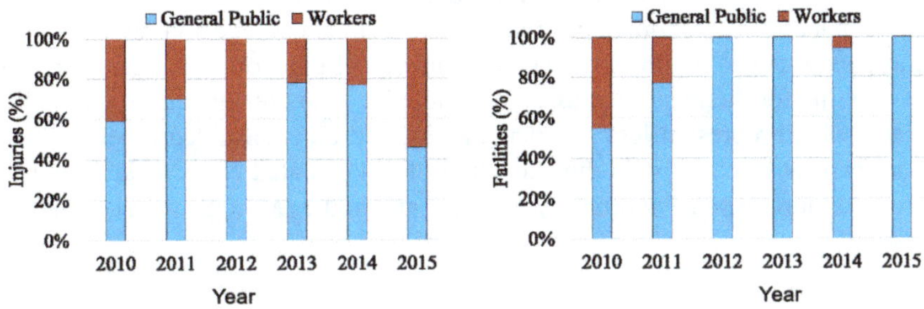

Figure 5: Annual percentage of workers and general public injured in distribution pipelines accidents.

public is larger affected than workers; only in 2012 one single event caused a higher number of injured workers.

Figure 6 focuses on the relationships between the pipe design variables and the injury rate. As a result, Classes 3 and 4, which are the most populated areas, have higher injury rates, as well as smaller diameter pipelines; all MAOP groups follow a similar trend over the time,

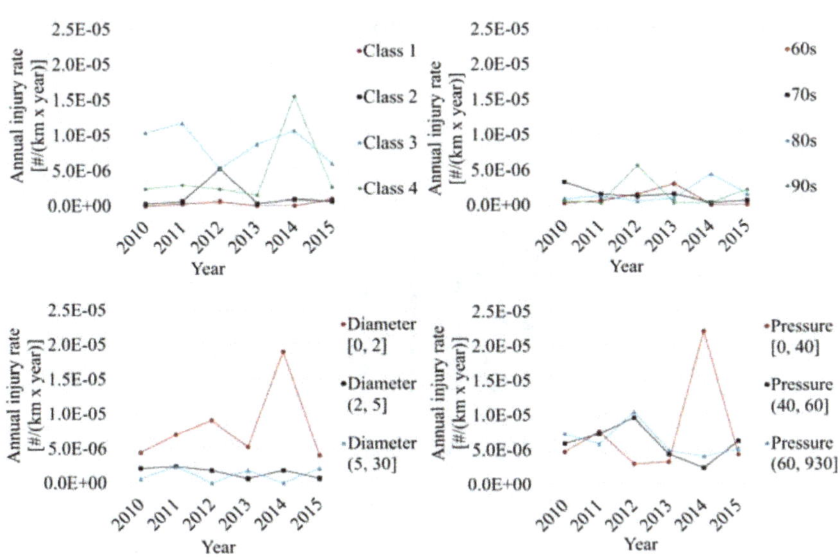

Figure 6: Annual number of injuries and fatalities for design variables for distribution pipelines.

with the exception of low pressure pipelines in 2014. However, this peak seems consistent with the injury rate and the other design characteristics, since in 2014 the number of injuries is higher than other years.

3.4 Property damage

The total cost of gas pipelines does not consider the cost of gas lost while hazardous material pipelines include the cost of product lost, as reported in section 2.1. Figure 7 shows that gas transmission pipeline accidents generally resulted in the most expensive annual average property damages.

The total property damage includes estimated cost of public property damage, estimated cost of intentional and unintentional gas released, operators' damage and repairs, emergency response, and other costs. Figure 8 shows the trends resulting from the regression analyses. The regression line for the location class and installation years results quite steady, although lower classes, which represent less populated areas, and older installed pipelines generally lead to more expensive costs, with the exception of few isolated cases. Additionally, larger diameters cause more expensive property damage, and higher MAOPs result in higher costs in most cases.

3.5 Spilled volume

The volume of product released is measured in cubic feet for gas pipelines and in barrels for hazardous liquid pipelines. The distribution pipeline database does not provide this information and gathering lines does not contain a valuable number of releases to be considered as relevant in this section. From the environmental point of view, hazardous material releases are the most dangerous due to the toxic chemicals that they contain and they can lead to environmental impacts; for this reason, this section focuses on hazardous liquid pipelines.

The hazardous liquid pipeline incidents database provides information on unintentional and intentional release of product; this study only considers the unintentional quantity of material released.

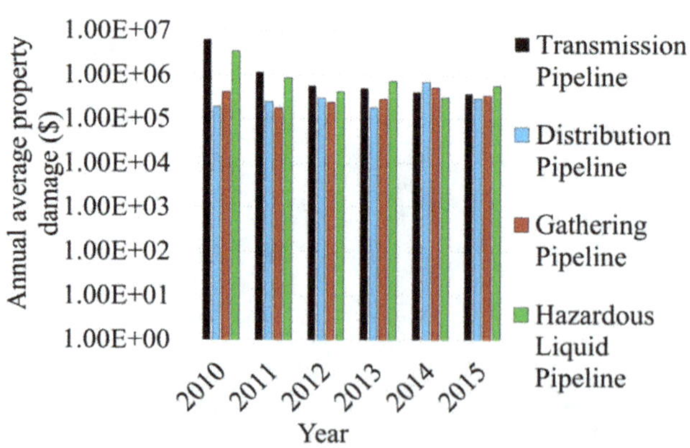

Figure 7: Annual average property damage per pipeline type.

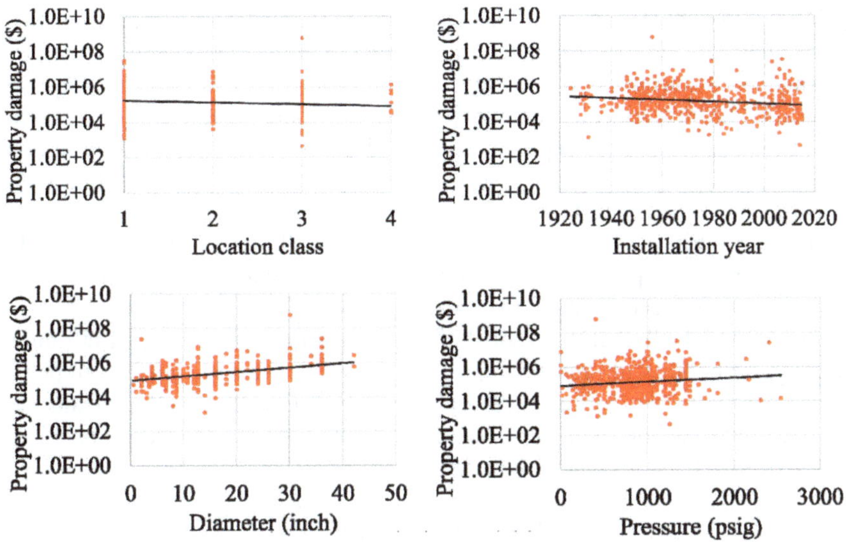

Figure 8: Property damage per design variables for transmission pipelines.

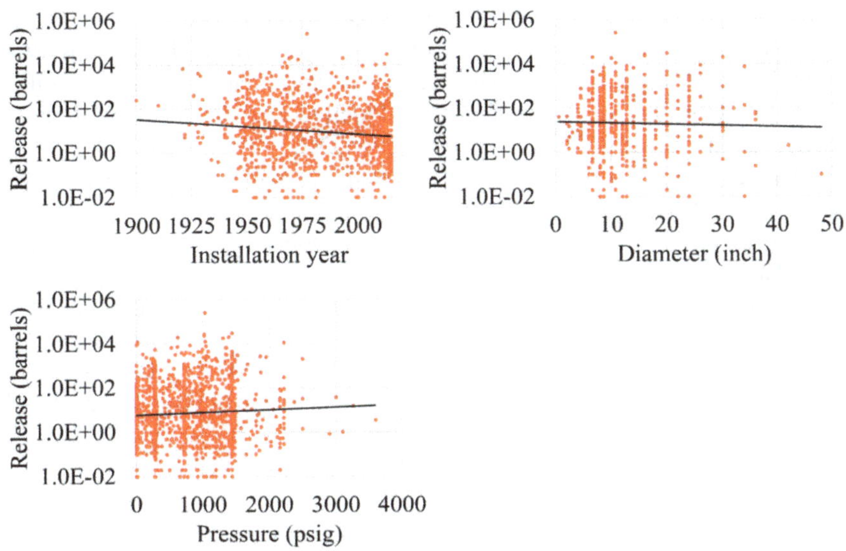

Figure 9: Unintentional released volume per design variables for hazardous liquid pipelines.

Figure 9 shows the scatter plot for the pipe design variables and the volumes of product released for hazardous liquid lines along with the regression line. Overall, product releases decrese with recent installed pipelines; smaller diameters and low and medium pressure pipelines result in most accidents, although the volume released increase with higher pressure pipelines.

4 CONCLUSION

This paper presents a statistical analysis of oil and gas pipeline failure event rates and consequences. Major pipeline incident events often result in ignitions, injuries, fatalities, property damage and release of hazardous materials. Pipeline design variables, such as class location, installation year, pipe diameter and maximum allowable operating pressure, were used as explanatory variables to estimate accident consequences. The results show that recently installed hazardous liquid pipelines of large diameters and high pressures have the largest ignition rates. Distribution pipelines of small diameters and located in high population density areas cause larger injury rates, especially affecting general public rather than workers. Furthermore, older installed transmission pipelines that fail in sparsely populated areas generally result in increased property damages. Finally, older installed hazardous liquid pipelines of smaller pipe diameters and higher maximum allowable operating pressures cause larger product releases in most of cases.

Based on this statistical analysis, probabilistic models may be developed in future works to calculate the probability of ignition, injury, property damage and spilled volume that may result in case of an accident. Moreover, future developments in this study will consider more pipeline design variables and will include other variables associated with pipeline failures, such as the type of release and the failure cause.

ACKNOWLEDGEMENTS

The authors gratefully acknowledge the financial support provided by NSERC.

REFERENCES

[1] Key world energy statistics; International Energy Agency, available at https://www.iea.org/publications/freepublications/publication/key-world-energy-statistics.html

[2] U.S. Department of Transportation (DOT), Pipeline and Hazardous Materials Safety Administration (PHMSA) Web Site, Washington DC.

[3] U.S. Department of Transportation (DOT), Pipeline and Hazardous Materials Safety Administration (PHMSA) Web Site, Washington DC.

[4] U.S. Government Publishing Office (GPO), Electronic Code of Federal Regulations (CFR), Title 49.

[5] Welded and seamless wrought steel pipe; The American Society of Mechanical Engineers (ASME), ASME B36.10M-2004.

[6] Lam, C. & Zhou, W., Statistical analyses of incidents on onshore gas transmission pipelines based on PHMSA database. *International Journal of Pressure Vessels and Piping*, **145**, pp. 29–40, 2016.
https://doi.org/10.1016/j.ijpvp.2016.06.003

VARIATION OF THE EXPLOSION RISK IN A HYBRID COLLECTOR DURING REVAMPING OPERATIONS

S. COPELLI[1], M. RABONI[2], M. RAGAZZI[3], E. C. RADA[3], V. TORRETTA[4]
[1]Department of Science and High Technology, Insubria University, Italy.
[2]Department of Industrial Engineering, Carlo Cattaneo University, Italy.
[3]Department of Civil, Environmental and Mechanical Engineering, University of Trento, Italy.
[4]Department of Theoretical and Applied Sciences, Insubria University, Italy.

ABSTRACT

Dusts removal from gaseous streams is a very common operation carried out in a number of industrial plants, e.g. those ones dedicated to the manufacturing of cement and steel, in order to meet the more increasingly stringent law requirements. Therefore, in the last decades, such industries have been to face more and more often the necessity to improve their de-dusting systems inspired by the principle of the 'maximum safety technologically feasible', but also pushed by the increased penetration of environmental issues into the public opinion. Unfortunately, the budget dedicated for all these improvements is often very low because de-dusting is, in the major part of the practical cases, an operation having a low influence on production performances; therefore, it is seen more as a 'necessary cost' than a profit generator. The unavoidable consequence is that plant managers investments are dedicated to other more fruitful processes than de-dusting, addressing the low budget available for gas stream purification to the so-called revamping strategies: that is, the reuse of existing de-dusting plants, either by enhancing their efficiencies through the use of various technical tricks or by adding another pieces of equipment, rather than install new and more technologically advanced plants. This solution is considered very interesting in terms of profit since the costs for the decommissioning of the obsolete equipments does not exist (in fact the old plant remains). Obviously, all the economic benefits derived from these strategies are counterbalanced by a series of technical disadvantages. In particular, a systematic risk assessment of safety of the whole new configuration of the revamped plant is necessary. In fact, such an evaluation cannot be limited to a safe design of the new installed equipment, but it has to be extended to the already existing equipment also considering the impact that changes in process conditions (induced by the new equipment) can have on the whole plant.

In this work, the relevant problem of an explosion in the de-dusting section of a cement plant, due to the establishment of an overpressure inside the apparatus, has been considered. Particularly, using fault tree analysis, it has been evaluated the changing in the overall risk (considering, for simplicity, 1 year of mission time) of explosion referring to a hybrid-like collector realized by introducing a Fabric Filter (FF) downstream with respect to an Electrostatic Precipitator (ESP). Results have shown that a chain of failures in the FF section may affect relevantly the explosion risk in the collector leading to the unavoidable need for the introduction of mitigation actions into the system.
Keywords: cement plants, explosions, hybrid collectors, revamping, risk assessment, safety.

1 INTRODUCTION

One of the most important mineral industries in the world, accounting only in the United States for an annual production capacity greater than 73 million tons, is cement manufacturing [1]. In Europe, this industry is not so developed and, during the last years, a relevant decrease in the consumption of cement has been registered. As the sake of example, in 2012, there has been a decrease in the manufacturing of cement of about 22% (with respect to the previous year) and this negative trend is actually continuing [2].

As a general consideration, cement industry is very capital intensive. Energy costs can account for up to 40% of the total cost of cement manufacturing. Currently, over 90% of the installed capacity uses coal as primary fuel for the kiln burner. Unfortunately, the use of coal as fuel is very dangerous because of the possible triggering of fire and explosions during cement

DOI: 10.2495/SAFE-V7-N2-113-125

manufacturing. In the last years, an innovative solution was proposed: the use of the Solid Recovered Fuel obtained from Municipal Solid Waste in order to substitute a big part of the coal [3, 4].

Cement plants are recognized as a source of undesirable airborne emissions. In an attempt to moderate the effects of such compounds on the environment and neighboring population, governments legislate maximum permissible limits for a multitude of emission types (such as sulfur dioxide, nitrogen oxides, mercury and dusts/particulate matter).

Focusing on dust emissions, which have become in the last years the most studied problem in terms of urban pollution (independently of the source: industrial processes, traffic or cooking or domestic heating), it is possible to notice that they are not as frequently regulated as SO_2, NO_x or mercury [5–8]. Anyway, the most stringent regulations covered are in the EU, specifically in Germany and Austria.

Dust removal from a gas stream is a common and quite cheap operation required in many industrial processes but, in the last decades, more and more stringent law requirements have forced this common operation to become very expensive. Budgets for introducing improvements (or new units) in the already existing de-dusting equipment play a fundamental role in the general asset management strategy: in fact, de-dusting, being often an operation having low influence on production performances, is unavoidably seen more as a ‘necessary cost’ than a profit generator. Companies’ trend, especially in the actual economic scenario, is to reuse existing de-dusting plants, enhancing their efficiencies rather than install new plants that would involve costs for the decommissioning of obsolete equipments. Unfortunately, the obvious economic benefits of retrofits are counterbalanced by technical disadvantages and possible safety-related problems. In particular, a systematic assessment of safety issues, not limited to the safety-oriented design of new machines, but extended in the evaluation of the impact that changes in process conditions (induced by the new equipments) can have on the existing ones, must be done.

Cement plants usually employ two main dust and particulate matter abatement technologies to help them meet emissions limits: ElectroStatic Precipitators (ESP) and Fabric Filters (FF). ESPs use an induced electrostatic charge to attract dust to charged surfaces. While efficient on start-up and mechanically reliable, ESPs become less effective as dust builds up on the charged surfaces, causing time lost to maintenance and cleaning and potentially causing emissions to rise above permitted levels. Fabric Filters use natural or synthetic fiber-based bags to physically remove dust from the gas-stream. With typically 4 to 12 bags working in parallel, baghouses maintain steady operation for longer than ESPs, although care must be taken with respect to the potential for damaged bags. If a bag breaks mid-operation, the potential exists for a massive emission of dust. This can have severe adverse effects for local residents, water courses and may involve financial penalties for the producer.

In this work, the relevant problem of the triggering of an explosion in the de-dusting section of a cement plant has been analyzed. Particularly, it has been calculated the changing in the overall risk of explosion (considering 1 year of mission time for the plant) referring to a relevant re-vamping case study where a hybrid-like collector has been realized by introducing a Fabric Filter downstream with respect to an already installed ElectroStatic Precipitator. The results of the analysis have shown that failures in the FF section may affect relevantly the risk of an explosion occurrence in the collector itself leading to the unavoidable need for the introduction of mitigation actions into the system.

2 DESCRIPTION OF THE HYBRID-LIKE COLLECTOR

Hybrid collectors are abatement systems that have been designed specifically to take into account all the factors affecting their operational efficiency. Because of this fact, they have a very high collection efficiency [9, 10], and they represent the best technology currently available in order to fully comply with the limit values laid down by specific authorizations issued by the competent bodies.

The hybrid collection system treated in this case study has been realized by a retrofit operation consisting of an addition of a fabric filter downstream an already installed electrostatic precipitation section.

Given the particular system design, the amount of dust emitted into the atmosphere is considerably lower than that allowed by the strict environmental standards in force.

Hot gases (about 300°C), coming from the pre-heating section of the kiln, are partly used to dry and ventilate both the coal and the crude feed mills but, while the gases of the coal mill are filtered independently, the dusty gases used in the kiln are mixed with the gas coming from the pre-heater in a tower (called 'conditioner') where they are cooled (till about 160°C) and humidified with water spray; the amount of dosed water is regulated by a computer as a function of the gas temperature.

Then, the conditioned gases enter the electrostatic section of the hybrid filter, whose design is such as to allow a very low-gas velocity (in order to favor the separation of the dusts from the gaseous stream). The operating principle of an electrostatic precipitator is based on the effect of ionization of a neutral gas (corona effect) when it passes through a strong electric field formed by issue (negative polarity) and precipitation (or collector) electrodes (positive polarity) [11]. Thanks to the very high applied voltage (in this case, 65 kV), the issue electrodes emit electrons by photoelectric effect and charge the dust particles of which the gas to be de-dusted is rich; in this way, the charged dust particles will be attracted by the precipitation electrodes (or plates) [12]. The dust settles on the plates so that, periodically, they have to be 'shocked' in order to be kept clean and fully operative. Particularly, an electrostatic precipitator can remove dust particles as small as 1 μm with an efficiency exceeding 99% [13, 14].

Immediately then, the partially de-dusted gases pass through another section: the fabric filter.

A fan located downstream of the fabric filter aspires the air flow to be de-dusted, causing its entry in the lower part of the filter; hence, the gas going up meets the rows of bags and passes from the outside to the inside of their tissue cylinder.

The particles suspended in the gas are retained on the outside of the bags thanks to both the structure of the fabric fibers and a layer of particulate laying outside the fabric of each bag (which is capable of retaining even the finer particles).

The collapse of each bag on itself, given by the outside-inside air flow, is prevented by a basket that keeps roughly the shape of a tube.

The de-dusted air inside of each bag is now aspired and removed by the filter in the upper part of the chamber to be sent to the chimney, always thanks to the fan cited above. Approximately, every 8 min, in order to avoid the clogging of the filtering surface because of an excessive deposit of dusts, an air jet at high speed is blown inside of each bag: such a jet, impacting against the filtering surface, causes a sudden expansion of the bag and allows for the detaching of the crust of dust that has been formed on it. This cleaning system is called 'pulsed jet' and has the advantage of not having to exclude the bags from the filtration system during cleaning operations; the air jet is provided instantly through solenoid valves from dedicated storage tanks, whose operating pressure is about 6 bar, and acquires a high speed through the venturi ducts placed on the basket of each bag. To maintain continuous filtration during the cleaning operation, the air is fed to a row of bags at a time, allowing the other rows to continue to carry out their activity. The detached material falls into the hoppers at the bottom

Figure 1: Sketch of the case study hybrid collector. (1) Final sections of the fabric filters; (2) hoppers; (3) de-dusted air duct; (4) fan; (5) bottom of the chimney. The ESPs are located behind the fabric filters.

of the filtering chamber and, then, it is extracted by means of a worm screw and returned to the kiln by means of conveyor belts.

All these constitutive parts of the plant have been reported in Fig. 1, which shows a simple sketch of the de-dusting unit.

3 RISK ASSESSMENT IN CEMENT PLANTS

3.1 General risk assessment procedure

When talking about Quantitative Risk Analysis (QRA), we referred to as a set of methodologies and structured procedures having the aim of determining (both qualification and quantification) the so-called 'risk function', which is a fundamental tool in a decision-making process [15].

The estimate of the risk arising from the exercise of a particular industrial activity is generally performed with a methodology that, starting from statistical data derived from the history of accidents actually occurred in plants similar to the one under study, will complete a series of specific analyses, related to a particular facility and the site where it is installed, in order to obtain a sufficiently accurate estimate of the risk [16]. In the case of mature and widespread technologies, the statistics of accidents can be so broad as to allow the direct realistic estimate of the risk or, at least, the probability of occurrence of conceivable accidents. The determination of the consequences depends very much on the site, for what concerns the aspects of meteorology and hydrology, population distribution, etc; therefore, it is generally necessary an adaptation of the statistical data to the particular case study. If you do not have a statistic sufficiently complete to be referred to, an analytical methodology to perform a risk assessment on a theoretical basis can be employed.

The risk in the performance of an industrial activity results from the use of potentially hazardous materials. The starting point of a risk analysis is always the identification of hazardous substances and the process that they undergo in the system under study. This procedure is usually referred to as hazard identification.

The protection system (more or less sophisticated) of the plant comes into play at this point, to prevent or, at least, limit the possible effects of an accidental event. Therefore, the adverse event occurs only if you have, at the same time, the failure of the process system and the failure of the protections that the system provided to confine the incident.

Considering the physical and chemical phenomena involved in an accident, you are able to evaluate their effect on individuals, economic systems, social systems, etc.., through:

(a) an identification of the events that contribute to the risk;
(b) an estimate of the probability of occurrence of such events and their consequences;
(c) a determination of the risk function and its use for decision making.

To develop such an analysis, a thorough knowledge of the system and the industrial process implemented in it has to be acquired for the search of those failures which may constitute the source of accidents (initiating events).

The risk analyst must then develop a model of the system that allows the identification of the possible states of the latter arising from each initiating event. To estimate the probability of occurrence of each state of the system, we can make use of appropriate techniques such as fault tree analysis.

The next step is the determination of all accidental scenarios associated with each degraded state of the system (sequence of events), possibly 'filtered' by the effect of the available protection systems.

At this point, in order to identify the consequences for each category of accidents, it is necessary to develop an environmental impact model, which describes the magnitude of the consequences of each scenario. In this phase, the possible implementation of an external emergency plan can take a decisive role for the mitigation of the consequences of the accident.

Finally, after determining the probability of occurrence of each accidental event, the integration of all results, (defined as an estimate of the magnitude and frequency of occurrence) on the entire spectrum of the categories of accidental scenarios, allows for the determination of the function associated with the risk linked to the activity in question [17]. Normally, on the basis of this risk function, decision-makers (political or technical, depending on the stage at which the study is actually applied) take the appropriate decisions.

3.2 Fault tree analysis for QRA [18, 19]

Fault tree analysis is a technique suitable for both determining the credible modes of occurrence of an undesired event (called, top event), caused by a complex concatenation of other events (qualitative analysis), and estimating the frequency of occurrence of undesired event on the basis of frequency of occurrence of the events that cause it (quantitative analysis). It is a deductive methodology particularly suitable for the analysis of complex systems whose development can be easily decomposed into a succession of more simple events, and therefore it is well suited to the analysis of industrial installations.

Once all the systems (and their parts) have been identified, it is necessary to continue the analysis of the chain of systems, subsystems, equipment, etc., up to the failure of the individual components for which we are in possession of sufficient information on the probabilities of failure to be used in the evaluation.

These are combined through logical operations (and / or), also called logic gates, going from bottom to up, until you get to determine the probability of the top event. The fault tree analysis allows you to treat the contribution to risk arising from human error, as well as other causes of failure common to several subsystems. Clearly, it is possible to use fault trees also for qualitative analysis. In this case, there is the advantage of highlighting the main causes of an accident (e.g. whether failures of equipment or human error) so that you can focus on preventive measures to reduce the probability of the accident itself.

The representation of a fault tree uses some symbols with standardized meanings; the main graphic symbols are summarized in Table 1.

3.2.1 Quantitative risk analysis

The quantification of a fault tree has essentially the aim of quantifying the reliability *R(t)* of a system, that is the probability that, in correspondence of an allotted time interval (also called mission time, t), a system performs properly the function for which it was built. From this definition, it is clear that the reliability of a system is inversely proportional to the mission time: the longer the time of the mission, the smaller results to be the reliability of the system (or it is more likely that the system fails).

It is important to note that the quantification of the fault tree is an operation valid for orders of magnitude; in other words, the results cannot be compared on the basis of small differences. For example, 1.5 and 4.1 are essentially the same number while 3.5 to 0.12 are two numbers significantly different. It follows that no unnecessarily high accuracy is required in the input data (i.e. in the estimate of the probabilities of occurrence of the primary events).

Table 1: Graphic symbols for fault tree analysis.

<table>
<tr><th>Graphic Symbol</th><th>Meaning</th></tr>
<tr><td>
</td><td>Primary events: these events, for one reason or another, are not further investigated; if you want to perform a quantitative analysis to estimate the frequency of occurrence of the top event you need to know the frequency of occurrence of these primary events, for example, estimating it through a historical analysis.</td></tr>
<tr><td>
</td><td>Intermediate events: these are events that occur before or after another event and represent the cause of the next event; they are connected to the events preceding or following them by gates.</td></tr>
<tr><td>
</td><td>OR-gate: so that the output of the gate takes place, it is sufficient that one of the inputs to the gate itself (which may be any number greater than 1) occurs.</td></tr>
<tr><td>
</td><td>AND-gate: so that the output of the gate happens, it is necessary that all the inputs to the gate itself (which may be any number greater than 1) occur.</td></tr>
<tr><td>
</td><td>INHIBIT-gate: represent a scenario in which the output event occurs if all input events occur and an additional conditional event (typically an event external to the configuration represented by the fault tree) also occurs.</td></tr>
</table>

3.2.2 Primary events quantification

For a system whose modes of failure can be represented in a fault tree, the information contained in *R(t)* are usually summed in the following functions: the unavailability q(t), which is the probability that the system is not able to perform its function (because it is broken) at time t, and the frequency of occurrence *W(t)*, which is the number of times that the system is not expected to be able to perform its function in its mission time.

To quantify a fault tree is first necessary to calculate these quantities for all primary events that have been identified in the construction of the fault tree itself.

If the component involved in the primary event is not repairable, the component is no longer able to perform its function at time t if it is spoiled before the same time t; its unavailability therefore coincides with its unreliability, namely [20]:

$$q(t) = \lambda t \tag{1}$$

where λ is the failure rate, that is, the frequency with which a system fails (or the fraction of the components that fail per unit of time, y^{-1}) and t is time, y.

If the component is repairable, its unavailability no longer coincides with its unreliability, since, even if the component had spoiled before time t, may still perform its function at time t in the case where it has been repaired. In this case [20], it is possible to demonstrate that the unavailability reaches an asymptotic value equal to:

$$q(t) = \frac{\lambda T_D}{1 + \lambda T_D} \approx \lambda T_D = \frac{\lambda}{\mu} \tag{2}$$

where T_D is the mean repair time, that is the time necessary to repair the component, and $\mu = 1/T_D$ is the repair rate, that is, the frequency with which a component can be repaired. For practical purposes, the unavailability can be approximated by the probability that the component has failed in the time interval T_D previous than the instant of time considered (therefore, resulting failure at time t).

3.2.3 Minimal cut sets and top event quantification [20]

A Minimal Cut Set (MCS) is the minimum combination of primary events which is necessary and sufficient to ensure the occurrence of the top event; in equivalent terms, the top event occurs if all the events in a MCS occur simultaneously. Then, the unavailability of a MCS is the probability that, at time t, all the system components are not able to perform their function; therefore, this probability coincides with the probability of occurrence of the top event due to the considered MCS. Since the individual events involved in a MCS are all independent, the overall probability that all events occur simultaneously is given by the following relation:

$$q(t) = \prod_{i=1}^{n} q_i(t) \tag{3}$$

where n is the order of the MCS (the number of events that involves the MCS) and $q_i(t)$ is the unavailability of each component of the MCS.

The frequency of occurrence of the MCS, which coincides with the frequency of occurrence of the top event in the mission time *(T)* because of the MCS considered, is then calculated as:

$$W(t) = \int_o^T \left[\sum_{i=1}^{n} \left(\left(\prod_{\substack{j=1 \\ j \neq i}}^{n} q_j(t) \right) \lambda_i \right) \right] dt \tag{4}$$

A top event can occur if anyone of the MCS that can cause it occurs. The likelihood of the top event in a given interval of time can thus be calculated as the union (logical OR) of the probability of occurrence of each MCS that can cause the top event itself.

4 CASE STUDY: DE-DUSTING SYSTEM REVAMPING

In accordance with the purposes of the present work, we have focused on the relevant problem of the triggering of explosions in the de-dusting section of a cement plant. As it is widely known, cement plants can suffer of explosion like problems only in the case of an accidental development of an overpressure inside process equipment. Therefore, much more serious explosions, such as dust explosions, cannot be triggered inside these plants because cement powder cannot burn as it does not contain any combustible group.

Using fault tree analysis, it has been evaluated the changing in the overall risk (considering one year of mission time) of a physical explosion referring to the case study where a hybrid-like collector is realized by introducing a Fabric Filter (FF) downstream with respect to an already installed ElectroStatic Precipitator (ESP). This revamping operation is capable of guaranteeing an increase in the de-pulverization efficiency of more than 0.5–1.5%; therefore, it is able to comply with the new environmental laws on dust emissions into the atmosphere.

Thus, we have proceeded with the drafting of the fault trees concerning the top event 'occurrence of an explosion in the de-dusting section' both before (ESP) and after (HYBRID FILTER = ESP + FF) the revamping operation.

For what concerns the layout of the de-dusting section (before and after) and its functioning, reference has been made to the plant sketch reported in Figure 1.

Figure 2 shows the graphical representation of fault tree realized for the plant configuration before the retrofit operation, which is considering the de-dusting section of the plant as constituted by the ESP (and its related instrumentation) only. Observing the tree, it can be derived clearly how an explosion of the ESP can be triggered as a consequence of different chains of events (all intended as failures).

In particular, the top event takes place in the case where both of the following final intermediate events occurs:

- E10: development of an overpressure higher than 6 bar inside the ESP chamber;
- E20: failure of the rupture disk installed on the ESP.

It is important to state that, within the scheme of the fault tree, all the protective devices present into the plant section object of the analysis have been considered.

Performing the MCS analysis, it is possible to observe that there are 4 MCS: 3 of order 4 and 1 of order 5. Among the MCS of order 4, we have all the combinations of events in which, due to the presence of external factors causing the blockage of the chimney (EXT), the feed to the de-dusting unit (FEED), the absence of a scheduled and correct maintenance activity (MANABS), the blockage of the rupture disk installed on the ESP (SCRBLOC), the absence of a human action as a consequence of the sounding of the high pressure alarm (HE) and the breakdown of the ESP pressure indicator (PI), a pressure overcoming the maximum threshold value can be reached causing a physical explosion of the ESP chamber.

For what concerns the MCS of order 5, it involves the following combination of events: fan breakdown (FAN), absence of a scheduled and correct maintenance activity (MANABS), block of the rupture disk installed on the ESP (SCRBLOC), pressure interlock (PHL) and temperature controller failure (TIC) on the water spray feeding line that constitutes the cooling system available for the hot dusty gases exiting the kiln and addressed to the de-dusting system.

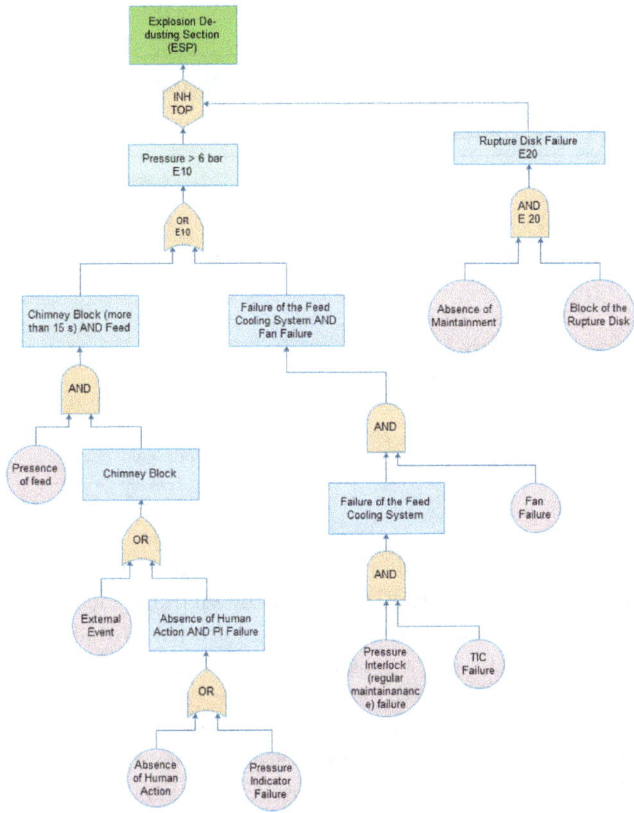

Figure 2: Fault tree for the top event 'explosion in the de-dusting section' evaluated before the plant retrofit.

In order to determine all failure rates and unavailabilities, Dossier Ambiente manuals [21] have been used. Table 2 contains all the data necessary for the quantification of the fault tree.

Quantifying the fault tree using a simple numerical probability calculation involving the computation of all MCS, a value of the probability of occurrence of the top event equal to 5.6 10^{-8} y^{-1} has been found. Such a value can be located in the acceptability zone for what concern both individual and societal (<100 inhabitants) risk. Therefore, just from this analysis, the system does not require to be integrated with redundant protection devices in order to improve the reliability of the de-dusting section.

Successively, we have proceeded with the generation of the fault tree concerning the top event 'occurrence of an explosion in the de-dusting section', now constituted by a fabric filter installed after the exit of the already existing ESP (that is, considering a revamping operation).

In this case, the generated fault tree is too wide to be reported here but it is possible to briefly summarized the main results arising from its detailed analysis.

Particularly, 7 different minimal cut sets can be observed: 3 of order 3, 3 of order 4 and 1 of order 5.

Among the MCS of order 3, quite dangerous, we have all the combinations of undesired events and failures in which, due to the absence of a regular and correct maintenance activity

Table 2: Database for failure rates, repair rates and unavailabilities. Case study before the plant revamping.

Primary event code	**Failure rate [y^{-1}]**	**Repair rate [y^{-1}]**	**Unavailability [-]**
EXT	–	–	$1 \cdot 10^{-7}$
FEED	–	–	$8 \cdot 10^{-1}$
MANABS	–	–	$2 \cdot 10^{-3}$
SCRBLOC	–	–	$1 \cdot 10^{-2}$
HE	–	–	$2 \cdot 10^{-3}$
PI	$3 \cdot 10^{-1}$	200	$1.5 \cdot 10^{-3}$
FAN	–	–	$1 \cdot 10^{-2}$
PHL	–	–	$1 \cdot 10^{-3}$
TIC	$1 \cdot 10^{-1}$	200	$5 \cdot 10^{-4}$

on the rupture disks of both the fabric filter and the electrostatic precipitator (MANABS), the blockage of the rupture disks (SCRBLOC), the breakdown of the cleaning air timing system (TEMP), the absence of cleaning air (NOAIR) and the failure in closed position of the exiting valve of the fabric filter (VALCLO), a pressure overcoming the maximum threshold value can be reached causing a physical explosion of the hybrid-like collector.

Among the MCS of order 4, we have all the combinations of events previously determined for the ESP unit alone (even if with some little variants): blockage of the chimney (EXT), presence of feed to the de-dusting unit (FEED), absence of maintenance activity (MANABS), block of the installed rupture disks (SCRBLOC), absence of human actions even if a high-pressure alarm is sounding (HE) and breakdown of the hybrid filter differential pressure indicator (PI). Finally, regarding the MCS of order 5, we can observe that it is the same involved in the plant configuration before the retrofit operation (this is because it involves only failures of equipment located upstream of the ESP and the fan).

As previously done, to determine all failure rates and unavailabilities, Dossier Ambiente manuals [21] have been used. Table 3 contains all the data necessary for the quantification of the new fault tree.

From a qualitative point of view, the new criticality that arises from the introduction of a FF downstream to the ESP is a blockage of the gaseous flux inside the FF. This can be due, as the sake of example, to the absence of air into the FF cleaning system: not removed dusts on the filter bags can increase pressure drops inside the equipment so that no gaseous flux can pass anymore. Therefore, the pressure into the FF and, consequently, into the ESP can only increase if there is the failure of the protective systems (rupture disks).

Quantifying the fault tree using a classical numerical probability calculation involving the computation of all MCS, a value of the probability of occurrence of the top event equal to 3.8 10^{-6} y^{-1} has been found. Such a value is located in the non-acceptability zone for what concern the societal risk (<100 inhabitants) and at the boundary of the acceptability region if an individual risk is considered.

This result confirm that a revamping operation performed without taking into account properly the different operating conditions under which the new revamped plant will operate can be very dangerous.

Table 3: Database for failure rates, repair rates and unavailabilities. Case study after the plant retrofit.

Primary event code	Failure rate [y^{-1}]	Repair rate [y^{-1}]	Unavailability [-]
EXT	–	–	$1{\cdot}10^{-7}$
FEED	–	–	$8{\cdot}10^{-1}$
MANABS	–	–	$2{\cdot}10^{-3}$
SCRBLOC	–	–	$1{\cdot}10^{-2}$
HE	–	–	$2{\cdot}10^{-3}$
PI	$3{\cdot}10^{-1}$	200	$1.5{\cdot}10^{-3}$
FAN	–	–	$1{\cdot}10^{-2}$
PHL	–	–	$1{\cdot}10^{-3}$
TIC	$1{\cdot}10^{-1}$	200	$5{\cdot}10^{-4}$
NOAIR	–	–	0.1
TEMP	–	–	0.05
VALCLO	–	–	0.05

5 CONCLUSIONS

In this paper the frequency of the relevant undesired event 'explosion in the de-dusting section of a cement plant' has been calculated for two different plant configurations: precisely, before and after the performing of an upgrade operation aimed at reducing the atmospheric emission of dusts and particulate matter (addition of a fabric filter downstream an electrostatic precipitator).

Analyzing the fault trees generated for the previously mentioned top event before and after the revamping operation, it has been showed that the probabilities of occurrence considering a mission time equal to 1 year are, respectively, 5.6 10^{-8} y^{-1} and 3.8 10^{-6} y^{-1}.

Referring to the most standard and widely accepted threshold value for the risk acceptability in industrial areas (which is 1.0 10^{-6} y^{-1} – reference: individual risk), the risk in the first plant configuration (before the retrofit operation) can be considered acceptable and the system does not require to be integrated with redundant protection devices in order to improve the reliability of the de-dusting section. On the contrary, in the revamped configuration, the risk falls within the non-acceptability zone; therefore, the system requires to be integrated with redundant protection devices in order to improve the reliability of the revamped de-dusting section. Particularly, the critical feature is precisely the introduction of the new apparatus, the fabric filter, which introduces a potential hazard before not present: that is, blocking of the air flux, consequent pressurization of the whole de-dusting line and, finally, explosion.

This simple analysis confirms that each process or pieces of equipment change must be fully evaluated from the safety and reliability viewpoint before being introduced into an already existing plant because it can imply the increase of the risk function. Particularly, in this case, there is an increase of the risk function of about two orders of magnitude: this locates the new configuration of the plant into a non-acceptability risk zone, even if the revamping operation was intended to a better integration of the plant with the environment.

REFERENCES

[1] Explosions in Cement Plants. Available at: www.fike.com/

[2] Available at: http://www.greenreport.it/

[3] Rada, E.C., Squazardo, S., Ionescu, G. & Badea, *A.*, Economic viability of SRF co-combustion in cement factory. *UPB Scientific Bulletin, Serie D*, **76**(3), pp. 199–206, 2014.

[4] Rada, E.C., Ragazzi, M., Ionescu, G., Merler, G., Moedinger, F., Raboni, M. & Torretta, V., Municipal solid waste treatment by integrated solutions: energy and environmental balances. *Energy Procedia*, **50**, pp. 1037–1044, 2014. https://doi.org/10.1016/j.egypro.2014.06.123

[5] Copelli, S., Rada, E.C., Ragazzi, M., Torretta V. & Schiavon, M., Reliability analysis and assessment of the explosion risk in a hybrid collector. *WIT Transactions on The Built Environment*, **151**, pp. 239–250, 2015. https://doi.org/10.2495/safe150211

[6] Torretta, V., Rada, E.C., Panaitescu, V. & Apostol, T., Some considerations on particulate generated by traffic. *UPB Scientific Bulletin, Series D,* **74**(4), pp. 241–248, 2012.

[7] Pei, B., Cui, H., Liu, H. & Yan, N. Chemical characteristics of fine particulate matter emitted from commercial cooking. *Frontiers of Environmental Science and Engineering*, **10**(3), pp. 559–568, 2016. https://doi.org/10.1007/s11783-016-0829-y

[8] Fuller, G.W., Tremper, A.H., Baker, T.D., Yttri, K.E. & Butterfield, D., Contribution of wood burning to PM10 in London. *Atmospheric Environment*, **87**, pp. 87–94, 2014. https://doi.org/10.1016/j.atmosenv.2013.12.037

[9] Gutierrez Ortiz, F.J., Navarrete, B., Cañadas, L. & Salvador, L., A technical assessment of a particle hybrid collector in a pilot plant. *Chemical Engineering Journal,* **127**, pp. 131–142, 2007. https://doi.org/10.1016/j.cej.2006.09.015

[10] Long, Z. & Yao, O., Numerical simulation of the flow and the collection mechanism inside a scale hybrid particulate collector. *Powder Technology,* **215–216**, pp. 26–37, 2012. https://doi.org/10.1016/j.powtec.2011.08.045

[11] Böttner, C.U., The role of the space charge density in particulate processes in the example of the electrostatic precipitator. *Powder Technology,* **135–136**, pp. 285–294, 2003. https://doi.org/10.1016/j.powtec.2003.08.020

[12] Neimarlija, N., Demirdžić, I. & Mrzaferija, S., Finite volume method for calculation of electrostatic fields in electrostatic precipitators. *Journal of Electrostatics,* **67**, pp. 37–47, 2009.

[13] Cooperman, G., A unified efficiency theory for electrostatic precipitator. *Atmospheric Environment,* **18**, pp. 277–285, 1984. https://doi.org/10.1016/j.elstat.2008.10.007

[14] Varonos, A.A., Anagnostopoulos, J.S. & Bergeles, G.C., Prediction of the cleaning efficiency of an electrostatic precipitator. *Journal of Electrostatics,* **55**, pp. 111–133, 2002. https://doi.org/10.1016/s0304-3886(01)00187-5

[15] Piccinini, N. & Demichela, M., Risk based decision-making in plant design. *Canadian Journal of Chemical Engineering,* **86**, pp. 316–322, 2008. https://doi.org/10.1002/cjce.20064

[16] Tixier, J., Dusserre, G., Salvi, O. & Gaston, D., Review of 62 risk analysis methodologies of industrial plants. *Journal of Loss Prevention in the Process Industries,* **15**, pp. 291–303, 2002.
https://doi.org/10.1016/s0950-4230(02)00008-6
[17] Piccinini, N. & Ciarambino, I., Operability analysis de-voted to the development of logic trees. *Reliability Engineering & System Safety,* **55**, pp. 227–241, 1997.
https://doi.org/10.1016/s0951-8320(96)00111-1
[18] CCPS, *Guidelines for Chemical Process Quantitative Risk Analysis*, Wiley, 2nd ed., 1999.
[19] Vaseley, W.E., Goldberg, F.F., Roberts, N.H. & Haasl, D.F., *Fault Tree Handbook*, U.S. Nuclear Regulatory Commission Washington D.C., 1981.
[20] Rota, R. & Nano, G., *Introduzione alla affidabilità e sicurezza nell'industria di processo,* Pitagora Editrice, 2007.
[21] Dossier Ambiente, http://www.amblav.it/dossier.aspx

NUMERICAL AND EXPERIMENTAL SIMULATION OF SPONTANEOUS COMBUSTION OF COAL

YU ZHIJIN & WEN HU
Faculty of Safety Science and Engineering, Xi'an University of Science and Technology, China.

ABSTRACT

In this paper, our main aims are to make up for the defects of large-scale coal spontaneous heating tests on trial methods. An unsteady-state numerical model is developed to simulate self-heating in a large-scale coal spontaneous heating reactor with experimental conditions. The self-heating process of coal is reproduced successfully, and the main characteristics of nonlinear heating and windward movement of the fire source are obtained during the whole process. By comparison, the simulated and experimental results of the Tianchi coal samples show a good agreement with the temperature rise features and fire source positions. This confirms the validity of the model. Based on the model, a sensitivity analysis is constructed to investigate the influence of the external environment on coal spontaneous heating. The main contents include (1) heat dissipation surroundings; (2) air quantity; (3) air direction. The simulated conclusions contribute to optimizing the experimental parameters and obtaining the shortest spontaneous combustion period. In addition, the model is found to be of great significance for predicting the fire source temperature and position in practical coal mine conditions.

Keywords: numerical simulation, temperature rising feature, fire source position, experimental parameters, spontaneous combustion period

1 INTRODUCTION

Spontaneous combustion of coal is a critical mining hazard that all coal-producing countries have to face [1, 2]. Accurate prediction of when and where the spontaneous combustion will occur is crucial to reduce the harm caused by spontaneous combustion of coal. It is well known that the oxidizing properties of coal are the essential reason for spontaneous combustion. However, in coal mines, whether or not a fire, caused by the coal's self-heating, occurs will depend on the external environment [3]. Experimental simulation is undoubtedly the most effective way to study the formation characteristics and the evolution law of coal self-heating. However, as there are extremely complex internal and external factors, a small-scale experimental method can better reveal the thermodynamics characteristics, temperature variation, and law of production for indicator gases, although it cannot monitor the position of the fire source due to passive heating or altering the external environment [4–8]. A large-scale test of the spontaneous heating of coal can well simulate the field conditions and investigate the coal oxidation and exothermic features [9]. Because of this, many large-scale experiments have been proposed under actual mine conditions and used in an attempt to predict the self-heating tendencies of coals. However, because of the long experimental period, tremendous workload, for the same coal sample, it is impossible to change the conditions and repeat experiments. Therefore, establishing numerical models is crucial to study coal's spontaneous combustion process under various conditions.

Numerous models were used to research the process of the spontaneous combustion of coal in mined-out areas, coal heaps or bunkers [10, 11]. Taraba *et al.* revealed the characteristic of fire source versus time in coal heaps by numerical simulation [12]. Yuan *et al.* numerically investigated the spontaneous combustion process of coal in a large-scale coal chamber and focused on the low-temperature oxidation reaction rate [13]. Some studies also discussed the influence of simulated parameters on the rising temperature feature [14, 15]. However,

DOI: 10.2495/SAFE-V7-N2-126-136

correlative studies on the characteristics of the fire source position and its influencing factors combined with various heat dissipation conditions were not found.

Firstly, the self-heating process of broken or crushed coals in a reactor is reproduced successfully by a numerical method. Then, through the model transformation, the influence of heat dissipation conditions, air quantity, and air direction on temperature and the position of the self-igniting fire source were evaluated, and the results were thoroughly analyzed.

2 NUMERICAL STUDY

Before the model is established, the following assumptions are presented:

1. Coal is homogeneous and isotropic;
2. The temperature of coal and air is equal at the same location;
3. The effects of water phase change, absorption, and desorption on temperature and volume of coal are deliberately ignored;
4. When the air outflow from the preheating water layer, its temperature is identical to the water temperature at this moment.

2.1 Mathematical model

The mathematical model of the spontaneous combustion of coal included gas flow, mass transport and energy conservation. Here, the governing equations are established as follow [16–18].

1. Gas flow

   The Darcy-Brinkman equation, which is suggested to describe the gas flow behavior for highly permeable media, can be written as Eq. (1)

   $$-\nabla p = \frac{\mu}{k}\upsilon - \mu_e \nabla^2 \upsilon, \qquad (1)$$

   where p is the gas pressure, Pa; $\nabla$is the Hamilton symbol; υ is velocity, m/s; μ is dynamic viscosity, Pa·s; *k is* permeability, m^2; and μ_e is effective viscosity, Pa·s.
2. Mass transport

   In this part, both convective and diffusive oxygen transport are considered; these are reported in Eq. (2)

   $$\varepsilon \frac{\partial c}{\partial t} + \upsilon \nabla c = D\nabla^2 c - r(1-\varepsilon), \qquad (2)$$

   where ε is porosity; *c* is oxygen concentration, mol/m^3; *t* is time, s; *D* is the diffusion coefficient of the gas component, m^2/s; *r* is the oxygen-consumption rate of coal, mol/(m^3·s); this is defined as the Arrhenius equation [19].

   $$r = \frac{c}{c_0} A \exp(-E/RT), \qquad (3)$$

   where c_0 is initial oxygen concentration, mol/m^3; *A* is the pre-exponential factor, mol/(m^3·s); E is the apparent activation energy with a unit of kJ/mol; R is the universal gas constant.

3. Energy conservation
   The equivalent energy conservation equation includes convective and conservative processes; it is formulated as Eq. (4)

$$\left[(\varepsilon\rho_g C_g)+(1-\varepsilon)\rho_s C_s\right]\frac{\partial T}{\partial t}+(\upsilon\nabla)(\rho_g C_g \mathrm{T})=\lambda_s\nabla^2 T+(1-\varepsilon)r\Delta H, \tag{4}$$

   where ρ_g, ρ_s is density of gas and coal, kg/m^3; ρ_g can be described as $\rho_g = Mp/TR$, M is the average molecular weight of the gas, T is temperature of the coal, K; C_g and C_s represent the heat capacity of the gas and coal, J/(kg·°C); T is temperature, °C; $\lambda_e = \lambda_g\varepsilon+\lambda_s(1-\varepsilon)$, λ_g, λ_s is the heat conductivity coefficient of gas and coal, W/(m·°C); and ΔH is heat of reaction, J/mol.

2.2 Simulated background

The simulated background of a large-scale experimental furnace is shown in Figure 1. The external and inner diameter of the cylindrical reactor was 100 and 60 cm, respectively, the height of coal loading was 150–170 cm and the largest coal was 1.5 tons. There was a 10–20 cm air buffer zone on both the top and bottom of furnace, correspondingly. In this reactor, automatic temperature measurement including thermocouples arranged in eight vertical layers of seven probes, the spacing between the layers of the monitoring point was 20 cm, the lowest layer was 5 cm from the bottom, and the radial spacings were 20, 40, and 60 cm, respectively. Gas samples including CO, O_2, CO_2 and CH_4 were collected manually or automatically using beam tubes which were arranged in the central axis of the furnace.

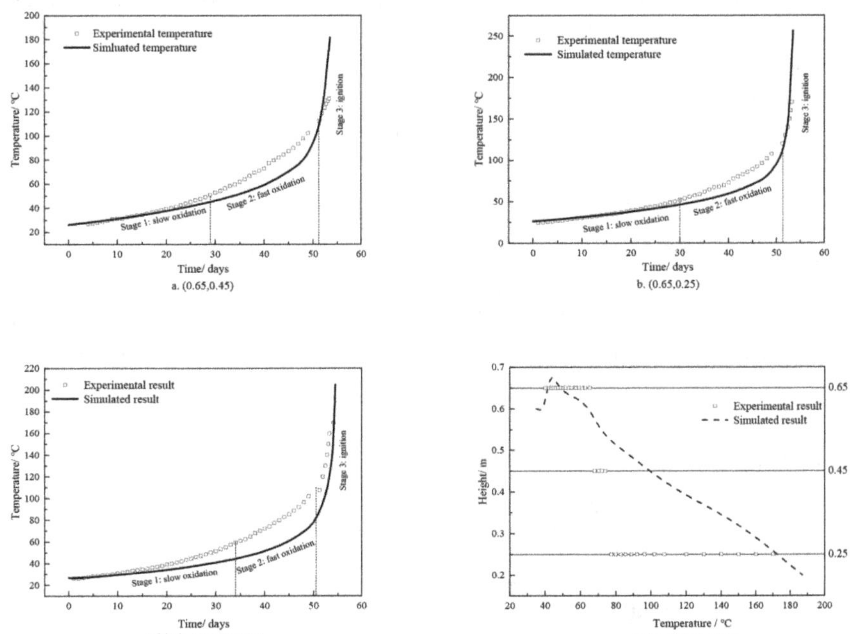

Figure 1: Large-scale coal spontaneous combustion experimental platform and its two-dimensional geometric model.

Table 1: Experimental conditions.

Reference particle size (mm)	Height (cm)	Mass (kg)	Volume (m^3)	Lump coal density (g mm^{-3})	Bulk density (g mm^{-3})	Porosity	Initial temperature (°C)
2.14	140	1415.13	1.58	1.40×10^{-3}	0.89×10^{-3}	0.36	26.2

Table 2: Proximate analyses of the coal sample.

Coal sample	Moisture (wt.%)	Volatiles (wt.%)	Fixed carbon (wt.%)	Ash (wt.%)
Tianchi	8.2	42.00	58.75	5.21

Gas chromatography was used to analyze the gas composition and concentration. The detailed structure and process of the experimental apparatus are described in the literature [9]. The geometric model is established at the ratio of 1:1. The experimental conditions and proximate analysis of the Tianchi coal sample are shown in Table 1 and Table 2, respectively.

2.3 Procedure for solution

The solved parameters are obtained either from the experimental results, including porosity, oxygen-consumption rate and heat of reaction, or from the related research results [9, 20]. The boundary conditions (Table 3) are set as far as possible according to the heat preserving circumstances of the experimental furnace. Then a finite element method is adopted for solving the governing equations.

Note: Q is the air quantity of the inlet, with air flow moving from the bottom to the top. The initial air quantity is 0.1 m^3/h. However, in view of with the temperature rising, the requirement of air volume significantly increases. Therefore, to obtain the shortest spontaneous combustion period, the air quantity increased to 0.2 m^3/h when the maximum temperature reached 60°C, 0.3 m^3/h when the maximum temperature reaches 110°C. Beyond 170°C, the air quantity is fixed at 0.4 m^3/h. T_1 is the temperature of the preheating water layer. When the minimum temperature of the coal close to the water layer is below 100°C, T_1 is equal to the coal temperature close to the water layer. Otherwise, T_1 is fixed at 100°C. h is the heat transfer coefficient of the wall which is 0.15 W/(m^2·°C) in the model; n represents the normal component.

Table 3: Boundary conditions of the model.

Physical field	Boundary conditions		
	Air inlet	Air outlet	Wall
Flow	$Q = 0.1–0.4 m^3/h$	$p = 0$	$-nv = 0$
O_2 concentration	$c = 9.375 mol/m^3$		$-n(-D\nabla c+vc) = 0$
Temperature	$T = T_1$	$-n(\lambda_{eq}\nabla T) = 0$	$-n(\lambda_{eq}\nabla T) = h(T_1\text{-}T)$

2.4 Results and validation

The test for spontaneous heating of Tianchi coal finished when the coal temperature reached 170°C [20]. During the experimental procedure, the central fire source zone always varied along the central axis. This is because the air flow along the central axis leads to a sufficient oxygen supply at this location. The comparison between the simulated and experimental results is shown in Figure 2.

The experimental results were compared to the simulated results in Figure 2a–d; this showed the same rule on temperature rise features at the same place in Figure 2a–c. There was an obvious nonlinear heating characteristic during the whole process of spontaneous combustion. The highest temperature reached by Tianchi coal was 170°C when the experiment had been underway for 54.1 days, contrasted with 53.1 days for the simulated results in Figure 2c. As is reported in Figure 2a–c, during the period when the temperature was between 50 and 100°C, the simulated results are always lower than the measured results. The influence of moisture on heat release is neglected in the model, which is the main reason for this phenomenon [21]. Figure 2d showed the variations of the maximum temperature postions, it illustrated that when the temperature of the fire source was more than 50°C, as the temperature rises, the position of the fire source changes correspondingly, and the tendency is to move towards the wind side. The significant increase of oxygen-consumption rate at high temperature stage should be responsible for this phenomenon. Finally, combustion always takes place at the face. Moreover, as can be seen in Figure 2d that there was a slight deviation in fire source positions between the experimental and the simulated results at the same temperature range. However, taking into account the distance between the adjacent measuring points (0.25, 0.45 and 0.65 cm height in the reactor), which is 20 cm in the experimental installation, also the locations of high temperature points could only be embodied by fixed sensing points in a certain space. Therefore, it demonstrated that simulated results can accurately reflect the positions of fire source at different temperature stages.

In a word, despite the existence of the deviation, the verified numerical methods can make up for the defects of the experiment and predict the temperatures and the positions of the spontaneous fire source of coal at various temperature stages in a large-scale spontaneous heating reactor.

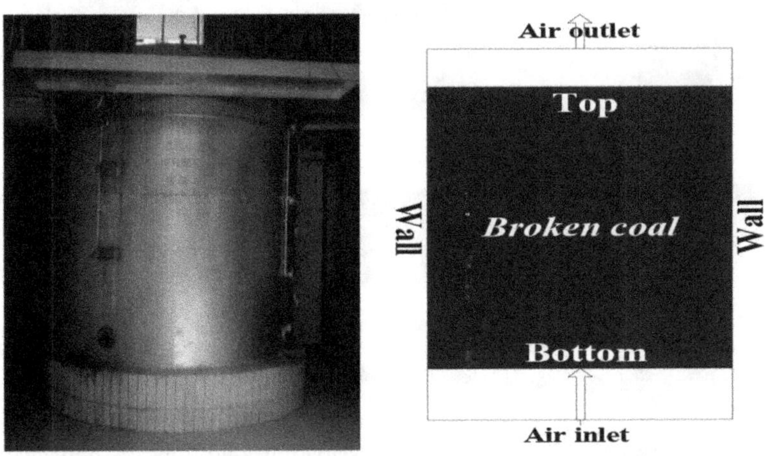

Figure 2: Comparison of simulated and experimental results.

3 SELF-HEATING SUSCEPTIBILITY ANALYSES

3.1 Effect of the conditions of heat dissipation

The spontaneous combustion of coal is closely related to the complicated external environment. Whether the heat from low-temperature oxidation accumulates or not is primarily dependent on the heat dissipation to the surroundings [20, 22]. However, the experimental equipment only provides a closed system and results in heat dissipation only by forced ventilation. These conditions are different from the actual coal mine environment. In this section, the boundary conditions in the heat transfer equations are modified to simulate various conditions of heat dissipation. The atmospheric environment which the parameters h and T_1 in Table 3 would be changed into 10 W/(m^2·°C) for h and fixed temperature 20°C for T_1 is used to contrast with the experimental conditions when the highest temperature reached 100, 200 and 300 °C; the respective temperature profiles are presented in Figure 3.

Figure 3 indicated that the conditions of heat dissipation were a significant factor in fire source positions. At the same temperature period, with a stronger heat dissipation capacity, the center fire source is formed at a location further from the air surface under the same air quantity, and the high temperature zone becomes more concentrated. In worse heat dissipation conditions, another distinct feature is that the temperature gradient is reduced and the high temperature zone is amplified due to a radial spread occurring simultaneously. Both the thermal conduction characteristics of coal and the buoyancy effect are responsible for the above-mentioned phenomenon. The rising coal temperature characteristics in experimental conditions and at atmospheric temperature are shown in Figure 4.

Figure 4 showed that, in the atmospheric environment, the spontaneous combustion period of coal was clearly extended. But a similar heating rate of coal will be obtained in both the atmospheric environment and the experimental conditions when the maximum temperature exceeds 100°C. The numerical and experimental results proposed a common rule that there

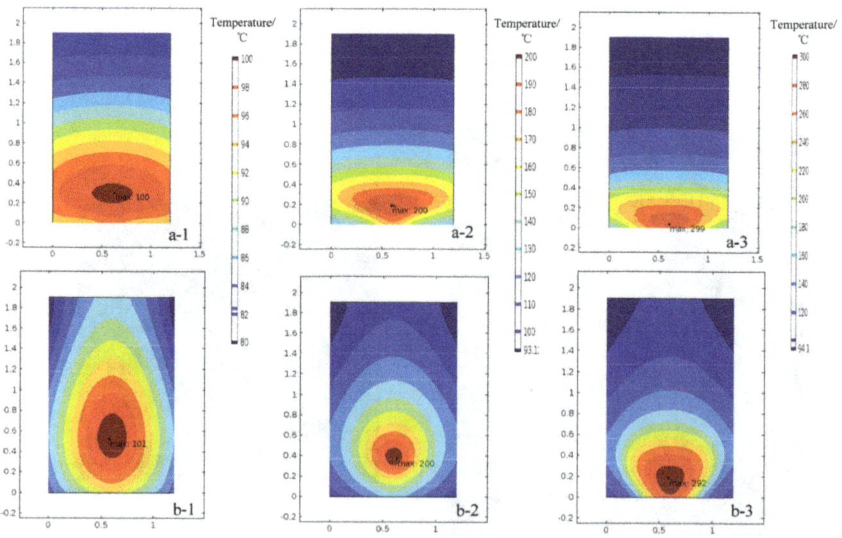

Figure 3: The temperature profile in various heat dissipation conditions; (a) under experimental conditions and (b) under atmospheric temperature.

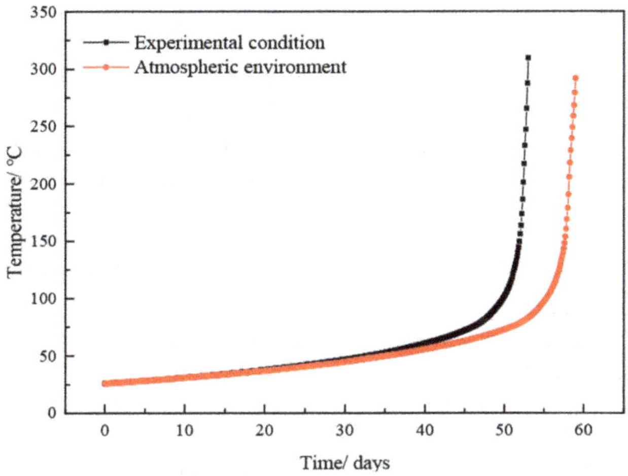

Figure 4: The characteristics of rising coal temperature in various heat dissipation conditions.

was a high demand for the conditions of heat accumulation at the low-temperature stage. In other words, as a result of increased heat exchange, the heat, which came from low-temperature oxidation, was unable to accumulate to make the temperature rise, which reduces the probability of spontaneous combustion of the coal. However, heat dissipation conditions have little impact on the temperature of the fire source at the high temperature stage.

3.2 Air quantity and direction

The air quantity was set as $0.5Q$ and $2Q$ in gas flow equations respectively; the temperature profiles on various air quantities are shown in Figure 5.

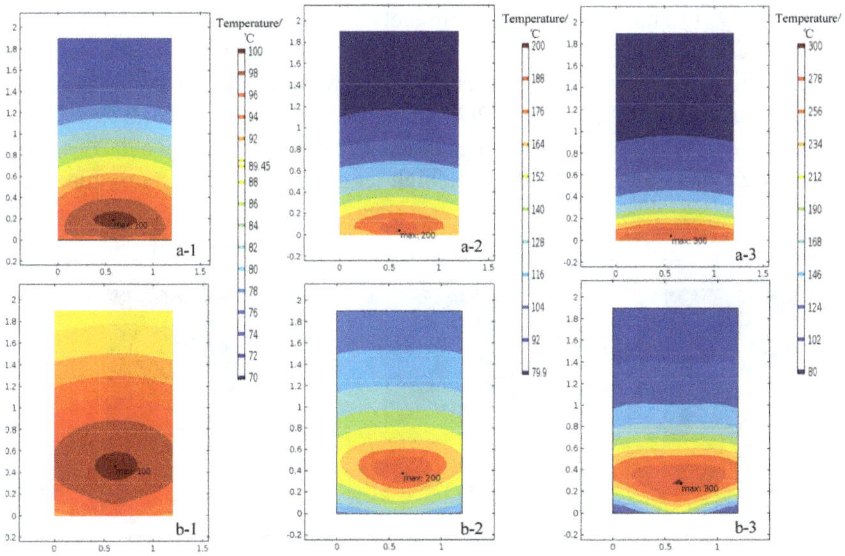

Figure 5: The temperature profile in various air quantities; (a) $0.5Q$ and (b) $2Q$.

Comparing Figure 5 with Figure 3a, at the same temperature stage, changing the air quantity would lead to different fire source locations. In Figure 5a the temperature was about 200°C when the high temperature zone gradually moved to the air surface; a similar case was 300°C in Figure 3a. This indicated that the greater the air quantity, the higher the temperature when the fire source arrived at the air surface. Such movement of fire source can be explained by oxygen starvation when there is a serious increase in the coal's oxygen consumption.

Meanwhile, as the air quantity increased, the coal temperature gradient decreased, apparently at an identical temperature stage. This means that an enhanced diffusion of air would increase the proportion of fire source zones significantly. The above results illustrated that at various temperature periods, air quantity was another critical factor affecting the fire source position.

To study the relationship between air direction and fire source position, the air inlet position is set at the middle or edge in the boundary conditions respectively. The temperature profile from simulated results can be seen in Figure 6.

Comparing Figure 6 with Figure 3a revealed that, when the air inlet was set at the border, because of less oxygen consumption in the low-temperature oxidation process, the driving force of the airflow may cause a slight shift of the high temperature zone in the reverse direction of the air inlet. With the temperature increase, the air direction did not have a significant impact on fire source distribution. However, above 200°C, the wind direction evidently promoted the radial diffusion of the fire source zone, and a larger fire area could be seen in Figure 6. The reason for this was that different air inlet positions brought various air leak intensities at the same position and then affected the development of the fire source position.

The fire source temperature in various air quantities during the coal's self-heating is shown in Figure 7a, meanwhile Figure 7b presents the fire source temperature difference for two air directions (middle subtract boundary) at the same time.

Figure 7a showed that the improvement in air quantity within certain limits will accelerate the occurrence of spontaneous combustion. However, a significant increase of air quantity would result in the heat dissipated completely, and then coal temperature remains stable throughout. When the highest temperature was below 70–80°C, the coal heating rate for various air quantities (0.5Q, Q and 2Q) was approximately equal. Once above 70–80°C, the coal's oxygen requirement increased significantly; There was an obvious influence on the heating rate of the fire source if the air quantity changed at this stage. Meanwhile, the temperature difference for various air quantities (0.5Q, Q and 2Q) increased remarkably. The above phenomenon indicated that the main reasons that improving air quantity will make the spontaneous combustion period shorten were obviously accelerating the high-temperature oxidation reaction of coal, and

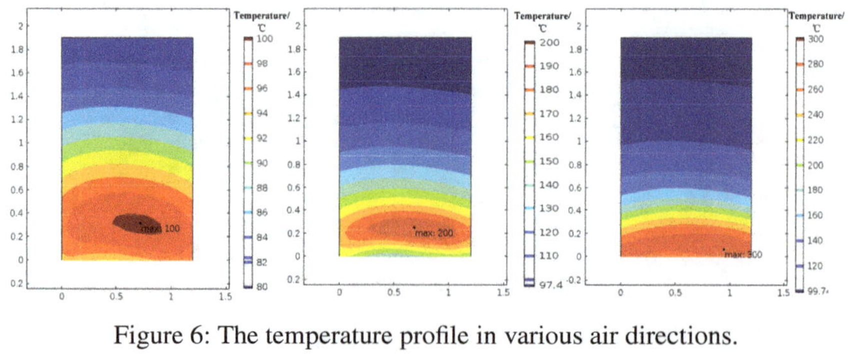

Figure 6: The temperature profile in various air directions.

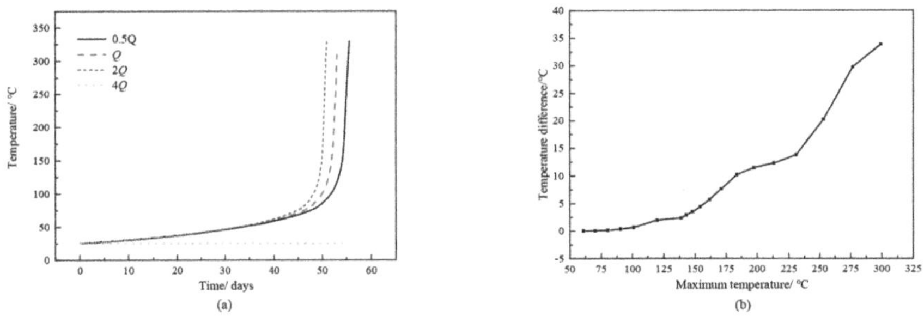

Figure 7: (a) The characteristics of coal self-heating in various air quantities. (b) Fire source temperature difference for two air directions.

shortening the period from the critical temperature to the combustion temperature. Furthermore, in this study, the optimum air quantity for the coal's self-heating test could be in the range of 2*Q*–4*Q*.

Based on experimental conditions, at the same reaction time, the temperature at the middle air inlet position was always higher than at the sides during the computational process. In Figure 7b, the temperature difference between the middle and the sides was remarkably increased when the fire source temperature was above 100°C. The above phenomenon indicated not only that the spontaneous combustion period of the middle was shorter than that of the sides, but it also revealed that the middle air inlet will generate more heat in the same period of time in consequence of better contact between the coal particle and oxygen; however, as the result of the total heat from oxidation was small, the influence of the air inlet position on the spontaneous-heating process of the coal was not obvious.

4 CONCLUSIONS

A numerical method is presented to reproduce the spontaneous-heating process of coal in a large-scale experimental reactor. This approach has been validated by comparison between simulated and experimental results of the Tianchi coal sample. The model is capable of predicting fire source temperature and position accurately at different temperature stages. Simultaneously, the model solves the problem that it is impossible to change the experimental conditions to perform repetitive experiments for the same coal, and it can also reduce the experimental workload and investigate spontaneous combustion of coal based on a more complex environment.

In addition, the model can be applied to simulate the more probable external environment. Firstly, we can summarize from simulated results that the stronger heat dissipation surroundings and the more air quantity, the higher the temperature when the fire source arrives at the air surface; the greater the temperature gradient, the more obvious fire source zone will focus on the central axis area. Thus, the air direction will have a slight impact on the fire source position during the low temperature oxidation process, and the air leakage conditions have a main influence on the heating rate of the high temperature stage. Finally, the increased heat dissipation will significantly prolong the process of low-temperature oxidation.

These conclusions can help to predict the fire source position and temperature in the actual coal mine environment.

ACKNOWLEDGEMENT

The authors are grateful for the support from the Key Projects of the National Natural Science Foundation of China (No. 5113-4019), and the National Natural Science Foundation of China (No. 5140-4195).

REFERENCES

[1] Song, Z. & Kuenzer, C., Coal fires in China over the last decade: a comprehensive review. *International Journal of Coal Geology,* **133**, pp. 72–99, 2014. https://doi.org/10.1016/j.coal.2014.09.004

[2] Taraba, B. & Michalec, Z., Effect of longwall face advance rate on spontaneous heating process in the gob area – CFD modelling. *Fuel*, **90**, pp. 2790–2797, 2011. https://doi.org/10.1016/j.fuel.2011.03.033

[3] Song, Z., Zhu, H., Jia, G. & He, C., Comprehensive evaluation on self-ignition risks of coal stockpiles using fuzzy AHP approaches. *Journal of Loss Prevention in the Process Industries*, **32**, pp. 78–94, 2014. https://doi.org/10.1016/j.jlp.2014.08.002

[4] Yuan, L. & Smith, A.C., Experimental study on CO and CO_2 emissions from spontaneous heating of coals at varying temperatures and O_2 concentrations. *Journal of Loss Prevention in the Process Industries*, **26**, pp. 1321–1327, 2013. https://doi.org/10.1016/j.jlp.2013.08.002

[5] Qin, Y., Liu, W., Yang, C., Fan, Z., Wang, L. & Jia, G., Experimental study on oxygen consumption rate of residual coal in goaf. *Safety Science*, **50**, pp. 787–791, 2012. https://doi.org/10.1016/j.ssci.2011.08.033

[6] Arisoy, A. & Beamish, B., Reaction kinetics of coal oxidation at low temperatures. *Fuel*, **159**, pp. 412–417, 2015. https://doi.org/10.1016/j.fuel.2015.06.054

[7] Taraba, B. & Pavelek, Z., Investigation of the spontaneous combustion susceptibility of coal using the pulse flow calorimetric method: 25 years of experience. *Fuel*, **125**, pp. 101–105, 2014. https://doi.org/10.1016/j.fuel.2014.02.024

[8] Wang, D., Dou, G., Zhong, X., Xin, H. & Qin, B., An experimental approach to selecting chemical inhibitors to retard the spontaneous combustion of coal. *Fuel*, **117**, pp. 218–223, 2014. https://doi.org/10.1016/j.fuel.2013.09.070

[9] Deng, J., Xiao, Y., Li, Q., Lu, J. & Wen, H., Experimental studies of spontaneous combustion and anaerobic cooling of coal. *Fuel*, **157**, pp. 261–269, 2015. https://doi.org/10.1016/j.fuel.2015.04.063

[10] Xia, T., Zhou, F., Liu, J., Kang, J. & Gao, F., A fully coupled hydro-thermo-mechanical model for the spontaneous combustion of underground coal seams. *Fuel,* **125**, pp. 106–115, 2014. https://doi.org/10.1016/j.fuel.2014.02.023

[11] Zhu, H.-q., Song, Z.-y., Tan, B. & Hao, Y.-z., Numerical investigation and theoretical prediction of self-ignition characteristics of coarse coal stockpiles, *Journal of Loss Prevention in the Process Industries*, **26**, pp. 236–244, 2013. https://doi.org/10.1016/j.jlp.2012.11.006

[12] Taraba, B., Michalec, Z., Michalcová, V., Blejcha, T., Bojko, M. & Kozubková, M., CFD simulations of the effect of wind on the spontaneous heating of coal stockpiles. *Fuel*, **118**, pp. 107–112, 2014.
https://doi.org/10.1016/j.fuel.2013.10.064
[13] Yuan, L. & Smith, A. C., CFD modeling of spontaneous heating in a large-scale coal chamber. *Journal of Loss Prevention in the Process Industries*, **22**, pp. 426–433, 2009.
https://doi.org/10.1016/j.jlp.2009.02.016
[14] Krajčová, M., Jelemenský, L., Kiša, M. & Markoš, J., Model predictions on self-heating and prevention of stockpiled coals. *Journal of Loss Prevention in the Process Industries*, **17**, pp. 205–216, 2004.
https://doi.org/10.1016/j.jlp.2004.02.002
[15] Zhu, W. C., Wei, C. H., Liu, J., Qu, H. Y. & Elsworth, D., A model of coal–gas interaction under variable temperatures. *International Journal of Coal Geology*, **86,** pp. 213–221, 2011.
https://doi.org/10.1016/j.coal.2011.01.011
[16] Saboorian-Jooybari, H. & Pourafshary, P., Significance of non-Darcy flow effect in fractured tight reservoirs. *Journal of Natural Gas Science and Engineering,* **24**, pp. 132–143, 2015.
https://doi.org/10.1016/j.jngse.2015.03.003
[17] Wessling, S., Kessels, W., Schmidt, M. & Krause, U., Investigating dynamic underground coal fires by means of numerical simulation. *Geophysical Journal International*, **172**, pp. 439–454, 2008.
https://doi.org/10.1111/j.1365-246x.2007.03568.x
[18] Song, Z.-y., Zhu, H.-q., Tan, B., Wang, H.-y. & Qin, X.-f., Numerical study on effects of air leakages from abandoned galleries on hill-side coal fires. *Fire Safety Journal*, **69**, pp. 99–110, 2014.
https://doi.org/10.1016/j.firesaf.2014.08.011
[19] Zhang, J., Kuenzer, C., Tetzlaff, A., Oertel, D., Zhukov, B. & Wagner, W., Thermal characteristics of coal fires 2: Results of measurements on simulated coal fires. *Journal of Applied Geophysics,* **63**, pp. 135–147, 2007.
https://doi.org/10.1016/j.jappgeo.2007.08.003
[20] Hu. W., *Study on Experimental and Numerical Simulation of Coal Self-ignition Process*, Xi'an University of Science & Technology, pp. 40–64, 2003.
[21] Lohrer, C., Schmidt, M. & Krause, U., A study on the influence of liquid water and water vapour on the self-ignition of lignite coal-experiments and numerical simulations. *Journal of Loss Prevention in the Process Industries*, **18**, pp. 167–177, 2005.
https://doi.org/10.1016/j.jlp.2005.03.006
[22] Jianming, W., *Investigation and Application of New Techniques and Air-leak Sealants for Coal Spontaneous Combustion Prevention and Extinguishment*, Tai yuan University of Technology, pp. 44–50, 2008.

ASSESSING VULNERABILITIES AS A STEP TOWARD CLIMATE CHANGE INDUCED HAZARD PREPAREDNESS

HARDY PUNDT, ANDREA HEILMANN & MARTIN SCHEINERT
Harz University of Applied Sciences, Wernigerode, Germany.

ABSTRACT

Increasingly, the consequences of climate change are recognized not only on a national, but also on the regional and local levels. More and more local administrations ask if and which measures should be implemented to be prepared concerning climate change induced hazards, such as flooding, soil erosion, or drought and heat periods in rural and/or urban environments.

Within the framework of a project carried out between 2013 and 2016, a local climate change adaptation strategy has been developed in a pilot region in middle Europe. Taking into account as many stakeholders, or actors from different sectors as possible, measures to adapt to climate change were defined based on the previous assessment of specific vulnerabilities. However, vulnerability assessment has been supported by the analysis of vast amount of spatial datasets using online geographic information services that were implemented as part of the project. Based on such technologies, as well as a web-based open forum, actors and the public were enabled to participate actively in the vulnerability assessment and especially concerning the definition of climate change adaptation measures. The participation process under explicit consideration of diverse relevant actors has lead to improved acceptance, and therefore more sustainable decisions about measures.

The benefits resulting from using open participation tools, including geographic information technologies and communication support, to identify and evaluate vulnerabilities will be discussed. This is linked to the goals of a follow-up project that starts in 2017, called 'BebeR', in which a special focus is on soil erosion due to increasing heavy rainfall events accompanied by flooding. The computer based support during the prioritization and implementation of measures to mitigate potential threats will be considered and conclusions be drawn.

Keywords: climate change adaptation, GIS, hazard preparedness, participation, vulnerability.

1 INTRODUCTION

Soil is an important natural resource. The erosion of soil consequently leads to a severe loss of this resource, thus having consequences for the economic prosperity of regions, as well as for natural sustainability. Besides that, soil erosion can lead to the destruction of infrastructures and hazardous situations for people. Within the framework of the project 'BebeR', the development of an adapted methodology to assess soil erosion and the consequences of adaptation measures is envisaged. The methodology will be tested and improved by different concerned parties.

The development, assessment and implementation of measures to minimize the erosion of soils require different steps, among others the identification of areas that are vulnerable to soil erosion. Soil erosion can be divided into surface erosion and riverbed erosion and caused either by wind or by water. To achieve a solid assessment of vulnerability, different methods and tools are needed that lead to reliable results on which future measures can be grounded. Until recently, the assessment of soil erosion risks on regional and national scales has been done based on the application of empirical models like the Universal Soil Loss Equation (USLE) and the Revised Universal Soil Loss Equation (RUSLE), often in combination with remote sensing and geographic information systems (GIS) [1]. However, the estimation of the potential loss of soil is only one aspect within a more complex framework that is needed to (a) identify vulnerabilities, (b) define measures, based on the vulnerability assessment and

DOI: 10.2495/SAFE-V7-N2-137-146

(c) evaluate whether such measures are effective. To achieve usable results in the areas (a) to (c), the integration of different methods and tools is necessary. Within the project 'Klimpass–Aktiv' (2013–2016) a methodology has been developed that enables local and regional authorities to generate specific, tailored to their needs, vulnerability maps as a prerequisite for the definition of measures to mitigate potential consequences of climate change. The methodology for the development of such maps considers vulnerability assessment as a process of continuous improvement. The methodology builds a basis for the project 'BebeR' that is aimed at the adaptation of existing, and the development of modified or new approaches that explicitly take into account experiences as well as expectations of all relevant actors from different concerned sectors like public authorities, agriculture, nature protection and water management. Therefore, the goal of 'BebeR' is to demonstrate, how planning processes aiming at the minimization of soil erosion can be optimized under consideration of requirements resulting from climate change. The project is focusing on rural, mountainous areas in which soil erosion can be a severe problem due to high gradients and slopes. The main envisaged result of the project, a practical and sustainable methodology to reduce soil erosion in view of future climate change induced weather events, should be transferable to other regions, thus making 'BebeR' a pilot project in this field of research and development.

2 VULNERABILITY ASSESSMENT

2.1 Methodology of vulnerability assessment

Vulnerability assessment is a complex task and requires the analysis of at least three factors: sensitivity, exposure and adaptive capacity, as shown in Figure 1. Whereas the sensitivity analysis requires insight into the environmental and socio-economic characteristics in a affected region, the exposure deals with the current and projected climate data in the region and its river catchments. Climate data are provided by simulation models such as WETTREG, REMO or STAR [2]. Using GIS, the layers (climate as well as socio-economic data) can be integrated easily aiming at making transparent statements about the situation of a region. The adaptive capacity, however, deals with the local and regional possibilities to

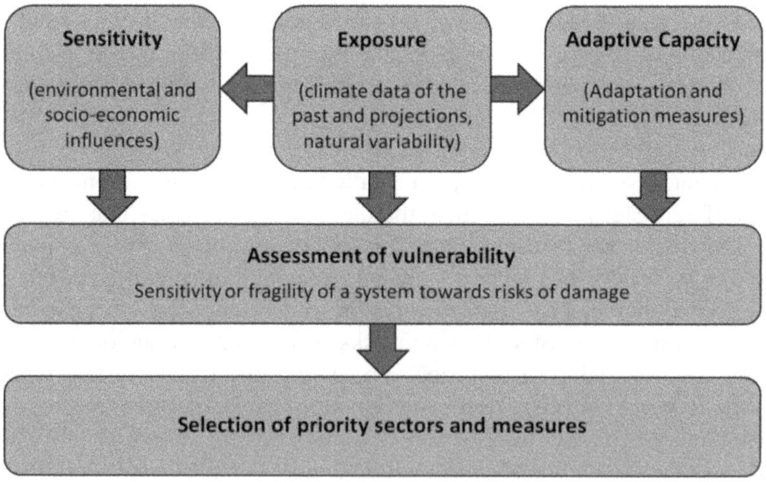

Figure 1: Determination of vulnerable areas (Source: [3], p. 10, modified).

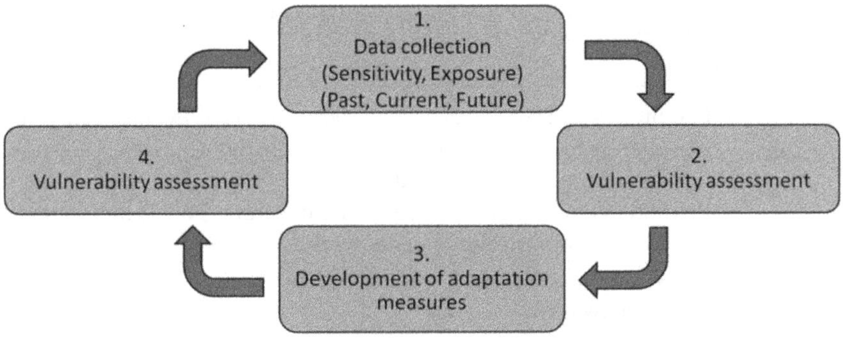

Figure 2: Step-by-step approach for vulnerability assessment.

implement adaptation measures concerning specific factors, in this case soil erosion. Different perspectives and experiences of the concerned actors have to be considered. The assessment of vulnerability, carried out under consideration of these three elements (sensitivity, exposure and adaptive capacity), leads to a solid grounding for the identification of the sectors that are prone to take action, and the definition of concrete measures to be implemented in the future in order to be prepared in case of hazardous situations.

A step-by-step procedure to determine vulnerability is presented in Figure 2. During data collection (step 1) the available data are compiled. Sensitivity and exposure are closely related. By using simulation models the effect of both, climate data as well as land use, agricultural technologies and/or new artificial elements (like storm water sewer) on soil erosion can be analysed. For the purpose of testing and model calibration, historical data are used. This will also lead to first assumptions concerning vulnerabilities (step 2). In order to assess the influence of adaptation measures (for instance on soil erosion), these have to be determined at first. The development of adaptation measures needs indispensably cooperation and cross-sector networking (step 3). The measures influence the adaptation capacity and finally the vulnerability assessment (step 4). Since data describing sensitivity as well as exposure have to be updated periodically and adaptation capacity is changing depending on planned measures it is essentially visible that vulnerability assessment is a process of continuous improvement.

2.2 Assessment of soil erosion vulnerability

When soil erosion is in the focus of vulnerability assessment, the RUSLE (Revised Universal Soil Loss Equation) is often used to predict the potential loss of soil [4]. The equation is based on the Universal Soil Loss Equation (USLE); however, the RUSLE model is a more adequate approach within the framework of the case study, which is presented in section 3. USLE enables users to predict soil erosion with acceptable accuracy [5], whereas RUSLE is more accurate in forecasting soil erosion. It is an evolution of the USLE equation, as it considers upstream areas contributing to downstream surface runoff. Thus, it is a model with better predictive ability. To assess the vulnerability of soils concerning erosion in river basins the RUSLE has been used many times and the usage of GIS to support vulnerability assessment has been documented in diverse publications (e.g. see Refs. [6–10]).

Within the framework of the above mentioned project 'Klimpass–Aktiv' vulnerability assessment has been carried out initially on a qualitative basis. The usage of different spatial

Table 1: Data sets required for the assessment of soil vulnerability.

Data category	Examples
Exposure	• Parameters measured during weather stations in the past (e.g. air temperature, precipitation totals and sunshine duration) • regional averages of these parameters • Predictions on a regional level using different models and scenarios • other observations (e.g. phenological data)
Sensitivity	• population and population development • stream network • land use (e.g. agriculture, forestry, infrastructure) • applied technologies (e.g. in agriculture) • critical infrastructure (water supply)
Adaptation capacity	• adaptation measure (implemented and planned) • Information, Awareness • Planning processes, co-operation, networking, etc. • Incentives

data layers and their integration using GIS led to vulnerability maps that enable users to get insight into areas that are threatened by harmful phenomena, e.g. flooding, or erosion. RUSLE has not been used so far but is seen as essential in the follow-up project 'BebeR'. Following the elements of Figure 1, and besides the geographic description of the region, the following data are required for the assessment of soil erosion vulnerability (Table 1).

During 'BebeR', the GIS-based intersection of different data layers will be combined with an adapted, model driven simulation approach to improve the identification of vulnerabilities to soil erosion in the areas under investigation. The envisaged methodology uses models to simulate and predict soil erosion (including RUSLE) based on scenarios. It integrates climate data resulting from future predictions (using results from models such as WETTREG, REMO, STAR) as well as data summarized in Table 1. Some insights into components of the methodology are described in [11, 12]; however, an effective integration into a methodological, stepwise and practical approach is foreseen as one achievement in 'BebeR'. The steps that will be carried out to end up with a concrete assessment of soil erosion vulnerability are mentioned below. It is important to note that the inclusion of the relevant actors is vital to achieve sustainable evaluations about the threats. Some components of the methodology to be developed are as follows:

a. Modification of an existing model to calculate the discharge in small rivers, called SMINF, including the RUSLE. For the latter, a GIS-based version is available (called ABAGis, as described by Reinstorf and Köhn [11]).
b. Generation of a sequential model chain (including SMINF and ABAGis) for the analysis and estimation of surface- and river-bed erosion.
c. Parameterization, calibration and validation of the models; sensitivity analysis of the models under usage.
d. Investigation of the erosion behaviour in the test river basins (current situation, current sensitivity).
e. Development of scenarios concerning soils and land use under consideration of climate change (exposure).

f. Analysis of the erosion behaviour in the test river basins under explicit consideration of climate change predictions; using the adapted soil erosion equations.
g. Simulation of impacts caused by planned adaptation measures (adaptation capacity).
h. Generation of vulnerability maps using GIS, thus representing different possible climate scenarios as well adaptation measures (which result in different soil erosion rates).

A main focus with respect to vulnerability assessment should be brought to the co-operation of the relevant stakeholders to identify a sustainable and economic set of measures particularly concerning the aspects d to h. By using the same data, it may happen that the assessment of vulnerability and the potential adaptation measures differ. In order to increase the adaptation capacity and the hazard preparedness the usage of vulnerability maps requires training of users. This is another reason for intensive networking cooperation and exchange of experiences.

3 CASE STUDY FOR SOIL EROSION ASSESSMENT SUPPORTED BY INTERACTIVE VULNERABILITY MAPS AND WEB TOOLS

3.1 Description of the model region

The pilot region for the development of vulnerability assessment tools is the administrative district of Mansfeld-Südharz. This district is situated in the south of the German federal state of Saxony-Anhalt and covers an area of nearly 1450 km^2. Land use in the rural areas is dominated by agriculture and forestry.

The administrational authorities of Mansfeld-Südharz have been working on local adaptation measures since 2011. The work was divided into two phases. The goal of the first phase ('Klimpass', 2011–2012) included the compilation of a local adaptation strategy. Within the second project phase ('Klimpass–Aktiv', 2013–2016), the advancement, implementation, and evaluation of specific adaptation measures were envisaged. The development of interactive vulnerability maps and the networking of relevant actors was a key element. Seven fields of action were identified as being of particular importance within the model region:

1. Reduction of soil erosion in agriculturally used areas
2. Change of the management of (small) rivers
3. Concept to dealing with invasive neophytes
4. Adaptation of forest management
5. Improvement of disaster management (especially in conjunction with extreme weather events)
6. Change of administrational methods to include climate change adaptation in urban planning.
7. Adaptation of water provision and wastewater removal.

Soil erosion and the management of small rivers are closely connected. Two river catchments have been selected as exemplary 'pilots' within the 'BebeR' project. The catchment of the 'Regenbeek' represents an area with partly high riverbed erosion, which is caused by strong current after snow melting periods or heavy rainfall. This kind of erosion leads to the destruction of the riverbanks and an unusually deepening of the riverbed. The soil or mud flows to underlying villages or cities pollutes, and potentially destroys, existing infrastructures or buildings (Fig. 3). The river 'Vietsbach' instead, represents an area with predominant surface erosion. Heavy rainfall can lead to anomalous bank erosion. This is accompanied by extensive erosion of surrounding arable land.

3.2 Development of vulnerability maps and the usage of web services

The participation of relevant actors from different sectors and administrative levels is a prerequisite for a decision making process aimed at the implementation of sustainable adaptation measures.

Therefore, web services or electronic workspaces for networking across governmental and non-governmental bodies, and the public are required, as Brennan *et al.* mentioned [13]. These tools help to overcome diverse barriers, thus allowing stakeholders to:

- access the information necessary,
- provide their own input and
- engage in a constructive discourse with other actors.

Figure 4 shows the interplay of components that are supported by GIS- and web services within the 'BebeR' project. It includes three main elements: simulation, participation and

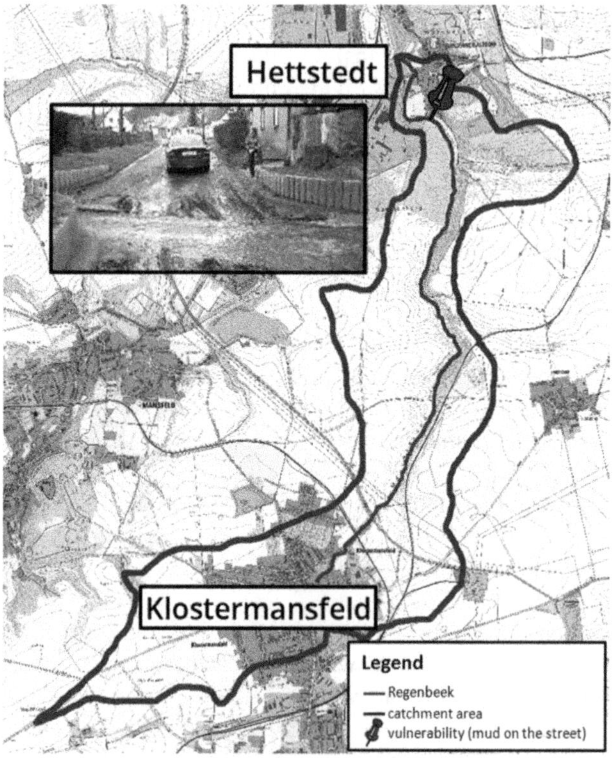

Figure 3: River catchment 'Regenbeek' and hazardous situation caused by erosion.

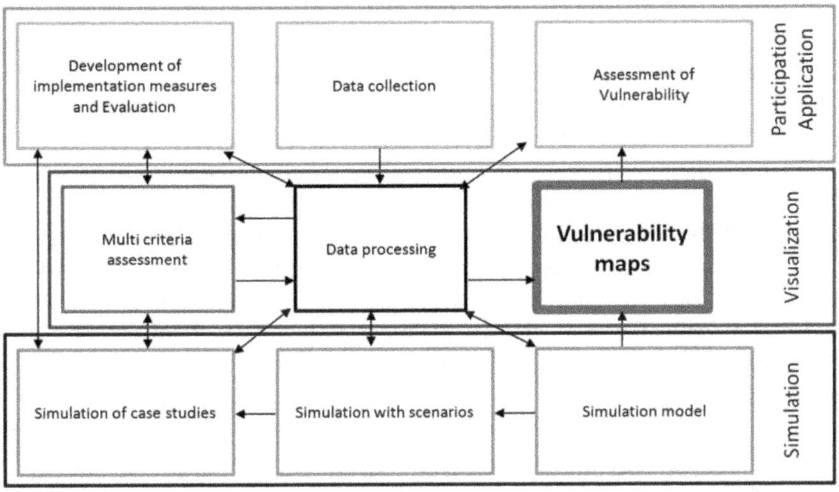

Figure 4: Interplay of components, supported by computer-based GIS- and web-services, as envisaged in the project 'BebeR'.

visualization. The simulation tools will be adapted to the pilot river systems. The goal is to simulate the erosion behavior of the past and calibrate the models based on historical events. Based on this, scenarios will be developed which should help to get insight into potential changes concerning the future erosion behavior under explicit consideration of climate change induced weather events. Scenarios consider different climate projections whereas the simulation of case studies also takes into account possible adaptation measures. Simulation utilizes data that are gathered either from official sources or from stakeholders involved. Besides data collection the web tool should also encourage an open discussion about potential measures as well as about the vulnerability assessment. In such a way, the tool enables active participation of different relevant actors and the proper consideration of their perspectives concerning soil erosion problems.

This methodology promotes a multidisciplinary, cross-sector and transparent approach supported by the mentioned computer tools. Aiming at evaluating and prioritizing adaptation measures the methodology will lead to more comprehensible decisions due to the inclusion of possibly all relevant stakeholders and the improved opportunities for their interaction via web-based services. The decisions will lead presumably to different options envisaging a possibly optimal adaptation to climate change. The occurring alternatives must be prioritized using a transparent methodology. A multi-criteria evaluation of the different outcomes can therefore contribute to a generally accepted prioritization.

The tool 'PRIMATE', a participative and multi-criteria evaluation method, has been used in the foregoing project. It proofed to be suitable for such evaluation purposes. Within the new project, PRIMATE will be used comprehensively and in close cooperation with the actors to achieve results that are accepted by at least most of the included parties [14].

Finally, the documentation is aimed at providing a methodology that can help stakeholders in other regions to act equally or at least similarly. The documentation includes a guide to support the analysis of spatial data layers and the development of vulnerability maps. The latter can be produced by an interactive, GIS-based online mapping system that has been implemented during the pioneer project 'Klimpass–Aktiv'.

3.3 Participation of stakeholders and the public

The actors' involvement in the development of adaptation measures to improve climate change induced hazard preparedness will vary with their roles. These can be distinguished into stakeholders involved (like representatives from authorities, forestry or agriculture) and the public. Stakeholders involved work according to a specific schedule and have designated tasks to fulfill (Fig. 4 – participation/application).

The public should be addressed in order to provide local knowledge, but foremost to increase its awareness. It is possible that important pieces of knowledge can complete the database. It is often the case that local citizens have experiences about facts and processes that can contribute to assess more critically alternative adaptation measures. The involvement of people in scientific and decision-making processes is nowadays often called 'Citizen science', which in such a sense offers great opportunities for climate adaptation [15].

The public should also be included to increase its willingness to implement measures for self-protection (e.g. defense of threats to their estates regarding mud flows from near slopes, or flooding from close rivers).

The online web services support broad participation: users can get insight into the environmental situation of the pilot region; they can also produce maps interactively. Such maps can be published via the project website and be discussed and modified in cooperation with other stakeholders. As an add-on, a discussion forum will be available via which users can critically review planned adaptation measures. The forum is also important to include the public and guarantee more transparency in view of the future scenarios. Citizens can also argue for other or possibly more suitable decisions taking into account that citizens sometimes have important knowledge about local conditions, which are prone to be ignored by administrational units. Apart from the web based participation tools, both stakeholders and the public will be informed using traditional communication instruments like newspapers, workshops and exhibitions.

4 CONCLUSIONS

The project 'BebeR', started in January 2017, is based on comprehensive experiences with a cooperative project management approach and online GI-services developed in the foregoing pioneer project 'Klimpass–Aktiv' (2013–2016). During the latter, concrete measures to adapt to climate change were determined and are nowadays under implementation. In 'BebeR', the specific problem of soil erosion is in the focus. The assessment of soil erosion vulnerability requires comprehensive datasets from different sectors, rich expertise from specialists, as well as local knowledge from citizens. Only the proper consideration and integration of data, knowledge and practical experiences will result in adaptation measures that are widely accepted. During this project, running until 2019, a methodology will be developed that enables the development, evaluation and prioritization of measures to mitigate hazardous soil erosion. An online and interactive GIS tool will be used to analyse spatial data and visualize relevant data layers aiming at assessing vulnerabilities and presenting threatened areas. Also, scenarios where and which alternative adaptation measures can be implemented will be presented cartographically using the GI-services. The cooperative weighting of such alternative scenarios including not only the stakeholders, but also the public, will be accompanied by using adequate evaluation methods, supported by an online discussion forum. In such a sense, different computer based tools will be used to enable a participative and cooperative approach: the modeling and simulation program, the GI-services, and the online forum. Vulnerability maps, and the forum, will support the critical discussion between project participants, thus envisaging an open dispute about alternative adaptation measures. This contributes to more transparency. The evaluation of adaptation measures will ensure a prioritization that is accepted by most stakeholders. In such a way, the methodology that is envisaged should contribute to minimizing surface and riverbed erosion sustainably or at least to being prepared in case of hazardous events.

REFERENCES

[1] Sotiropoulou, A.M., Alexandridis, T., Bilas, G., Karapetsas, N., Tzellou, A., Silleos, N. & Misopolinos, N., A user friendly GIS model for the estimation of erosion risk in agricultural land using the USLE. In: *Proceeding of the International Conference on Information and Communication Technologies for Sustainable Agri-production and Environment*, eds M. Salampasis & A. Matopoulos, pp. 795–801, 2011.

[2] Wegehenkel, M., Heinrich, U., Jochheim, H., Kersebaum K.C. & Röber, K., Evaluation of three different regional climate change scenarios for the application of a water balance model in a mesoscale catchment in Northeast Germany. *Advances in Geosciences*, **27**, pp. 57–64, 2010.
https://doi.org/10.5194/adgeo-27-57-2010

[3] Schmidt, C., Klimaanpassung auf Regionaler Ebene am Beispiel der Region Westsachsen und Oberlausitz-Niederschlesien. In: Lehr- und Forschungsgebiet Landschaftsplanung der Techn. Univ. Dresden eds, Klimaanpassungsstrategien in der Landschafts- und Raumplanung, Band 2.

[4] NSERL (National Soil Erosion Laboratory, Purdue University Indiana), 2017. Revised Universal Soil Loss Equation, Version 2 (RUSLE2), available at: http://fargo.nserl.purdue.edu/rusle2_dataweb/RUSLE2_Index.htm (accessed 23 January 2017).

[5] Beskow, S., Mello, C.R. & Norton, L.D., Soil erosion prediction in the Grande River Basin, Brazil using distributed modeling. *CATENA*, **79**(1), pp. 49–59, 2009.
https://doi.org/10.1016/j.catena.2009.05.010

[6] Biswas, H., Raizada1, A., Mandal, D., Kumar, S., Srinivas, S. & Mishra, P.K., Identification of areas vulnerable to soil erosion risk in India using GIS methods. *Solid Earth*, **6**(4), 1247–1257, 2015.
https://doi.org/10.5194/se-6-1247-2015

[7] Durães, M.F. & Mello, C.R., Groundwater recharge behavior based on surface runoff hydrographs in two basins of the Minas Gerais State. *Revista Ambiente & Água*, **8**(2), pp. 1–10, 2013.
https://doi.org/10.4136/ambi-agua.1127

[8] Mello, C.R., Viola, M.R., Beskow, S. & Norton, L.D., Multivariate models for annual rainfall erosivity in Brazil. *Geoderma*, **202–203**, pp. 88–102, 2013.
https://doi.org/10.1016/j.geoderma.2013.03.009

[9] Routschek, A., Auswirkungen des Klimawandels auf die Bodenerosion durch Wasser, Schriftenreihe des LfULG, Heft 29.

[10] Pradhan, B., Chaudhari, A., Adinarayana, J. & Buchroithner, M.F., Soil erosion assessment and its correlation with landslide events using remote sensing data and GIS: a case study at Penang Island, Malaysia. *Environmental Monitoring and Assessment*, **184**(2), pp. 715–727, 2012.
https://doi.org/10.1007/s10661-011-1996-8

[11] Reinstorf, F. & Köhn, J., Ursachenermittlung und Maßnahmenplanung zur Vermeidung von Bodenerosion im Einzugsgebiet des Regenbeeks, Landkreis Mansfeld-Südharz, Project Report, Univ. Appl. Sciences Magdeburg-Stendal, 2015.

[12] Scheinert, M., Pundt, H. & Heilmann, A., Climate change adaptation and interactive participation of stakeholders - first Results of the Project "KLIMPASS-AKTIV". In: J. Marx Gómez, M. Sonnenschein, U. Vogel, A. Winter, B. Rapp & N. Giesen, eds. EnviroInfo 2014, *28th Internatonal Conference on Informatics for Environmental Protection*. BIS-Verlag, Oldenburg, pp. 181–188, 2014.

[13] Brennan, J., Heilmann, A. & Pundt, H., An information systems approach to developing adaptation strategies. *Proceeding of the European, Mediterranean & Middle Eastern Conference on Information Systems 2012*, EMCIS2012, Munich, Germany, pp. 231–241, 2012.

[14] Scheuer, S., Haase, D. & Meyer, V., Exploring multicriteria flood vulnerability by integrating economic, social and ecological dimensions of flood risk and coping capacity - from a starting point view towards an end point view of vulnerability. *Natural Hazards*, **58**(2), pp. 731–751.
https://doi.org/10.1007/s11069-010-9666-7

[15] Available at: www.buergerschaffenwissen.de: Green Paper Citizen Science Strategy 2020 for Germany, 2016.

FLOOD ADAPTATION BY INFORMAL SETTLERS IN KATHMANDU AND THEIR FEAR OF EVICTION

NEERAJ DANGOL & JENNIFER DAY
Faculty of Architecture, Building and Planning, The University of Melbourne, Australia.

ABSTRACT

Informal settlements in Kathmandu are increasing in size and number. The housing demand in the city is rising due to the population growth and rural–urban migration, resulting in rising housing price. The high cost of housing means it is difficult for the low-income group to afford necessary housing. The government has not addressed the necessity of affordable housing for the low-income group. In this situation, some people of low-income build their dwellings on public land without legal title. This phenomenon has added to the number of informal settlers in the city. Most of the informal settlements in Kathmandu are located in the floodplains of rivers, putting them at flood risk. Annual monsoon season flood incidents in recent years demonstrate how these riverbank informal settlements are at risk. Informal settlers need to take initiatives themselves to reduce their flood risk as the government assistance is absent. Moreover, the government considers the informal settlements unlawful as they are built on public land without any authority. Therefore, there is always the possibility of their eviction by the government. The study investigates how informal settlers perceive their fear of eviction and how the fear influences their flood adaptation. The study pursues the qualitative approach to understand and analyse the informal settlers' fear of eviction and their flood adaptation. Semi-structured interviews were conducted in 41 houses from three informal settlements situated along the largest river of Kathmandu called the Bagmati River. The flood adaptive measures implemented in these houses were also identified. The study finds that the informal settlers can be encouraged for the flood adaptation by reducing their fear of eviction.

Keywords: fear of eviction, flood adaptation, informal settlement, Kathmandu.

1 INFORMAL SETTLEMENT

Defining informal settlements is complex in a universally accepted manner [1]. However, UN-Habitat has proposed the operational definition of a slum [2]. According the definition, a slum household is defined as a household lacking one or more of the following: improved water; improved sanitation; sufficient living area; durable housing (location and permanency of structure); and secure tenure. Among these indicators, some are more distinctly applicable than others to characterise the informal settlements in Kathmandu. For example, considering permanency of house, many houses in informal settlements can be considered not permanent because they are constructed using sacks, bamboo, plastic sheet, wood, metal sheet, mud and other temporary and semi-permanent materials (Lumanti, 2008). Some houses combine temporary and permanent material constructed with low-quality workmanship. Location wise, most of informal settlements in Kathmandu are situated at hazardous sites such as flood prone riverbanks and landslide prone steep slopes. Therefore, hazardous location and lack of permanency of houses are visible characteristics of informal settlements in Kathmandu.

Another indicator which is clearly applicable to characterise the informal settlements in Kathmandu is tenure security. All the informal settlements in Kathmandu are situated in public lands without legal title [3]. The informal settlements the study is investigating are situated at the riverbanks, which are the public lands. Informal settlers in Kathmandu are called 'Sukumbasi', which literally means 'landless people' in Nepali. Therefore, informal

DOI: 10.2495/SAFE-V7-N2-147-156

settlements or Sukumbasi settlements in Kathmandu are distinguishable based on the lack of legal tenure. Because of the lack of legality, these settlements are not recognised by the administration and they do not have the formal address.

Informal settlements are increasing in size and number in Kathmandu. The survey conducted in 2008 by Lumanti, a non-governmental organisation working with the informal settlements, shows that the informal settlements in Kathmandu expanded from 17 smaller to 40 larger settlements from 1985 to 2008 [3]. In 2008, the population of informal settlers was 12,726 as per the survey. In 2013, two informal settlers' organisations conducted another survey of informal settlements in Kathmandu. Their survey shows 24,021 settlers were living in 46 informal settlements in Kathmandu in 2013. This means the population grew almost double in 5 years period.

2 STUDY AREA AND METHOD

Among the riverbank informal settlements in Kathmandu, the study selects settlements indicated by red dots in the following Figure 1 as its study area. The red oval groups seven settlements, which are located close to each other. These seven settlements are considered in a cluster, which is named as Shantinagar taken from the name of the area. Other two settlements are Bansighat and Shankhamul. These settlements are selected as the study area because they are situated along the largest river of the city, the Bagmati River, which has the larger population of informal settlers than other rivers; and these settlements are within the boundary of Kathmandu Metropolitan City (KMC).

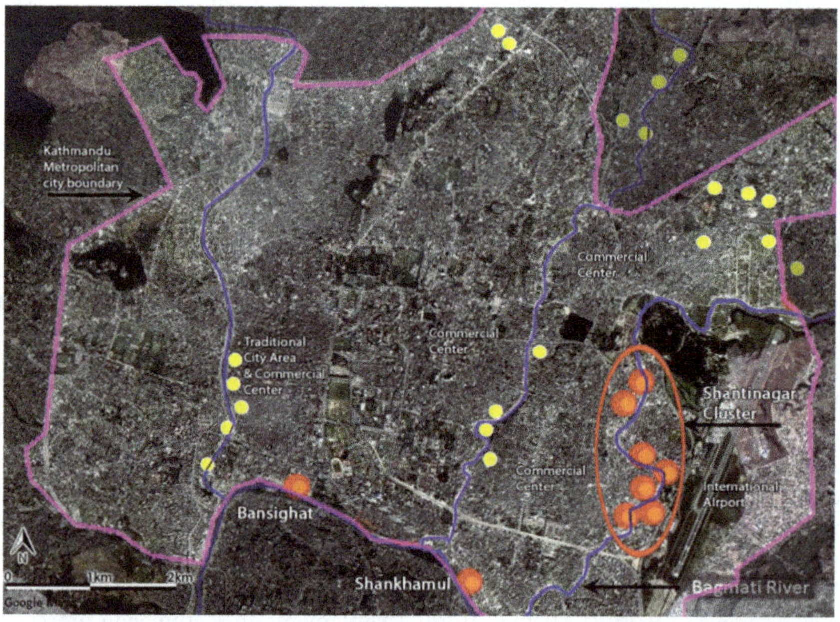

Figure 1: Location of informal settlements in Kathmandu where red area represents selected study area, and yellow dots represent other informal settlements. Blue lines are rivers and the purple line shows the boundary of Kathmandu Metropolitan City *(Source: Satellite image from Google Map).*

Table 1: Total number of houses in the settlements of study area and sample size for semi-structured interviews.

Informal Settlements	No. of houses in settlements	No. of sample houses
Shanti Nagar Cluster	693	27
Bansighat	180	7
Sankhamul	105	7
Total	837	41

(Source: Nepal Basti Basobas Samrachhyan Samaj and Nepal Mahila Ekta Samaj, 2013. They are informal settlers' organisations)

The study pursues qualitative approach to understand informal settlers' fear of eviction and to identify their flood adaptive measures. The qualitative data are collected through semi-structured interviews and observations and analysed through thematic analysis. Purposive sampling is applied for to decide the sample houses for interviewing. The location and the physical features of houses were considered for the sampling purpose to obtain variety in the resulting sample for interviews. The gender balance of interviewees was maintained as far as possible. The unit of analysis is the housing unit, which is appropriate because the paper investigates the physical flood adaptation in informal settlers' houses. The households living in the housing units were approached for the data collection.

Regarding the sample size, the study intended to continue interviewing until theoretical saturation was achieved. Forty-one semi-structured interviews were conducted to achieve the theoretical saturation in all three settlements. The numbers of sample houses from each settlement for the interview are shown in the Table 1 below.

3 FLOOD IMPACTS

Findings from interviews and observations show many of informal settlers in the study area are affected by flood incidents. Most of flood incidents occurred during monsoon season from June to September. During this season Kathmandu receives major portion of its annual rainfall causing increase in water level in the rivers (Figs 2 and 3). The rivers also collect storm water from the city through drains contributing in river level rise.

Figure 2: The increased river level after a rainfall in Shantinagar *(Source: Author)*.

Figure 3: Houses adjacent to the raised water level in Shantinagar *(Source: Author)*.

Following flood impacts are observed in the study area and learned from respondents.

- The housing structures get damaged or affected by the flood. For examples, walls made up of wooden plank, corrugated metal sheet or other lightweight materials get washed away or damaged by floor water. The wall facing to the river gets more damages. In some cases, flood also damages walls made up of robust materials such as brick or cement block. In these cases, flood water washes away soil underneath of the house weakening the foundation or support below walls.
- Several respondents shared how the flood damages their belongings every year or more than once in a year. Damages are more when the flood occurs unexpectedly because they do not get time to move their stuff in safer locations. Many of them lost their furniture, bed, mattress, clothes, utensils, groceries and textbooks. Even if they manage to prevent of washing away of their stuff by closing the doors and windows, they all get wet and damaged. Some of them even shared how they could not save their food and stayed hungry in following days.
- Many of residents, whose house got flooded, had to spend nights outside for a few days. They manage to drain out flood water in a few hours by digging waterway (channel) and bucketing up; however, it takes a few days to clean the mess caused by the flood. They need to dry the stuff which are reusable and throw other damaged ones. Additionally they also have to guard their belongings which they put outside before they get caught by the flood. Some of them also shared they lost their belongings when they put outside.
- It takes time for residents to come back to their normal life after flood incidents as they need to rearrange and manage everything affected by the flood, as explained in the previous point. During this period, they miss their working days for around a week and kids miss their school.
- Some respondents shared their experience of about to get drowned in the flood. For example, a lady in Shantinagar settlement got stuck in her house in a flood in the night in 2015 when the water level reached up to her neck. She managed to get out from the house but she hurt her leg stepping on something sharp and had to have treatment for a month. Similarly, a fourteen year old girl was trapped in the flooded house (Fig. 4) in Bansighat settlement in 2016. It also occurred in the night time when all other family member except her managed to escape from the house. Later, the police came and helped her to get out using rope. She was sick for several days because of cold and fear. An elderly man in Shantinagar shared how he ran through the roof carrying a kid as it was hard to escape from the door. These are some of examples of how residents' lives are at flood risk.
- Many residents mostly kids get cold, fever, itching skin, diarrhoea after flood incidents. The river itself is polluted because city sewerage gets mixed into the river without any treatment. Moreover, the sewerage running through the settlement also overflows because of river flood (Fig. 5). Contact with the polluted water causes waterborne and skin diseases. Similarly, getting wet in the flood causes cold and fever. Additionally, sleepless nights outside the house and anxieties make flood affected residents health more vulnerable.
- When the flood water enters into the sewer pipes in the settlements, the sewer water springs up blowing the manhole lids (Fig. 5). This worsens the flood in the settlement by adding sewer water. When flood water enters into the sewer pipes, it also causes a backflow of sewer water through toilets in the houses.

Figure 4. Houses get flooded in Bansighat *(Source: an interviewee).*

Figure 5. Flooded road in Shantinagar and stones are put on the manhole cover to prevent overflow *(Source: Author).*

4 FLOOD ADAPTATION

In defining adaptation, the literature focuses on a system's ability to cope. For example, Brooks [4], defines adaptation as 'adjustments in a system's behaviour and characteristics that enhance its ability to cope with external stresses.' Similarly, Pelling [5] defines adaptation as 'the process through which an actor is able to reflect upon and exact changes in those practices and underlying institutions that generate root and proximate causes of risks, frame capacity to cope and further rounds of adaptation of climate change.' Both definitions present system's or actor's ability to cope with external stresses or risks as an important aspect of adaptation. It is important to have adaptive capacity to implement the adaptive measures; however there is also the possibility that people may not implement the adaptive measures even they are able to do. The study identified the measures of flood adaptation implemented by informal settlers in their houses from observations and interviews. Following are the identified adaptive measures:

- Raising the ground floor (plinth) level by filling with soil, sand, stone and bricks. The ground floor level is raised depending on the past experience of the flood. Four feet to ten feet rise in the floor level was observed. When the floor level is raised, nine inch thick brick wall is constructed as the foundation or support to hold filled materials and superstructure.
- In the case of houses adjacent to the river, sandbags are piled up to protect the foundational brick wall below ground level. Piles of sandbags are supported by the bamboo and reed sticks (Fig. 6).
- Soling the floor with brick and finish the floor with cement plaster so that flooded water in the house can be drained out quickly. It also reduces seepage or dampness from the floor.
- Adding an upper floor (first floor) to protect to be safe from flood and to protect belongings.
- Raising riverside walkways by filling soil, sand and stones and supported by bamboo poles (Fig. 7).
- Changing the outer wall of the house from comparatively temporary materials to rubust ones. For example, residents who had wooden plank or bamboo wall change it to tin (CGI) wall to reduce flood water intrusion. Similarly, who had tin wall convert it into brick or cement block wall to reduce flood impacts. Changing the mud mortar brick wall of a house to cement mortar wall is also a way of changing construction material responding flood impacts.

Figure 6: walkway or land next to the river raised by pilling sandbags and filling sand and soil *(Source: Author).*

Figure 7: Houses made up by temporary materials raised and supported using piles of sandbags *(Source: Author).*

- Additional to the ground level rise, residents also raise the threshold of door or platform outside the house to prevent intrusion of flood water.
- Raising the toilet level or moving the toilet to upper floor to prevent backflow from the toilet.
- Plastering the lower level of outer wall to prevent seepage.
- Building boundary walls, if space is available to prevent from flood water.
- Raising furniture height by using bricks and stones.
- Changing CGI sheet at roof and using putting to stop leakage from the roof or use plastic sheet over the roof.
- Providing pipes to drain out flood water from or around the house.

5 FEAR OF EVICTION

Informal settlements in the study area are built on public land without any authority which is considered unlawful by the government of Nepal. The Public Crime and Punishment Act – 2027 (1970) [6] prohibits 'to trespass on any governmental or non-governmental office or anyone's building or land by committing riot; or to stay or remain there without any authority.' It clearly implies that occupying the public land is against the law and that the government has legal rights to take action against it. Consequently, the government evicted Thapathali settlement forcefully by bulldozing dwellings in May 2012 leaving 251 families nowhere to go and without any alternatives [7]. Similarly, the government has attempted several times to evict Shankhamul settlement [8] and imprisoned some of the settlers for weeks [9]. When a successful or attempted eviction occurs, it not only affects the settlement where the action occurs but also influences other informal settlers by raising their fear of eviction [10].

All respondents who participated in the interview accepted that they are living on public land and agreed that there is possibility of their eviction by the government any time. However, the study observed that they have different levels of fear of eviction even all of them have no legal title. Based on the analysis of interviews and observations, the study identified factors which affect the level of fear of eviction. These factors are community, recognition of the local administration, informal ownership and experiences of discrimination, humiliation and criminalisation.

The study found that the sense of being in the community affects the level of fear of eviction in an individual informal settler. More than two-third among 41 respondents mentioned that they have a large community with them and indicated that their belief on the solidarity of the community helps to console them when they feel the fear of eviction. For example, the researcher asked about fear of eviction, a widow living with her daughter in law and grandson said 'I always have fear of getting evicted, but I am not alone in this settlement.' She indicated that she is with community by saying she is not alone in the settlement and it is considerable that she mentioned being in community immediately to compensate her fear of eviction. Similarly, an elderly tailor who actively participates in protests against government actions towards informal settlers, emphasised on the strength of their community. He said 'We are not a few in number we have a large community and we won't keep quiet if the government uses the force to evict us without proper alternative. We are ready to face the bullet but won't leave the place.' This strong statement shows their confidence on the solidarity of the community, which supports them to minimise their fear of eviction.

More than half of respondents mentioned that recognition by the ward office (local administration) affect their fear of eviction. The administrative area of Kathmandu Metropolitan City is divided into 35 wards which are the lowest level of administrative units. Each ward has a ward office. The study observed less fear of eviction in informal settlers when the ward office recognises them. The recognition by the ward office is important for informal settlers because it gives them assurance that their residency in the ward is acknowledged. The ward office is the first place of governmental administration where residents can report their concerns. The ward office is responsible in identifying the residents in the ward and providing them a recommendation letter so that they can apply for citizenship card, for electricity connection, for water supply connection and for other various purposes. Informal settlers who received their citizenship card being local residents of Kathmandu with the help of the recommendation letter obtained from the ward office feel more secure about their residency in the settlement. For example, a respondent who got citizenship card proudly says 'I even got the citizenship card from here (referring the address of the informal settlement). I am now a resident of Kathmandu.' This indicates that they perceive having citizenship referring the address of informal settlement means they are accepted by the government as local residents of the place. This perception provides them confidence to claim that they belong to that place, which consequently diminishes their fear of eviction. In the same way, obtaining a migration certificate from the ward office also contributes in reduction of fear of eviction. Migration certificate shows the person is migrated to the place where they are living from their place of origin and it also proves that they do not own any property at the place of their origin.

The study identified that having sense of informal ownership of the house make informal settlers feel more secure. Informal settlers who have an identity card expressed higher level of sense of informal ownership. The identity card was prepared and distributed to informal settlers in 2004–2005 by Lumanti (NGO working on informal settlements) in coordination with Kathmandu Metropolitan City Office (KMC). Informal settlers living in the study area prior that time received the identity card. The purpose of the identity card was to keep record of informal settlers as mentioned by Lumanti. The informal settlers who have the identity card perceive it as an important document because it supports their recognition as genuine informal settlers who have been living there for more than ten years. A middle aged man, who is a construction worker, proudly shared that 'We have our identity card distributed 10 years ago. It has our pictures with this house in the background. We can use it to show that we are informal settlers living here in this house for many years.' He takes the identity card as an evidence to show that he owns the house and that he is living there for a long time. Similarly, the retired army said regarding owning the identity card, 'When Keshav Sthapit was Mayor; we got the identity card which shows we are living here. At that time, our lands were also demarcated and based on that the identity card includes the information....there was a program organized at nearby open area to distribute identity cards to us by then Mayor. It is actually temporary permission for us provided by KMC to live here.' He takes the identity card as an official document containing information about his family, land and house and acknowledge by then Mayor of the city. This makes him believe the document as the temporary permission of them to live in the settlement. These responses clearly show that they have sense of ownership of their land and house because they own the identity card. Therefore, owning the identity card encourage them to be less fearful to eviction.

Discriminative and humiliating behaviours by the government, society and neighbours, who have legal tenure, are identified as a contributing factor in increasing fear of eviction in informal settlers. Most of respondents share their experience of being discriminated in various forms which enhanced their fear of getting evicted. An example of discrimination by government was when the road passing between the informal settlement and the formal neighbourhood was expanded. The government took land only from the area of informal settlements. A lady who lost half of her land and a wall of her house during road expansion expresses her fear saying 'if the road gets expansion further we may lose our house because last time when the road was expanded they (the government) did not touch the formal land at the other side of road but took our land and a wall of our house. We are pushed closer to the river because of the road expansion.' This indicates how informal settlers view discriminative actions by the government on them because when a road is expanded elsewhere, generally lands from both sides are taken. Such experience makes them feel the government does not consider them as residents of the city which increases their fear of eviction.

The study found that informal settlers are criminalised by the government and formal residents for being informal settlers. Experiences of being criminalised make them guilty of being informal settlers. A middle aged lady who had such experience shared, 'When we were adding a floor in our house, people from nearby area (residents with legal title from adjacent neighbourhood) came to tell us now to do. They told us that we do not have legal title and permission to build, so we should stop it. When we did not stop, they brought police and took my husband to the police station. We already added floor at that time, so what to do? The police warned him not to do so anymore and let him go.' She also added to justify the addition of floor, 'Because of the river, we always have fear to get flooded or washed away. We cannot sleep in peace at night. So, adding a floor is necessary to protect ourselves and stuff.' Her expression shows she has fear to get punished for improving house and she is also afraid of formal residents who can complain against her. Likewise, a construction worker lady expressed her fear, 'If we try to improve our house, police van arrives. How we can feel this is ours and how can I can say I have right to live here?' In this way, experiences of being criminalised increase their fear to be punished which does not allow them to feel secure in their settlement. Informal settlers not only have fear from the state but also from formal residents living nearby neighbourhoods because they know formal residents do not like them and can complain against them to the police. Informal settlers feel that formal residents consider them as eyesore.

6 FEAR OF EVICTION AND FLOOD ADAPTATION

The study analysed and found that informal settlers who have higher level of fear of eviction do not invest for flood adaptation as much as they could do. They worried if they may lose everything if they are evicted next day. For example, a respondent said, 'We don't have money; we have to take loan to improve our house to make it safe from the flood. But if next day, we get evicted or we have to leave place, we will lose everything..... I have that fear..... government is not trustworthy, it can come any day and tell us to leave because this is not our land, this is public land. Then we will have to leave.' Therefore, they decide for adaptation carefully and do not invest at once. They prefer to invest gradually. For example, if they have to change the material of wall from tin to brick, they do one wall at a time. They may take one to two years to change all the walls.

They tend to choose a temporary adaptive measure although they need to repair or replace that measure time to time. For example, they raised the ground level at river edge by filling with sand and soil and support the filled level using piles of sandbags. These sandbags get damaged by the river and washed away and they have to be repaired or replaced time to time. They do not use stone wall to support the filled level because they hesitate to spend for long term by spending more. The reason is they are uncertain when they may get evicted by the government forcefully.

Around half of respondents indicated that the government need not to help them to protect them from the flood. They can protect themselves if the government assures them that they will not be evicted forcefully. A respondent representing this view said, 'If government allows us to live here and provide certain kind of tenure security, the government does not need to do anything for our safety from the flood. We can improve our houses by ourselves. We are ready to avoid our meal for six months and take loans to strengthen our house.'

7 CONCLUSION

The riverbank informal settlers take initiatives for the flood adaptation in absence of assistance from the government. However, they are afraid to invest for the adaptation because they have fear of getting evicted anytime. If the government can reduce this fear, informal settlers could enhance their flood adaptive measures. The fear of eviction can be reduced, if the government could assure informal settlers that they will not be displaced without any alternative provision. Such assurance can be provided by communicating with informal settlers, by issuing notice and by introducing and modifying the policies relevant to informal settlers.

ACKNOWLEDGEMENTS

Australian Government Research Training Program Scholarship, Riady Scholarship, The University of Melbourne, Anna Hurlimann and Richard Tomlinson.

REFERENCES

[1] Nuissl, H. & Heinrichs, D., Slums: perspectives on the definition, the appraisal and the management of an urban phenomenon. *DIE ERDE–Journal of the Geographical Society of Berlin*, **144**(2), pp. 105–116, 2013.

[2] Chapter 1: Development Context and the Millennium Agenda, The Challenge of Slums: Global Report on Human Settlements 2003 (Revised Chapter 1). Available at: http://unhabitat.org/wpcontent/uploads/2003/07/GRHS_2003_Chapter_01_Revised_2010.pdf (accessed 8 October 2016).

[3] Lumanti. *Status of squatter communities along the Bagmati River and its tributaries in the Kathmandu Valley.* Kathmandu: Lumanti Support Group for Shelter, 2008.

[4] Brooks, N., Vulnerability, risk and adaptation: A conceptual framework. *Tyndall Centre for Climate Change Research Working Paper*, **38**, pp. 1–16, 2003.

[5] Pelling, M., Adaptation *to climate change: from resilience to transformation*, Routledge, London, 2010.
[6] Public Crime and Punishment Act - 2027 (1970), (1970).
[7] Ghimire, B., Government evicts Thapathali squatters. *The Kathmandu Post*, 9 May 2012.
[8] Anjaria, J.S. & McFarlane, C., *Urban navigations: politics, space and the city in South Asia,* Routledge, 2011.
[9] Kathmandu Post. UN Park stalled due to squatters. *Kathmandu Post*, 17 November 2003.
[10] Paul, B.K., Fear of eviction: The case of slum and squatter dwellers in Dhaka, Bangladesh. *Urban Geography*, **27**(6), pp. 567–574, 2006. https://doi.org/10.2747/0272-3638.27.6.567

IN THE WAKE OF WINSTON – CLIMATE CHANGE, MOBILITY AND RESILIENCY IN FIJI

A. ROWAN GARD & JOELI VEITAYAKI
Centre for Pacific Studies, University of St Andrews, United Kingdom & School of Marine Studies, University of the South Pacific, Fiji.

ABSTRACT

When tropical cyclone Winston hit Fiji on 20 February 2016, it became the strongest storm recorded in the southern hemisphere and the second strongest storm ever recorded in the world. Forty-four people died as a result of the superstorm, and the ongoing governmental recovery plans have estimated nearly three billion Fijian dollars (£1.1 billion) in damages. Fijian Prime Minister Frank Bainimarama, in a national address following Winston, stated 'almost no part of our nation has been left unscarred'. This paper explores the adaptation measures for extreme weather and climate change in Fiji following Winston, as well as the author's experiences both as a survivor of Winston and as a volunteer aid-relief worker living in Suva, the Fijian capital, at the time of the superstorm. Notably, Winston struck 5 days after Fiji became the first nation to ratify the United Nations Climate deal agreed to by 195 nations in Paris in December 2015. Many Pacific island nations have played a vital role in leading the global dialogue on climate change. In the lead up to the Paris summit, Prime Minister Bainimarama, stressed that 'unless the world acts decisively in the coming weeks to begin addressing the greatest challenge of our age, then the Pacific, as we know it, is doomed.' This is undoubtedly true and a global response is desperately needed, but equally true is the importance for such a reaction to be a cultural and faith-integrated process across multiple scales, ranging from village and community to regional, in scope. One critical aspect of extreme weather and climate change response in the Pacific is migration. However, the impacts of migration include organizing, adaptation and urban poverty, a further source of vulnerability. Consequently, human mobility in response to extreme weather and climate change is far more complex than originally addressed and is deserving of deeper consideration.
Keywords: climate change, extreme weather, faith-based organisations (FBOs), fiji, mobility, resiliency, tropical cyclone Winston.

1 TROPICAL CYCLONE WINSTON

In the flickering candlelight I stare at the ceiling and wonder if the roof will peel away in the incredible winds of tropical cyclone Winston. On this particular Saturday night in Suva, the Fijian capital, the tall palm trees in the backyard crash to the ground in the devastating winds and it feels as if the world is attempting to invert itself, as Winston generates gusts of wind recorded at 200 mph and average wind speeds of 143 mph [1].

Winston was a Category 5 superstorm of unprecedented strength and destruction. Dozens of people were killed, buildings were stripped to their foundations and towns reduced to rubble, leaving many in abject poverty.

For months after Winston, families struggled to procure basic necessities, such as food and clean drinking water for their children [2]. My own neighbourhood of Suva Point was without electricity for one week after Winston. This situation was nothing when compared to rural communities which faced far worse difficulties, going weeks and months without electricity and running water [3]. Consider too, that at the time of this publication, more than one year after the event, some are still housed in tents [4]. It is clear that this is the beginning of what is to come for the Pacific.

DOI: 10.2495/SAFE-V7-N2-157-168

Figure 1: The largest Fijian island, Viti Levu, covers ~ 57% of the nation's land area and has the two largest urban centres (Suva, the capital, and Lautoka). It also hosts most of the major towns, including Nadi, the site of the international airport, and contains ~ 69% of the population (image courtesy of Google Maps; data from [5]).

2 RISING TIDES IN A CHANGING CLIMATE

Globally, more than 600 million people, or 8.6% of the world's population, live in coastal zones (zero to ten metres elevation) which comprises only 2% of the world's land area. Of these 600 million people more than half, ~ 360 million, live in urban centres. This amounts to 13% of the world's urban population living in the highest risk zones of sea-level rise [6].

Pacific Islands Countries and Territories (PICTs), especially the low-lying atoll nations, are presently experiencing the adverse effects of human-induced climate change. PICTs and their urban centres are vulnerable to extreme weather events, such as Winston.

The dense urban and economic centres of PICTs are located in coastal zones and are furthermore vulnerable to rising sea levels and other extreme events. Beyond tropical cyclones, these events include floods, tidal surges, droughts and landslides. Sea-level rise is commonly misunderstood as being a uniform process. However, it is a far more complex process caused by various factors, such as the formation of the earth's crust. Additionally, ocean temperature and salinity reflect regional variability, as well as the gravitational and rotational changes driven by the ice-ocean mass exchange [7–9]. Ultimately, this results in uneven rise and fall in overall sea-level rise, with some areas experiencing an amplified rise, especially in the equatorial region [10].

While these processes have occurred for millennia, humans have affected sea-level rise directly by way of manipulation of ground and surface water flows and indirectly through greenhouse gas emissions that increase global temperatures [11]. These warmer temperatures impact the mass and volume of seawater through increased terrestrial ice melt and warmed ocean temperatures, both of which compound rising global sea-levels.

3 STAYING AFLOAT IN FIJI'S ECONOMY

Winston severely affected the lives and livelihoods of 540,000 Fijians or 62% of the population of the island nation. An estimated 30,369 houses or 22% of households were destroyed or damaged, forcing some 131,000 people into temporary shelter. Further destruction of infrastructure included damage to 495 schools and 88 health facilities [12].

The estimated damage from the cyclone was £1.14 billion or nearly three billion Fijian dollars, roughly 31% of the gross domestic product (GDP). With such an extensive loss and the Fijian economy being based primarily on agriculture and tourism, it was unsurprising that the Fijian government issued a month-long state of emergency [13]. Early economic assessments indicated that agriculture (especially sugar that comprised one-third of industry activities) was seriously impacted [6]. Fiji Sugar Corporation Chairman, Abdul Khan, specified 'Our all up losses in the sugar cane industry add up to 163.35 million Fijian dollars. The loss of workers' livelihoods is 53.3 million Fijian dollars' [14]. Unsurprisingly, the tourism industry suffered high cancellation rates of 30%–40% immediately following Winston [15]. However, with recovery and reconstruction activities ongoing and the Government prioritising spending to address the emergency needs, the Fijian economy is anticipated to grow by 3.6% in 2017 and by 3.2% in 2018 [12].

National statistics indicate that 31% of Fiji's population lived below the poverty line prior to the devastating effects of Winston [16]. However, in urban areas, such as Suva, the statistics reflect a drop to 19% of the population, when compared with 43% in rural areas living below the poverty line. Yet, the statistics belie a more complex reality. I argue that indigenous Fijian or I Taukei lifeways be contemplated and reflected upon when seeking to understand poverty in Fiji pre- and post-Winston. Traditionally, Fijian land is held communally and all members of a village have access to support, and feed themselves from various horticultural and maritime practices. This more nuanced consideration can be summed up by Papuan New Guinean writer Martyn Namorong [17] as he discusses traditional Melanesian life and ideas of poverty: 'The environment was the source of their physical, spiritual and intellectual nourishment. While individuals had certain property rights, such as the ownership of personal artifacts of value, the land was owned communally... Is it poverty, if a rural [man] only wears astanget and does not own a laptop?... How do you define poverty and wealth in this present time when Melanesians live in two realities? We live in the reality of our ancestral land and in the reality of the modern State that exists on that land'.

Figure 2: The remains of a family home and bathroom on Qoma Island. Image courtesy of Rowan Gard.

Statistics indicate that although there are higher rates of poverty in rural areas, those communities are generally better placed to support themselves through traditional subsistence practices. Urban poverty is a pressing concern as individuals and families living in urban centres often lack the means to support themselves. Contributing to this are multiple factors, such as lack of education and skills training, and the harsh fact that 65% of those in full-time employment in Fiji are earning wages below the basic needs poverty line [18]. Subsequently, the international aid community's understanding and forward movement especially around climate change-related economic and mobility policies must take into account the pressing reality of PICT contemporary urban life, such as in Fiji.

4 AFTER THE STORM—ON THE GROUND RESPONSE

In the days and weeks following Winston, I volunteered with Fiji Red Cross and also joined a dozen colleagues at the University of the South Pacific (USP) in gathering and purchasing relief supplies for the fishing community on Qoma Island located northeast of Suva. The island was especially heavily hit and as we arrived via motor boat I stared at the palm trees, their fronds looked to be permanently bent, as if they were experiencing the catastrophe still.

After we had unloaded the relief supplies from the boats, which ranged drastically from freshly baked bread, to plastic sheeting, to a football for the children to play with, we were formally greeted by the village leaders on the concrete foundation slab that had once served as a community gathering space and village hall. We sat in a large circle and were invited to partake in a kava ceremony. Kava, also known as yaqona in Fiji, is a mild narcotic beverage known for its calming effects and is made from combining the root of a shrub, *Piper methysticum*, with water; kava is used in a range of political, religious and social ceremonies throughout the Pacific. I noted that even the large kava bowl had been damaged with one leg broken off and was propped up with a plastic cup.

Following our formal greeting, I wandered the community in a state of emotional numbness and spoke with the villagers and listened to their Winston or 'the big monster' survival stories. It was clear that the devastation was near-total on Qoma. Many homes and buildings had been stripped to concrete slabs, while other buildings, such as the Methodist Church, which had been the largest building on the island, had a roof that had collapsed and was unsafe to enter. As I surveyed the damage, I was especially touched by the story of one family that had sheltered together in the bathroom, which had partial concrete walls and thus was stronger than the rest of their home. As I spoke with the middle-aged father he shared with me that 'God kept us safe; he is always with us' and it was the love of his family that had enabled him to face the future. Drawing upon faith and family, he was already in the process of rebuilding his family's former home.

After we had spent the afternoon on Qoma, my USP colleagues and I piled back into the boats we arrived in for the short ride back to Viti Levu. On the drive back to Suva, we discussed ways in which the villagers on Qoma, and all of Fiji, could be made more resilient for future extreme weather events. In a previous fieldwork on Qoma, one colleague had encouraged the villagers to plant mangroves, as a means of stabilising their shoreline. Indeed, much of their shoreline was in far better shape than their neighbours, though the villagers' commercial seaweed gardens had been destroyed. They asked for assistance in securing more seaweed starters to replant their aqua gardens, as well as seeds for their terrestrial gardens.

The need to literally rebuild and replant is a powerful one for those who live through the trauma of an extreme weather event, and the need to positively plan for the future is vital. It occurred to me how markedly different sharing seeds and plant starts with a community was when compared to more traditional Western aid deployment, which often provided multiple rounds of canned Western food items for immediate consumption. This also reminded me of an observation my colleague Lucie Hazelgrove-Planel, who survived tropical cyclone Pam in Vanuatu noted: 'Does the food aid package as a whole unintentionally include a concept of adaptation to climate change which promotes a Western diet, lifestyle and economy?' [19]. The deeper connotations of this insight are well worth considering as Fiji continues reconstruction and future resiliency measures.

In a more immediate sense, help that comes as standard international aid isn't always necessarily helpful. As was expanded upon by Fe'iloakitau Kaho Tevi [20], a survivor of tropical cyclones Pam and Winston, and a political analyst based in Port Vila, Vanuatu: 'When a disaster happens...they send second hand clothes, they send electrical kitchen appliances to places that don't have electricity. They send high heels, to people who get them and don't have anything to do with them. After the disaster in both Vanuatu and Fiji there was food past its use by date that came. These gifts sit in containers on wharves, and we are not too sure what to do with them. Sometimes we have to sell the clothes, to get some money for the things we need".

In the days and weeks following Winston there were further issues around organising and deployment of relief assets. When speaking with friends, colleagues and community members working in relief deployment, there was more than one account shared of Fijian military personnel consuming and selling supplies intended for Winston victims. It was also known that representatives of the Fijian government had distributed faulty information associated with the national declaration of emergency and aid request, specifically some of the banking account numbers for donations had been altered so that funds were being syphoned into third-party accounts. Though this information is anecdotal, I do not doubt it. Some of these corruption charges were briefly acknowledged by the Fiji Independent Commission Against Corruption (FICAC) and the National Disaster Management Office (NDMO) in a press release in March [21]. In relation to this, it is a significant fact that the freedom of the academic and artistic communities as well as the press around issues of investigative journalism and critiques of the government in Fiji remain suppressed by censorship and full blackouts [22], as was the case in the 2014 elections [23]. There is little wonder then that I observed those seeking assistance beyond their own families and trusted community members, to be most comfortable going to their local churches, parishes and faith-based organisations.

5 THE BUOYANCY OF FAITH – THE ROLE OF FAITH-BASED ORGANISATIONS IN HUMANITARIAN AID

In the past, religion as well as the role of faith-based organisations (FBOs) have often been ignored by those involved with humanitarian aid and development. Sociologist Kurt Alan ver Beek [24] even went so far as to call it 'a development taboo' after completing a study that explored the void of research on the roles of religious organisations in development. However, in the past 20 years that stigma has lessened, and now the academic community and the wider international aid community have taken a 'religious turn' of sorts exploring in greater depth the role FBOs play in humanitarian aid [25].

Though FBOs have a long history of humanitarian work, ranging from Christian hospitals to Hindu meal programmes, a recent proliferation in religious charities, NGOs and other associations have flourished. Importantly, the World Bank has estimated that as high as 50% of all health and educational services globally are provided by FBOs [26] and further evidence for the growth of FBOs can be found in the dramatic increase in US government funding from 2001 to 2005, where in that time period an increase of 10.5% to 19.9% occurred [26]. The World Bank's 2000 study entitled *Voices of the Poor* concluded that many people in lower socioeconomic conditions demonstrated more confidence in religious organisations than other secular, governmental or military governing bodies [27].

Presently, there are some 33,500 international NGOs in the world with 3,183 being granted consultative status with the Economic and Social Council (ECOSOC) in the United Nations (UN) [28]. Embedded within Article 71, Chapter X of the UN's Charter, authority is given to 'make suitable arrangements for consultation with non-governmental organizations which are concerned with matters within its competence' [29]. This is advantageous as many FBOs often have deeper histories than other NGOs, which can in turn strengthen their connections with local communities. In times of crises, FBOs are able to build on established foundations of trust and expedite response in emergency situations. I observed this first-hand while in Fiji, in particular whilst engaging with the congregation of Saint Luke's Anglican Church and the aid they provided to the wider community following Winston. However, it should be noted that past relationships and pre-established sympathies might impact the FBOs' abilities to engage with the wider community in an unbiased and unpolitical manner given the competitive dynamics of religious organisations in some communities [30]. Nevertheless, an increased integration of FBOs at all levels of response to extreme weather and climate change may prove advantageous, as they have much to offer the communities in which they are embedded, as well as the wider humanitarian community.

6 STRINGS ATTACHED – THE POLITICS OF HUMANITARIAN AID

The international response to Cyclone Winston was significant not just from the pressing humanitarian relief aspect, but also the political influence exerted by various nations vying for Fiji's affections. Australia committed $35 million dollars as assistance to recovery, including supplies and health services to over 200,000 people. The HMAS *Canberra* and over 1,000 Australian Defence Force and humanitarian personnel were deployed with 341 tonnes of relief supplies; including 30,000 litres of drinking water; 10,000 hygiene kits; and 7,000 shelter kits; as well as undertaking repairs to seven primary schools, two medical centres and one secondary school [31].

Australia's Department of Foreign Affairs and Trade (DFAT), with the support of Foreign Minister Julie Bishop, have committed to supporting Fiji in a long-term recovery plan. Australia has prominently highlighted their humanitarian efforts with a spokesperson noting 'Australia has a longstanding friendship with Fiji and we acted swiftly to support the Fiji Government response to the devastation of Tropical Cyclone Winston. Australia remains the largest donor to the response effort and we will continue to be there for the long-haul to help Fiji recover' [32].

Political tensions following the Fijian coups in 2000, and again in 2006, and the slow governmental response to restoring democracy, have led to a strained relationship with the governments of Australia and New Zealand. In 2014, Australia lifted the last restrictions on the Fijian government. Fiji was also reinstated in 2014 into both the Commonwealth and Pacific Island Forum. These shifts likely enabled Australia and New Zealand to provide substantial aid following Winston and politically re-engage with the Fijian government after an approximate 15-year gap. Further, the New Zealand Prime Minister John Key, described the aid thus: 'New Zealand's response to Cyclone Winston is shaping up to be our largest humanitarian response in the Pacific' [33]. Although disheartening it may be to consider the political opportunities natural disasters yield, especially with an eye to the future of climate change in the Pacific, the international aid Fiji received following Winston is deserving of critical consideration.

Shortly after my arrival in Fiji in January of 2016, 20 shipping containers containing weapons arrived in the Port of Suva from Russia. It seemed likely that this shipment was intended for use by Fijian peacekeeping forces, given that 45 Fijian UN peacekeepers were captured by the al-Nusra Front in Syria in 2014 [34]. It has been suggested that the peacekeepers' capture was partially related to obsolete military equipment and technology [35]. The recent donation of arms to Fiji from Russia likely was noted by the international community, as it may well signify a greater geopolitical realignment by Fiji. Significantly, while still operating under coup sanctions by the international community, the Prime Minister Bainimarama fostered a 'Look North' political mind-set within Fiji and turned to other nations, such as Russia and China, less concerned with the state of democracy and human rights in Fiji. Political commentators have also noted that Russia is currently using 'chequebook diplomacy' in the Pacific. For example, in 2008, Nauru, Vanuatu and Tuvalu recognised Georgia, Abkhazia and South Ossetia following a Russian invasion and border reconfiguring, following international aid support from Russia [35]. And again in 2014, these PICTs along with Fiji and Tonga, recognised the annexation of Crimea [35].

While Australia and New Zealand continue to dominate aid relief in the South Pacific, Russia and China's continuing contributions are notably expanding. While discussing the Fijian economy one afternoon, a Fijian friend referred to the foreign powers presence in the islands as 'sharks in the water.' Consequently, the symbolic significance of aid warrants attention, especially as it reflects blood in the water and the evolving global political multipolarity and regional power potentially in flux.

7 BUILDING A BETTER TOMORROW – UNDERSTANDING MOBILITY AND INCOME DIVERSIFICATION

In an increasingly warming world frequently cited figures estimate that by 2050 the number of people forced to move primarily because of climate change will range between 200 million [36] and 1 billion [37]. Much of the literature to date on mobility in response to environmental degradation often assumes that migrants are a monolithic group making similar responses to environmental crises and moving to international destinations as yet unknown. This assumption is out of step with a more nuanced view of migration; a view in which migration is a vital adaptive strategy in response to environmental and socioeconomic mutability. If we are to work towards a more comprehensive view of mobility and migration, then understanding the duration, destination and composition of migration flows is essential for achieving a deeper knowledge of their impact on both origin and destination communities, as well as ways in which policy can successfully support individuals and communities on the move [38, 39].

Extreme weather events force people to leave their homes as they are no longer habitable. Yet, we know that most displaced people return to their homes as soon as possible, seeking to literally rebuild their lives [40–43]. Increasing levels of mobility in response to extreme weather events are likely, though the mobility may be cyclical in nature. For example, many Fijian colleagues who lived in Suva returned to their home villages following Winston to assist in the rebuilding of rural communities with familial connections. However, within weeks of Winston, many Fijians returned to the altered normality of a post-Winston world within their respective rural or urban spheres. Ultimately, mobility is a part of the solution to climate change and extreme weather events, not the problem.

Assumptions are often put forward that transnational migration is the major form of migration, when this is simply not true. Consider that in numerous conversations with Fijian friends, colleagues, relief workers and community members not a single person mentioned emigrating after Winston, even in lengthy discussions on climate change and the bleak future many PICTs faced with regard to increasing extreme weather events and rising sea level. Although, it should be noted that Fiji does have advantageous highland topography when compared with other lower-lying PICTs. However, contradictory perspectives put forward by American and European politicians around concerns of migrants from low-income nations arriving on wealthy shores in 'swarm(s)' [44] is oversimplified and incomplete, not to mention based on xenophobia and racism [45], especially when it comes to issues of environmental change [38]. In my experience, few Pacific Islanders wish to leave their island homes. Adaptation to limited-range and cyclical mobility within national boundaries reduces the necessity to move great distances away from impacted zones and enables those affected by extreme weather disruptions to return more quickly to their homes and communities.

According to the International Institute for Environment and Development 'There is growing evidence suggesting that mobility, in conjunction with income diversification, is an important strategy to reduce vulnerability to environmental and non-environmental risks—including economic shocks and social marginalization' [38]. Development and policy that support mobility and differentiates the variability within migration, is an important step forward. For example, the European Union (EU)-funded Pacific Climate Change and Migration (PCCM) project, in partnership with the International Labour Organisation (ILO) and the United Nations Development Programme (UNDP) has been recently implemented by the United Nations Economic and Social Commission for Asia and the Pacific (UNESCAP), and is positively supporting PICTs such as Kiribati, Nauru and Tuvalu, which are subject to future sea inundation [46]. The primary goals of the PCCM which commenced in 2014 are two-fold, (1). 'To increase protection of individuals and communities that are vulnerable to climate change displacement and migration through targeted national and regional policies; and (2) To increase labour mobility opportunities for Pacific Islanders, through well-managed labour migrations schemes' [46]. The former President of Kiribati, Anote Tong, speaking in regards to the realities of slow-onset climate change and the PCCM project noted: 'We are providing options for those that want to migrate now as a matter of choice, to do so as worthwhile citizens. We are providing training [and] up-skilling so that they can apply to different countries to seek migration status on merit' [47].

Attempting to predict human actions and agency in the advent of climate change and extreme weather events is tremendously challenging. It can be useful to extrapolate migration patterns from previous natural disasters, as a means of making analogous parallels. Studies [48–50] of migration patterns within wealthier, social-networked and better-educated groups point to these groups remaining mostly unaffected by environmental degradation, while younger groups with fewer dependents and possessions are increasingly likely to permanently resettle in new areas than those with land ownership in the impacted zones. Although this should not be confused with impoverished groups who have limited ability to invest in migration and are therefore less likely to employ mobility as an adaptation strategy. For example, 'Approximately 10,000 people across Kiribati, Nauru, and Tuvalu attempted to migrate between 2005 and 2015 but were unable to do so, primarily due to financial constraints' [39]. For those too poor to migrate who are dealing with both short-term environmental degradation and slow-onset climate change the ability to survive will be determined by the accessibility to local opportunities for income diversification. Simultaneously, those individuals and groups who are socioeconomically and politically marginalised will be least able to compete for the increasingly sought after resources and opportunities.

8 THE ANXIOUS EDGE OF NOW – CONCLUSIONS & FUTURE RESEARCH

The main purpose of this analysis was to give a first-hand account of the impacts of extreme weather in Fiji, the roles FBOs can play in community resiliency schemes, and to further elucidate the role of mobility within short- and long-term climate change situations. To that end, research that explores analogous parallels with the predicted impacts of climate change suggests that environmental degradation does not necessarily result in migration. Where mobility does occur, as with extreme weather events, it is often shorter in duration and distance, similar to what I experienced and observed in Fiji. My findings mirror the behaviour of previously researched episodes of drought in other regions of the world. However, in regards to slow onset climate change and rising sea levels, considerably less is known. The total PICTs populations required to move in the Pacific – most especially citizens of Kiribati, Nauru, Tuvalu and the Marshall Islands – will greatly depend on adaptation and resiliency initiatives and the support and planning of wider international strategies [39]. In the future, there is an opportunity to research in greater depth the correlations between environment, cyclical mobility and income diversification within national boundaries in the Pacific, as these types of trends and movement tend to go undocumented in national census data [38].

Finally, predicting the impact of climate change on population distribution and movement is fraught with difficulties. However, it seems unlikely that the political alarmist claims of a wave of millions of environmental refugees will occur. Further, a paradigm shift is need when understanding migration, as it is a part of the solution not the problem. Many migration management policies try to control the volume and types of population movement, when they would be wise in supporting more nuanced migration patterns that result from environmental degradation, economic growth or crisis, while simultaneously integrating appropriate cultural and faith precepts in an attempt to support individuals, families and communities on the move.

ACKNOWLEDGEMENTS

My research was generously made possible by Dr. Tony Crook and by funding from the University of St Andrews and was further supported by the European Consortium for Pacific Studies (ECOPAS) with funding from the European Union (EU). I owe tremendous debt to my co-author Dr. Joeli Vitayaki for his invaluable advice and assistance from my first days in Fiji through the present, and for generously and cheerfully sharing his illimitable knowledge of marine conservation. I thank too, the Happy Home Trust for their generous hospitality during my time in Fiji, providing me with sanctuary during tropical cyclone Winston. Finally, I am thankful to Dr. Elizabeth Minor, Laura Pond and Ricky Young for their friendship and contributions to this project.

REFERENCES

[1] Westcott, L., Tropical cyclone winston slams fiji with 200 MPH winds, kills 21. *Newsweek*, 22 February 2016, available at, http://europe.newsweek.com/fiji-tropical-cyclone-winston-photos-429070?rm=eu (accessed 23 January 2017)

[2] Bolatagici, L., Food supply runs short. *The Fiji Times*, 21 May 2015, available at http://www.fijitimes.com/story.aspx?id=354571 (accessed 15 December 2016).

[3] Round, S., Cry for more help from remote Fijian village. *Radio New Zealand*, 12 April 2016, availablea at http://www.radionz.co.nz/international/pacific-news/301293/cry-for-more-help-from-remote-fijian-village (accessed 27 November 2016).

[4] Round, S., Resilience amid the struggle—one year on from Cyclone Winston. *Radio New Zealand*, 20 February 2017, available at http://www.radionz.co.nz/international/pacific-news/324871/resilience-amid-the-struggle-one-year-on-from-cyclone-winston (accessed 20 February 2017)
[5] Fiji High Commision. Wellington 6011, New Zealand, available at http://www.fiji.org.nz/about-fiji/ (accessed 15 January 2017)
[6] McGranahan, G., Balk, D. & Anderson, B., The rising tide: assessing the risks of climate change and human settlements in low elevation coastal zones. *Environment and Urbanization,* **19**(1), pp. 17–37, 2007.
https://doi.org/10.1177/0956247807076960
[7] Farrell, W. & Clark, J., On postglacial sea level. *Geophysical Journal International,* **46**, pp. 647–667, 1976.
https://doi.org/10.1111/j.1365-246X.1976.tb01252.x
[8] Mitrovica, J., Tamisiea, M., Davis, J. & Milne, G., Recent mass balance of polar ice sheets inferred from patterns of global sea-level change. *Nature,* **409**, pp. 1026–1029, 2001.
https://doi.org/10.1038/35059054
[9] Mitrovica, J., Gomez, N. & Clark, P., The sea-level fingerprint of West Antarctic collapse. *Science,* **323**, p. 753, 2009.
https://doi.org/10.1126/science.1166510
[10] Tamisiea, M., Mitrovica, J., Davis, J. & Milne, G., Long wavelength sea level and solid surface perturbations driven by polar ice mass variations: fingerprinting Greenland and Antarctic ice sheet flux. *Space Science Rev,* **108**, pp. 81–93, 2003.
https://doi.org/10.1023/A:1026178014950
[11] NOAA (National Oceanic and Atmospheric Administration). U.S. Department of Commerce. 2016. Global ocean heat and salt content, available at www.nodc.noaa.gov/OC5/3M_HEAT_CONTENT (accessed 28 January 2017)
[12] Asian Development Bank, Fiji: Emergency Assistance for Recovery from Tropical Cyclone Winston. 2016. Available at: https://www.adb.org/projects/50181-001/main#project-pds (accessed 15 January 2017).
[13] The Fijian Government. Suva, Fiji, available at http://www.fiji.gov.fj/Media-Center/Press-Releases/FIJIAN-GOVERNMENT-ESTABLISHES-DISASTER-RELIEF-FUND.aspx (accessed 18 January 2017)
[14] Qalubau, P., Sugar Industry records $163.35 million loss. *Newswire,* 29 May 2016. Available at: https://www.newswire.com.fj/national/tc-winston/sugar-industry-records-163-35-million-loss/ (accessed 12 November 2016)
[15] Dateline Pacific. *Radio New Zealand*, 24 February 2016. Available at: http://www.radionz.co.nz/international/programmes/datelinepacific/audio/201790617/fiji's-tourism-industry-takes-big-hit-following-cyclone-winston (accessed 12 January 2017).
[16] Fiji Bureau of Statistics. Suva, Fiji. Avilable at: http://www.statsfiji.gov.fj/statistics/ (accessed 17 January 2017)
[17] Namorong, M., The melanesian way: the definitions of leadership and poverty are in direct contrast with western lifestyle. *The Melanesian Way*, 1 October 2016. Available at: https://melanesianwaypacific.blogspot.co.uk(accessed 12 February 2017).
[18] Barr, K., Economic growth for each one of us. *The Fiji Times*, 1 November 2014. Available at: http://www.fijitimes.com/story.aspx?id=284819 (accessed 18 December 2016)
[19] Hazelgrove-Planel, L., Cyclone pam from the field: adapting to climate change? *Anthropology Today,* **31**(6), 2015.
https://doi.org/10.1111/1467-8322.12213

[20] Anglican Communion News Service, Church to develop disaster strategy as Pacific islands disappear. *Episcopal News Service*, 11 May 2016. Available at: http://episcopaldigitalnetwork.com/ens/2016/05/11/church-to-develop-disaster-strategy-as-pacific-islands-disappear/ (accessed 18 February 2017)

[21] The Fiji Independent Commission Against Corruption (FICAC). *The Fiji Sun*. 5 March 2016. Available at: http://fijisun.com.fj/2016/03/05/ficac-on-lookout-for-corrupt-practices-on-the-back-of-cyclone-winston/ (accessed 17 November 2017).

[22] Simmons, M., No media freedom in Fiji while decree still in place, says Prasad. *Asia Pacific Report*, 5 May 2016. Available at: http://asiapacificreport.nz/2016/05/05/no-media-freedom-in-fiji-while-decree-still-in-place-says-prasad/ (accessed 18 February 2017).

[23] Agence France-Presse (AFP). Fiji imposes draconian media blackout ahead of this week's poll. *The Australian*, 15 September 2014. Available at: http://www.theaustralian.com.au/news/world/fiji-imposes-draconian-media-blackout-ahead-of-this-weeks-poll/news-story/a11a2823ff456f93c0f83ab024b3440d (accessed 14 January 2017).

[24] Alan ver Beek, K., Spirituality: a development taboo. *Development in Practice*, **10**(1), p. 31, 2000.

[25] Petersen, M., International religious NGOs at the UN: a study of a group of religious organizations. *The Journal of Humanitarian Assistance,* 17 November 2010.

[26] James, R., What is distinctive about FBOs?, The international NGO training and research centre (INTRAC). *Praxis Paper,* **22**, p. 7, 2009.

[27] Narayan, D., Chambers, R., Shan, M.K. & Petesch, P., *Voices of the Poor*. World Bank, 2000.

[28] Boli, J. & Brewington, D., Religious organizations in religion. *Globalizaiton and Culture*, p. 207, 2007.

[29] United Nations Charter, chapter X, Article 71, available at www.un.org, (accessed 2 February 2017)

[30] Newland, L., Fiji in Globalization and the Re-Shaping of Christianity in the Pacific Islands, pp. 383–385, 2006.

[31] Australian Government. Department of Foreign Affairs and Trade. Barton ACT 0221 Australia. Available at: http://dfat.gov.au/aid/topics/investment-priorities/building-resilience/humanitarian-preparedness-and-response/tc-winston/Pages/default.aspx (accessed 12 January 2017)

[32] Wyeth, G., After cyclone winston, focus on fiji's regional relationships. *The Diplomat Magazine,* 6 April 2016. Available at http://thediplomat.com/2016/04/after-cyclone-winston-focus-on-fijis-regional-relationships/ (accessed 10 December 2016).

[33] Cyclone Winston: NZ Government offers extra aid injection to help Fiji. *New Zealand Herald*, 1 March 2016. Available at http://www.nzherald.co.nz/nz/news/article.cfm?c_id=1&objectid=11598113

[34] ABC/Reuters. Fijian peacekeepers released by Syrian militant group Jabhat al-Nusra, two weeks after capture in Golan Heights: UN. *ABC News*. 12 September 2014. Available at: http://www.abc.net.au/news/2014-09-11/captured-fiji-peacekeepers-released-un-says/5737772 (accessed 17 January 2017).

[35] Madaus, R., The bear returns to the south pacific: Russia sends arms to Fiji. *The Diplomat Magazine*, 9 April 2016. Available at: http://thediplomat.com/2016/04/the-bear-returns-to-the-south-pacific-russia-sends-arms-to-fiji/ (accessed 15 November 2016).

[36] Myers, N., Environmental refugees: an emerging security issue. *Proceedings of the 13th Economic Forum*, May 2005.

[37] Christian Aid, Human Tide: The Real Migration Crisis. *Christian Aid Report*. May, 2007. Available at: https://www.christianaid.org.uk/Images/human-tide.pdf, (accessed 2 January 2017)
[38] Tacoli, C., Crisis or adaptation? migration and climate change in a context of high mobility. *Environment & Urbanization*, **21**(2), pp. 513–525, 2009. https://doi.org/10.1177/0956247809342182
[39] United Nations, Economic and social commission for asia and the pacific. pacific climate change and migration project. Suva, Fiji. Available at: http://www.unescap.org/subregional-office/pacific/pacific-climate-change-and-migration-project (accessed 5 February 2017)
[40] Perch-Nielsen, S. & Bättig, M., Exploring the link between climate change and migration. *Proceedings of the 6th Open Meeting of the Human Dimensions of Global Environmental Change Research Community*, October 2005.
[41] Raleigh, C., Jordan, L. & Salehyan, I., Assessing the impact of climate change on migration and conflict. Paper presented at *Social Dimensions of Climate Change*, Social Development Department, the World Bank, March 2008.
[42] GECHS, *Disaster Risk Reduction, Climate Change Adaptation and Human Security*, Report 2008:3, GECHS - Global Environmental Change and Human Security, University of Oslo, Oslo, 2008.
[43] Piguet, E., Climate change and forced migration. Research Paper No 153, New Issues in Refugee Research Series UNHCR, Geneva, 2008.
[44] Elgot, J., How David cameron's language on refugees has provoked anger. *The Guardian*, 27 January 2016. Available at: https://www.theguardian.com/uk-news/2016/jan/27/david-camerons-bunch-of-migrants-quip-is-latest-of-several-such-comments (accessed 12 January 2017)
[45] Yakushko, O., Xenophobia: understanding the roots and consequences of negative additudes toward immigrants. *Educational Psychology Papers and Publications*, University of Nebraska, Lincoln, 1 January 2009
[46] UNESCAP. United Nations. Economic and Social Commission for Asia and the Pacific. *Climate Change and Migration in the Pacific: Links, attitudes, and future scenarios in Nauru, Tuvalu, and Kiribati*, 2016, available at http://www.unescap.org/sites/default/files/PCCM%20-%20%20Survey%20Fact%20Sheet.pdf (accessed 20 November 2016).
[47] Singh, D., Analysis: Pacific climate change and migration project addresses climate induced migration. *The Fiji Sun*, 29 October 2015. Available at: http://fijisun.com.fj/2015/10/29/pacific-climate-change-and-migration-project-addresses-climate-induced-migration/ (accessed 16 January 2017)
[48] McLeman, R. & Smit, B., Climate change, migration and security, Commentary 86, Canadian Security Intelligence Service, Ottawa, 2004.
[49] Massey, D., Axinn, W. & Ghimire, D., Environmental change and out-migration: evidence from Nepal. *Population Studies Centre Research Report* 07–615, Institute for Social Research, University of Michigan, Ann Arbor, 2007.
[50] Morrissey, J., *Environmental Change and Forced Migration: A State of the Art Review*, Refugee Studies Centre, Oxford Department of International Development, University of Oxford, Oxford, 2009.

IMPACTS OF RAINFALL AND VULNERABILITIES IN THE METROPOLITAN REGION OF BAIXADA SANTISTA, BRAZIL

[1]FABIANO DE ARAUJO MOREIRA, [2]NÚRIA APARECIDA MIATTO RAMPAZO & [3]MARINA SÓRIA CASTELLANO
[1]Department of Geography, University of Campinas, Brazil.
[2]Department of Energy/Faculty of Mechanical Engineering, University of Campinas, Brazil.
[3]Center for Human Sciences, State University of Western Paraná, Brazil.

ABSTRACT

In altered environments, natural events such as rainfalls can bring a series of damages and their impacts are related not only to rainfall itself, but also to the socioeconomic aspects of the population. This paper aims to discuss the process of urbanization and occupation of the municipalities of the Metropolitan Region of Baixada Santista/RMBS (Brazil) and the vulnerability of the population considering the events triggered by rainfall in the decade of 2000. For this purpose, the events were collected in pre-existing electronic journals and official databases. The impacts were mapped based on the Social Vulnerability Index, provided by the Brazilian Federal Government's Planning, Development and Management Ministry. A total of 33 types of impacts related to rainfall across the RMBS were identified, with nearly 42,000 occurrences for the period. The event triggered by heavy rains with the highest number of deaths occurred in the municipality of Guarujá. During the whole period, the years of 2008 and 2009 stand out with the higher number of occurrences, mainly affecting the municipalities of Peruíbe and Mongaguá, which faced important rain episodes in the summers of both years. It was observed that there was no direct relationship between vulnerability and the number of occurrences related to rainfall, a tendency that differs from other similar studies, that emphasizes the most vulnerable population is the most affected by these types of events. This fact could be explained by the use of a municipal index, which have generalized specificities inside the urban area.

Keywords: Brazil, disasters, rainfall, urbanization, vulnerability.

1 INTRODUCTION

The Brazilian urbanization is associated with the generation of poverty since the accelerated and uncontrolled growth of the cities induces an urban crisis [1], as a large part of the population does not have access to essential services. The problem becomes even more serious with the fact that 83.8% of the country's population lives in urban centers.

One of the features of urbanization is the modification of the natural environment [2–7], which include the replacement of green areas, paving, densification of the population, waterproofing of soil, presence of pollutants, among others, which contribute to changes in the energy, thermal and water balances.

In these altered environments, natural phenomena such as rainfall can bring a series of problems. Their impacts may include direct effects, relating to the most visible consequences, such as the destruction of buildings and deaths; and indirect effects, which arise later and are more difficult to attribute to events, such as influence on the economy and social life [8].

The impacts associated with rainfall are not only related to precipitation itself, but also to the socioeconomic aspects of the population involved [9, 10]. In this way, the agent of the disaster cannot be considered independent of the social context in which it occurs: it must include characteristics of land occupation, planning and the dynamics of the affected society.

DOI: 10.2495/SAFE-V7-N2-169-179

This paper aims to discuss the process of urbanization and occupation of the municipalities of the Metropolitan Region of Baixada Santista/RMBS (Brazil) and the vulnerability of the population considering the events triggered by rainfall in the decade of 2000.

2 STUDY AREA

The RMBS is located on the coast of the state of São Paulo, Brazil (Fig. 1) and occupies an area of 2.818 km², englobing nine municipalities: Bertioga, Cubatão, Guarujá, Itanhaém, Mongaguá, Peruíbe, Praia Grande, Santos and São Vicente, totaling more than 1.6 million inhabitants, which corresponds to almost 4.0% of the total population of the state of São Paulo, the most economically important state in the country. Due to tourism, the area coexists with an expressive floating population, which can more than double in the high seasons. Although 67% of the region is vegetated, it presents a high degree of urbanization (over 97%) and population density.

In this region, confined between the coast and the Serra do Mar massif, there are mangrove areas, representing about 8.0% of the surface of the region, with the main local mangrove systems being those of the Estuarine Complex of Santos/São Vicente, on the Atapanhaú River, on the Bertioga Canal and at the mouth of the Itanhaém River. In addition, we can highlight the presence of the Atlantic Forest, representing about 40.0% of its surface, as well as 10.0% of the restinga system.

High rainfall volumes throughout the year and the fact that there is no significant difference in the distribution of precipitation between seasons shows the precipitation mechanisms on the coast are distinct from those in more continental areas [11]. This sector constantly registers frontal systems, sea breezes and the action of the South Atlantic Convergence Zone, mainly in

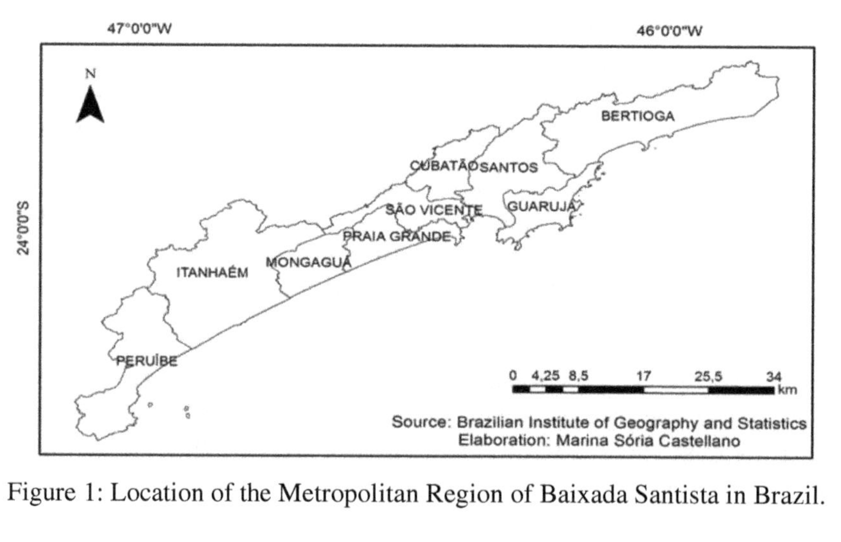

Figure 1: Location of the Metropolitan Region of Baixada Santista in Brazil.

the summer, as well as local precipitation intensified in some points by the orographic effect. The temperatures are constantly high, and all these factors causes a naturally fragile and unstable environment, promoting strong chemical weathering.

The singularities of the area in relation to its physical characteristics were determinant in the process of local urbanization and economic development, as well as in the emergence of susceptible areas and vulnerable populations. Being one of the oldest occupying areas of the country, its environment has been altered for several centuries, significantly impacting its original natural characteristics.

Its regional importance is evident with the presence of the largest port, petrochemical and steel pole in Latin America [12], with the port representing 25.1% of the total Brazilian trade balance in 2010 [13]. The area has vocations that are apparently disparate because of its economic potential (large port and industrial pole and tourism potential), which combine with conservation reserves and areas of legal protection, denoting extreme complexity in its spatial relationships [12].

The intense urbanization has significantly altered the physical environment, and the conservation of the natural areas close to the urbanized areas continues to be considered of little importance by the government. Young [14] emphasizes that the lack of conservation of natural areas in the RMBS is observed from the processes of environmental degradation and impacts related to urban activities.

Numerous anthropogenic actions contribute to the increase of environmental problems in the region: mangroves and estuarine waters are contaminated by the heavy industry effluents located in Cubatão, while toxic gases released by the industries promote erosive processes on the hillsides indirectly, by contributing for the loss of vegetation; continuous deforestation for the expansion of the urban area; the characteristics of the areas near the sea have been almost completely modified, losing the original areas of dunes and flooded valleys; contamination of soil and water by effluents and municipal solid waste; irregular occupations that impact the local characteristics of the soil or that increase the vulnerability of the population in relation to phenomena such as floods or landslides, among others [14].

The vulnerability of the population, in addition to the problems arising from the impact of human action on the environment, is also identified by activities that should be guaranteed by local governments, such as basic sanitation, garbage collection and sewerage, access to services such as health and public transportation, as well as problems related to the security sector. All these issues have arisen or were intensified due to the spatial segregation caused by the occupation of the city of Santos by the population with high or medium purchasing power, especially in the coast area with great verticalization which, through real estate speculation, expelled the less affluent population to distant and irregular regions, highlighting the municipalities neighboring Santos that lack more incisive policies on urban occupation.

Such facts illustrate contemporary, generalized aspects that characterize urban coastal occupation: the lack of planning and the difficulty of public policies to operate in an effective way and at the same speed as urban occupation. This makes the challenges for the planning of social and environmental public policies significant in RMBS.

3 METHODOLOGY

The main source of occurrences associated with rain episodes in the RMBS were the news published in electronic journals available on the internet and pre-existing databases, with information from Brazilian institutions such as the Institute for Technological Research (IPT), the Civil Defense (Disaster Records Database) [15] and Natural Disaster Database of the Institute for Meteorological Research (IPMet) [16].

The database was systematized in Microsoft Excel, organizing the information in the following columns: date, time, location, phenomenon, characteristic of the phenomenon (eg. rain duration), impacts and source. In turn, the impacts were subdivided into: environmental, material and/or economic damages; number of fatalities, homeless, displaced, missing and/or buried, sick or injured, and affected persons. It should be noted that the category 'environmental, material and/or economic damage' has been filled with a variety of related damages, from flooding to roadblocks or cracking of buildings, among others.

The mapping of the occurrences was made to understand where the problems associated with rains occurred more frequently and which were the most affected social strata. It should be noted that although the use of indexes offers limitations to the research, precisely because they do not represent the reality, they bring information that can characterize certain areas and communities, being a valuable instrument of analysis and comparison between different areas.

It is important to emphasize that the concept of vulnerability is quite complex and may be associated with many social factors. According to Cutter et al. [17], within the social sciences, there is a consensus that some key factors influence the vulnerability of a given group, including lack of access to resources, limited political representation, insufficient social capital, among others. Thus, the differences between the concepts of vulnerability in different academic researches are due to the selection of the variables and the weights given to each of them, generating several indexes associated to the theme, such as the Social Vulnerability Index (SoVI), developed by Cutter et al. [17], considered the academic standard for measuring social vulnerability.

In this work, it was used the Atlas of Social Vulnerability of the Institute of Applied Economic Research (IPEA), an important and respected public foundation, linked to the Ministry of Planning, Development and Management of the Brazilian federal government. The Atlas of Social Vulnerability is a platform for consulting the Social Vulnerability Index (IVS), composed of 16 indicators and calculated from the arithmetic mean of three sub-indices. According to the Atlas, "These dimensions correspond to sets of assets, resources or structures, whose access, absence or insufficiency indicate if the life standard of families is low, suggesting, in the limit, the non-access and non-observance of the social rights" (free translation from Portuguese) [18].

The indices are detailed as follows:

- IVS Urban Infrastructure, which addresses the conditions of access to basic sanitation and urban mobility, factors that significantly impact the well-being of the population.
- IVS Human Capital, which involves health and education issues, aspects that determine directly the current and future perspectives of social inclusion.
- IVS Income and Labor, which group's income and other indicators that may create a situation of income insecurity.

All indicators were calculated from the 2000 and 2010 Demographic Census of the Brazilian Institute of Geography and Statistics (IBGE) and normalized on a scale of zero to one, where zero corresponds to the ideal situation (less vulnerability) and one to the worse situation (greater vulnerability) (Fig. 2).

Areas with an IVS value between 0 and 0.200 are classified as areas of very low vulnerability; between 0.201 and 0.300, low vulnerability; between 0.301 to 0.400, medium vulnerability; between 0.401 and 0.500, high vulnerability and finally from 0.501 to one, very high vulnerability.

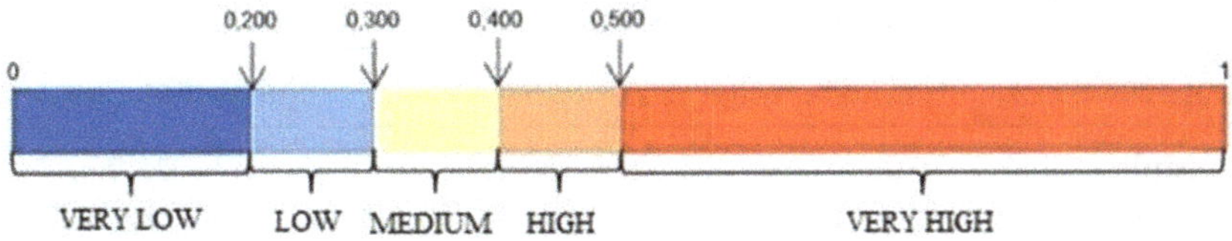

Figure 2: Vulnerability and classification values according to the Social Vulnerability Index (IVS) [18].

Only a few types of impacts were mapped, including: affected people, house collapse, fatal victims and total occurrences. The choice was made because the first two had the highest number of occurrences, the fatal victims were the most violent, and the total in order to give an overview of the RMBS situation.

4 RESULTS

A total of 33 types of rainfall impacts were identified throughout the RMBS, and they are listed in Table 1 according to the number of occurrences for each of them.

There were almost 42,000 occurrences, most notably the affected people, with more than half of the total recorded. However, it should be noted that the number of people affected by the events is certainly underestimated, since not all the news consulted provided this type of information, besides being impossible to know for sure the number of people who, in one way or another, were affected by the phenomenon. Moreover, there was no standard in the sources consulted: some considered the sum of the dead, wounded, homeless and displaced, while others included people without electricity, for example. Although it fades from the scope of this work, this fact illustrates the difficulty of providing a reliable picture of the real consequences of rainfall in the area.

Human losses and impacts may also be underestimated, because in some cases, the news only mentioned the event left people displaced or homeless - or even fatal victims - but did not specify that number.

In the analyzed period, 20 events with fatal victims were identified. Of the total (147 registered), 125 people lost their lives in a single event in the municipality of Guarujá, on April 4th, 2005, with landslides, vehicle damage and other impacts triggered by heavy rains. Four of the 20 events with fatal victims were caused by lightning, corresponding to 25.0% of the cases, totaling one victim in Cubatão, two in Itanhaém and one in São Vicente. All episodes of fatal victims caused by lightning occurred during the summer, the main season of lightning in Brazil, due to the warming caused by the Sun [19].

Regarding the temporal analysis, Figure 3 shows the total number of occurrences per year, for the entire RMBS. It is clear that the years 2008 and 2009 stood out, but mostly 2008, with a large number of cases. Among these, about 93.0% occurred only in the municipality of Peruíbe.

Figures 4–7 show the number of affected people, house collapses, fatal victims and total number of occurrences, respectively, by municipality, as well as their vulnerability ranges.

Even though the 'affected people' was the most numerous impact, it was concentrated in only two municipalities: Peruíbe and Itanhaém. Peruíbe stood out for the large number of cases, reaching just over 27,730, with 71.0% of these occurring in just one day, January 12th, 2008. The data provided by the Department of Water and Electric Energy [20] show a series

Table 1: Impacts related to rainfall in RMBS and their respective number of occurrences.

Type of impacts	Number of occurrences
Affected people	25.777
House collapse	7.575
Displaced people	4.150
Homeless people	2.830
Sick and injured	484
Falling trees	227
Fatal victims	147
Landslides	118
Damage to paving	104
Unroofing	102
Surface flood	69
Gradual floods	24
Overflow of rivers and streams	23
Interdicted properties	22
Traffic/Interdiction of routes	18
Interruption in transportation	16
Vehicle damage	14
Wall fall	14
Cutting off power and water supply	10
Block fall	9
Accidents	7
Coastal flood	7
Barrier fall	6
Traffic sign fall	5
Erosion/hole	4
Flooded houses	4
Sudden floods	4
Missing and/or buried persons	4
Light pole fall	4
Damage to bridges	2
River flood	2
Dam rupture	1
Total	41.783

of inconsistencies, not accusing a heavy rain on this day. However, data from the nearby municipality of Mongaguá, which also had some impacts in this episode, indicate a rainfall of 248.5 mm between January 11th and 12th, 2008, and of 23.2 mm between January 12th and 13th, 2008 [20]. The analysis of vulnerability did not show the expected trend - that the most vulnerable populations would present greater exposure to risk and, consequently, more occurrences - since both municipalities mentioned are characterized as areas of low vulnerability in the regional scenario.

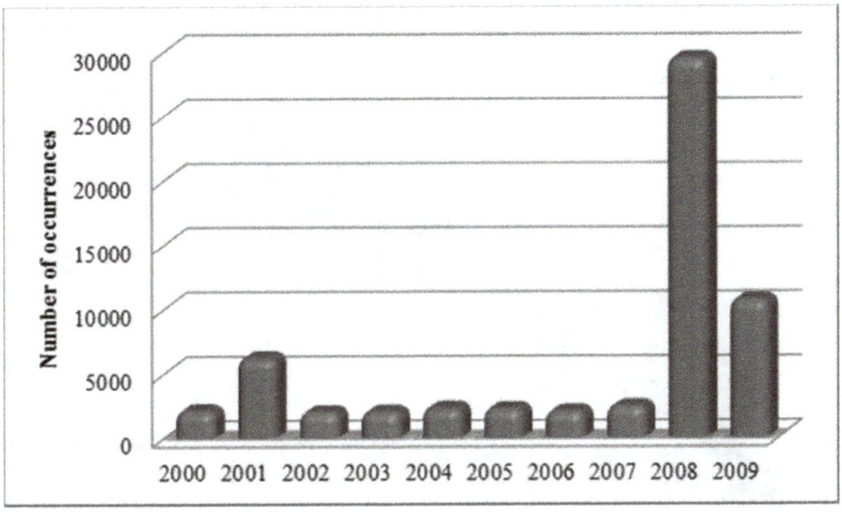

Figure 3: Total occurrences for all RMBS municipalities per year.

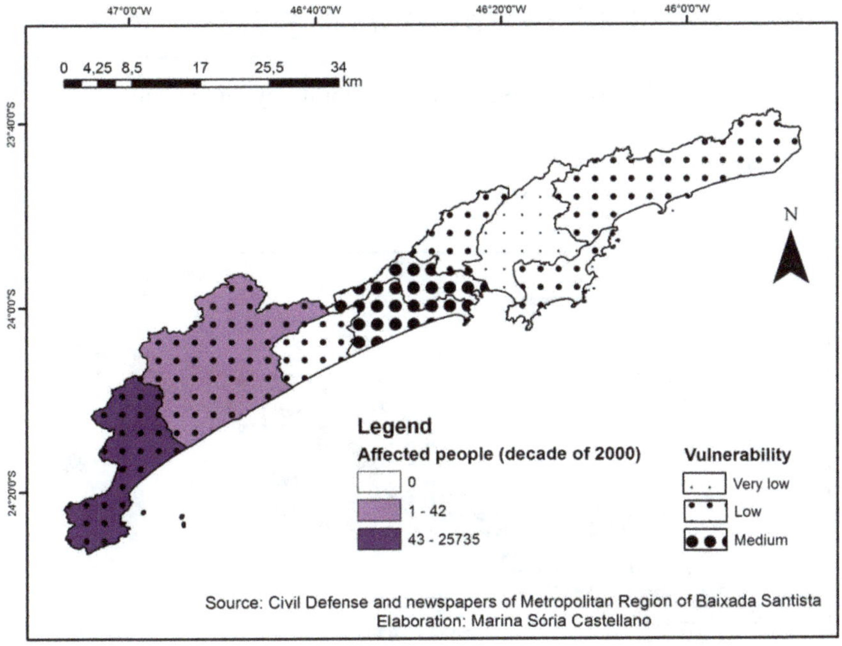

Figure 4: Affected persons and population vulnerability by municipality.

All municipalities of the RMBS presented house collapse, Peruíbe being the one with the highest number of occurrences, despite having presented low vulnerability in the period. As in the cases of affected people, a large part of the occurrences in this municipality (about 74.0%) occurred on January 12th, 2008. It was noted that among the municipalities with medium vulnerability, São Vicente stood out with 69 cases and Santos, the only one with very low vulnerability, had 24 cases.

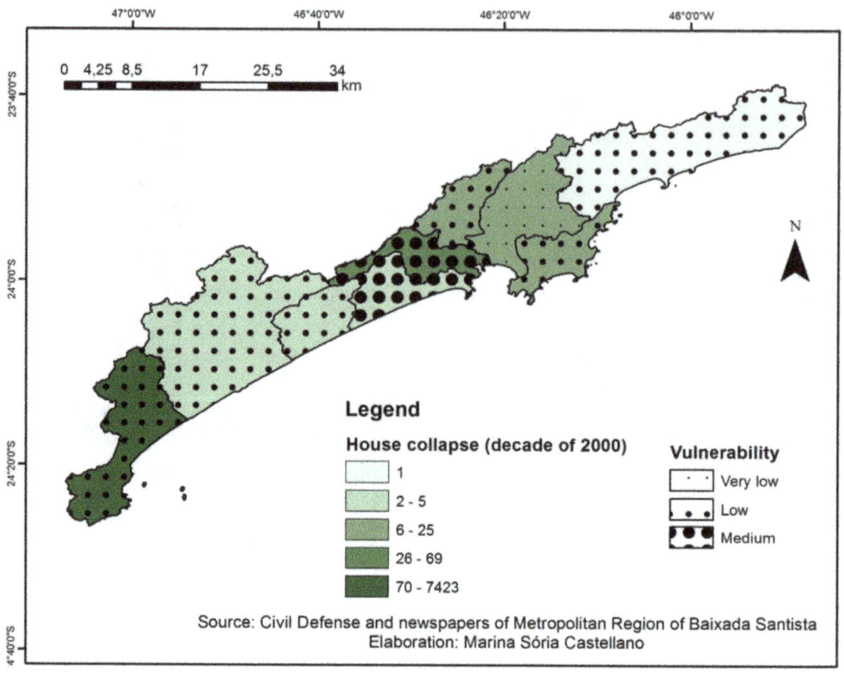

Figure 5: House collapses and vulnerability by municipality.

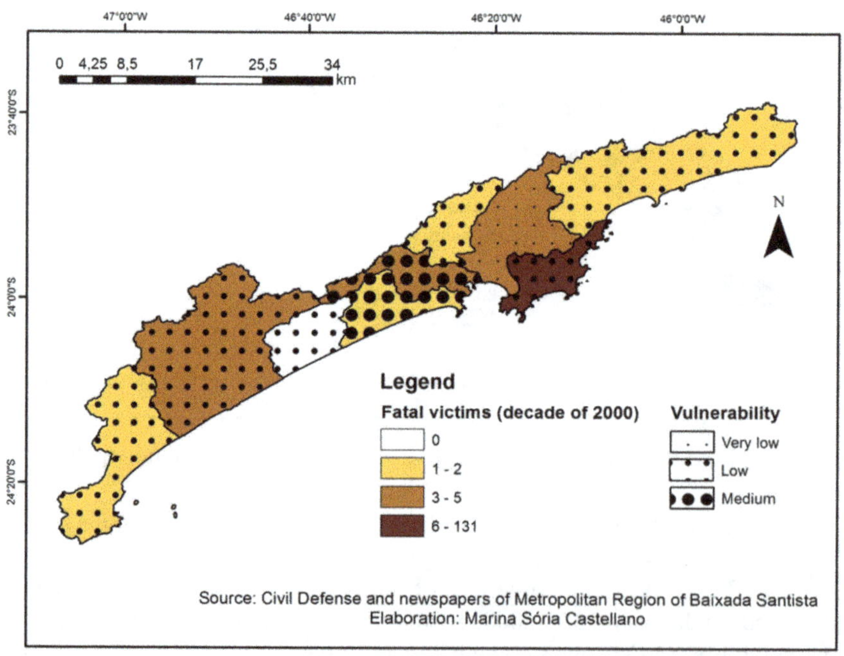

Figure 6: Fatal victims and population vulnerability by municipality.

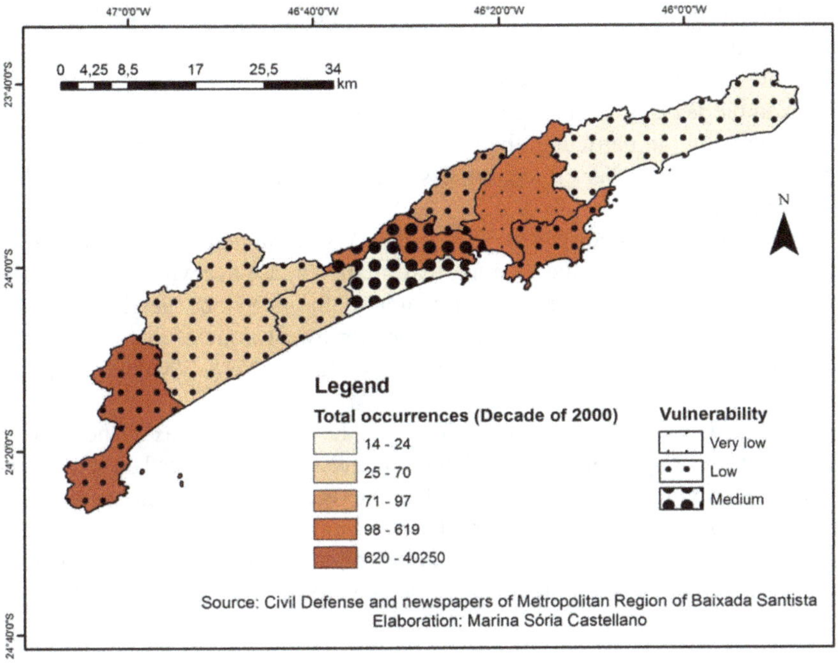

Figure 7: Total occurrences and vulnerability of the population by municipality.

With the exception of Mongaguá, all municipalities had cases of fatal victims. Santos, the only one with very low vulnerability, had three cases. Among the municipalities with low vulnerability, Guarujá stood out with 131 occurrences, 125 of which occurred in one day (April 4th, 2005). Among those with average vulnerability, São Vicente stood out with five cases.

Despite being classified as low vulnerability, Peruíbe was the municipality with the most occurrences: 40,250, especially in cases of house collapse, affected people, displaced people and homeless people. Most of the incidents were concentrated in 3 days: January 12th, 2001; January 12th, 2008; and February 25th, 2009 (all during the summer). Santos, the only one with very low vulnerability, presented an intermediate number of occurrences if compared to other municipalities (with emphasis on cases of tree falls and homeless people - not mapped in this work). Among those with low vulnerability, in addition to Peruíbe, it was noted that Guarujá presented a considerable number of cases (many records of homelessness and house collapse), and among those with a medium vulnerability population, São Vicente stood out (especially because of the many cases of house collapse and displaced people).

5 CONCLUSIONS

Some aspects should be taken into account from the spatialization and analysis of the data. At first glance, vulnerability classifications may seem strange, since there are no municipalities with high or very high vulnerability and most of them are classified as areas of low vulnerability. This was due to the fact that the index used in this study (IVS) was defined by the IPEA for several Brazilian municipalities - very heterogeneous areas - and not only for RMBS.

Another important point was that although the literature and knowledge of the area show that vulnerable populations tend to suffer more from rainfall, the mapping showed that there was no direct relationship between vulnerability and the number of cases related to rainfall, since municipalities with population in situation of greater vulnerability were not necessarily those with more occurrences and vice versa.

A clear example is Santos, the only one with very low vulnerability, but who presented many cases (including, in particular, the number of fatal victims). Peruíbe can also be cited, since it had the highest number of occurrences in the period, although it was classified as an area of low vulnerability.

Such trends can be explained by the use of the IVS, a municipal index, which probably generalizes and does not expose the specifics that smaller areas have within the municipalities, such as the neighborhoods. It is known, for example, that in Santos, neighborhoods such as São Bento, Nova Cintra, Marapé, Embaré and the Northwest Zone of the municipality have, historically, problems related to rainfall and are areas where extremely vulnerable populations inhabit.

Thus, it was noted that, for the purposes of this work, the municipal index did not accomplished the desired role, concealing essential particularities to understand how populations are exposed to risks associated with rainfall. This should be taken into consideration when using these data to formulate public policies aimed at reducing the vulnerability of populations at risk, and it is necessary to aggregate information available to other organizations or entities that have specific data of smaller areas within each municipality, such as some prefectures or civil defenses, for example.

ACKNOWLEDGEMENTS

This research was supported by Fundação de Amparo à Pesquisa do Estado de São Paulo (FAPESP – São Paulo State Research Support Foundation – Process no. 2013/24236-3 and 2014/14598-8), in connection with project "METROPOLE: An Integrated Framework to Analyse Local Decision Making and Adaptive Capacity to Large-Scale Environmental Change: Community Case Studies in Brazil, UK and the US" (Forum Belmont, Process no. G8MUREFU3 2201-040, Fapesp Process no. 2012/51876-0).

REFERENCES

[1] Santos, M., *A urbanização brasileira,* São Paulo: HUCITEC, 1993.

[2] Sant´Anna Neto, J.L., *Clima e organização do Espaço,* Boletim de Geografia. Universidade Estadual de Maringá, ano 16, no. 1, 1998.

[3] Brandão, A.M.de P.M., Clima urbano e enchentes na cidade do Rio de Janeiro. In: Guerra, A.J.T. & Cunha, S.B.da (eds), *Impactos ambientais urbanos no Brasil,* Rio de Janeiro: Bertrand Brasil, 2001.

[4] Yahn, A.G. & Giacomini, A.A.R.V., Recursos hídricos e saneamento. In: Fonseca, R.B., Davanzo, A.M.Q. & Negreiros, R.M.C. (eds), *Livro verde: desafios para a gestão da Região Metropolitana de Campinas*, Campinas: Unicamp/IE, 2002.

[5] Gonçalves, N.M.S., Impactos pluviais e desorganização do espaço urbano em Salvador. In: Monteiro, C.A.de F. & Mendonça, F., *Clima urbano,* São Paulo: Contexto, 2003.

[6] Tucci, C.E. & Braga, B. (eds), *Clima e recursos hídricos no Brasil,* Porto Alegre: ABRH, 2003.

[7] Nunes, L.H., Landslides in São Paulo, Brasil: an integrated historical perspective. *Pages News*, **19**(2), 2011.

[8] Smith, K., *Environmental hazards - assessing risk and reducing disaster*, Londres: Routledge, 2004.

[9] Mattedi, M.A., *As enchentes como tragédias anunciadas: impactos da problemática ambiental nas situações de emergência em Santa Catarina*, PhD thesis - Universidade Estadual de Campinas, 1999. (in Portuguese).

[10] Nunes, L.H., Mudanças climáticas, extremos atmosféricos e padrões de risco a desastres hidrometeorológicos. In: Hogan, D.J. & Marandola Jr., E. (eds), *População e mudança climática. Dimensões humanas das mudanças ambientais globais*, Campinas: Núcleo de estudos de População - NEPO, Unicamp; Brasília: UNFPA, 2009.

[11] Nunes, L.H., *Distribuição espaço-temporal da pluviosidade no Estado de São Paulo: tendências, variabilidade, processos intervenientes*, EPUSP, 1997.

[12] Zundt, C., Baixada Santista: uso, expansão e ocupação do solo, estruturação de rede urbana regional e metropolização. In: Cunha, J.M.P.da, *Novas Metrópoles Paulistas: População, vulnerabilidade e segregação*, Campinas: Núcleo de Estudos de População – Nepo. Unicamp, 2006.

[13] CODESP – Controle de segurança dos portos, http://www.codesp.com.br/

[14] Young, A.F., Mudanças climáticas: entre a coesão e a fragmentação dos assentamentos humanos, os conflitos e as transformações da paisagem na Baixada Santista. In: Hogan, D.J. & Marandola JR., E. (eds), *População e mudança climática: dimensões humanas das mudanças ambientais globais*, Campinas: Núcleo de Estudos de População – Nepo. Unicamp; Brasília: UNFPA, 2009.

[15] Sistema Integrado de Informações sobre Desastres (S2ID). Available at: http://150.162.127.14:8080/bdrd/bdrd.html. (Accessed July 2014).

[16] IPMET - Instituto de Pesquisas Meteorológicas. Available at: http://www.ipmet.unesp.br/index2.php?menu_esq1=&abre=ipmet_html/defesa_civil/index.php. (accessed 24 January, 2015).

[17] Cutter, S.L., Boruff, B.J. & Shirley, W.L., Social vulnerability to environmental hazards. *Social Science Quarterly*, **84**, pp. 242–261, 2003. https://doi.org/10.1111/1540-6237.8402002

[18] IPEA - Instituto de Pesquisa Econômica Aplicada. Available at: http://ivs.ipea.gov.br/ivs/

[19] Tavares, R., Clima, Tempo e Desastres. In: Tominaga, L.K., Santoro, J. & Amaral, R. (eds), *Desastres Naturais: conhecer para prevenir,* 1ªed. São Paulo: Instituto Geológico, **01**, pp. 111–146, 2009.

[20] Departamento de Águas e Energia Elétrica. Available at: http://www.hidrologia.daee.sp.gov.br/. (accessed 20 March, 2017).

INTEGRATING BUILDING EFFECT OF LOCAL GRID REFINEMENT IN URBAN FLOOD MODELLING

CHEN-JIA HUANG & MING-HSI HSU
Department of Bioenvironmental Systems Engineering, National Taiwan University, Taiwan

ABSTRACT

Flooding often causes the devastating impact of people and the tremendous economy loss, especially in developed cities. Recently, many efforts have been made in urban flood modelling, which provides a useful tool for disaster mitigation. The high resolution flood modelling in large floodplain remains computationally demanding. Therefore, the multi-scale approach provided a rapid solution for solving the complex flow pattern in local region where densely distributed buildings often affect the propagation of overland flow, as well as excluding interior spaces from flooding.

An urban inundation model including the coarse and fine grid resolutions was developed to represent the blockage effect of buildings. The proposed model firstly solves the non-inertia overland flow equations using a coarse main grid for the global domain, then the results are used as the boundary conditions for the fine grid computation. The local refinement technique reflects the details of hydrodynamics in areas near buildings with high resolution. The computing of global domain and all the local regions of the multi-scale inundation model are able to improve the efficiency.

A synthetic case was adopted for model validation. The simulation results of multi-scale model demonstrated the good accuracy. The model was further applied to the simulation of typhoon Morakot event (2009), which induced the worst flooding that ever happened in Taiwan in half a century. The simulation results showed that the proposed model can reveal the better representation of water depth in local refined urban area and the computing efficiency was significantly improved for the real-time flood forecasting.

Keywords: blockage effect, Multi-scale, Urban inundation model

1 INTRODUCTION

Flood is one of the devastating natural hazards in the world. The number of floods occurring has significantly increased in the past half century. Many efforts have been made to mitigate the impact of floods; however, modelling and forecasting floods is still a challenging issue. Recently, the rapid advance of remote sensing makes the more detailed data in resolution available for implementation in inundation modelling. These data, like Digital Elevation Models (DEMs), can enhance the topographic detail to obtain better accuracy in 2D flood modelling [1–4]. However, modelling large floodplains with these detailed topographic data still require extraordinary computing time.

The multigrid approach provided a rapid solution for solving the complex flow pattern in local area [5–7]. Ruge *et al.* [8] improved the accuracy of the shallow-water equation on a sphere using local refinement coupled with multigrid method without the use of global fine grids. Lamby *et al.* [9] developed a fully adaptive multiscale scheme using a local time stepping strategy at different resolution grids to increase the efficiency and to reduce the computational costs of numerical simulations in two-dimensional shallow water equations. Henonin *et al.* [10] developed a multigrid approach which considers the detailed elevation in fine grid to compute the water level of coarse gird in continuity equation and yield acceptable results at several flood events in Beijing City.

In urban flood modelling, densely distributed buildings usually exclude surface runoff from the interior spaces when flood occurs. To characterize the physical situation in urban area, fine-resolution data have proven valuable as the simplest approach to represent the

DOI: 10.2495/SAFE-V7-N2-180-189

building using its roof elevation, but it seems to be a be computationally costly solution in large scale floodplains modelling. The present paper aims to propose an inundation model using two different scale grids to simulate the detailed flow condition in local area and improve the computing efficiency in large basins.

2 METHODOLOGY

2.1 Two-Dimensional Inundation Model

To evaluate the flood hazard of a flood adaptation measure in the future, a 2D inundation model [11] was applied to solve the sophisticated flow propagation in urban area.

$$\frac{\partial(1-\alpha_0)d}{\partial t}+\frac{\partial(ud)}{\partial x}+\frac{\partial(vd)}{\partial y}=q \tag{1}$$

$$-\frac{\partial h}{\partial x}=u\left[\frac{n^2\sqrt{u^2+v^2}}{d^{4/3}}\right] \tag{2}$$

$$-\frac{\partial h}{\partial y}=v\left[\frac{n^2\sqrt{u^2+v^2}}{d^{4/3}}\right] \tag{3}$$

where eqn (1) is the continuity equation, and eqns (2) and (3) are momentum equations in Cartesian horizontal directions; α_0 is the building coverage ratio (BCR), d is the depth of flow [m]; h is the water stage [m]; u and v are the velocity component in x and y directions [m/s]; q is the source or the sink per unit area [m/s]; and n is Manning's roughness [s/m$^{1/3}$].

The building is represented as a square located in the center of the simulation cell (Fig. 1). The occupancy of building in one cell is defined as α_0 (BCR) such that the flux between two neighboring cells is not affected by the building and the surface runoff can only be allowed to distribute and move in the area outside the building, which covers the grid of ratio $(1-\alpha_0)$.

2.2 Local Grid-refined Method

The local grid-refined approach uses two different resolution grids to calculate the water depth and velocity. The coarse grid is used for simulation in global domain for the

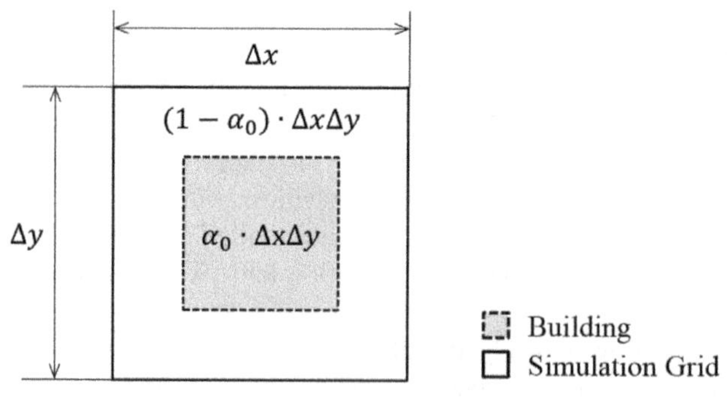

Figure 1: Schematic of simulation grid containing buildings.

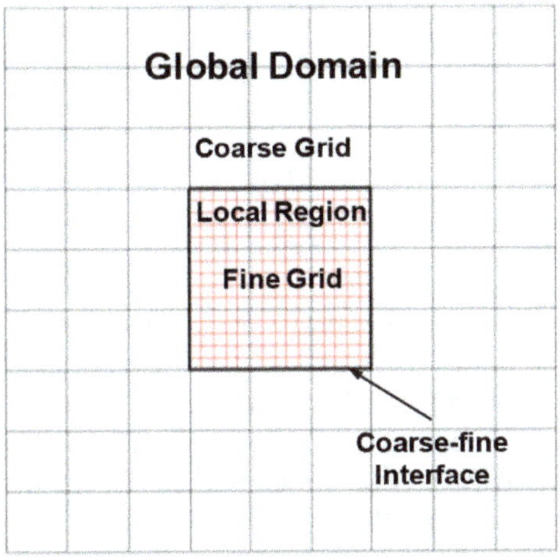

Figure 2: Relationship of coarse and fine grids.

consideration of computational efficiency (Fig. 2). The fine grid is then used in the specified vulnerable zones where the population and infrastructures concentrated to obtain a higher precision for hazard demand.

In local region of the specified zone, the boundary condition should be provided by the simulation results of global domain. In this study, we adopted the discharge as the boundary conditions for the local region. Therefore, the boundary conditions of the fine grid are determined by the fluxes between the coarse grids that are adjacent to the coarse-fine interface (Fig. 3). The flux between two coarse grids is calculated as below:

$$Q = d_c \times a \times U_c \tag{4}$$

where Q is the flowrate between two coarse grids [m^3/s]; d_c is the water depth of coarse grid at the interface [m]; a is the cell size of coarse grid; U_c is the flow velocity between the coarse cells, i.e. u or v in eqn (1) [m/s].

The fluxes crossing the interfaces of global domain and local regions are used as the boundary conditions for the fine grids. The flux between two coarse grids is redistributed to the fine grids (Fig. 4) that are adjacent to the same interface section. The discharge through the boundary of each fine grid is assumed constant during a coarse time step and can be expressed as following:

$$q_i = K_i Q, (i = 1 \sim p) \tag{5}$$

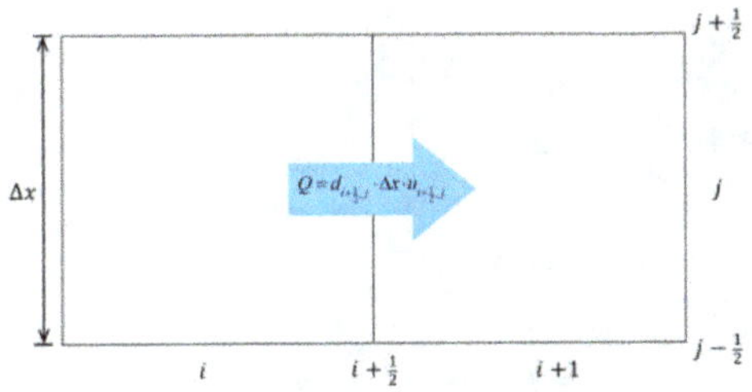

Figure 3: Flow rate of coarse grid.

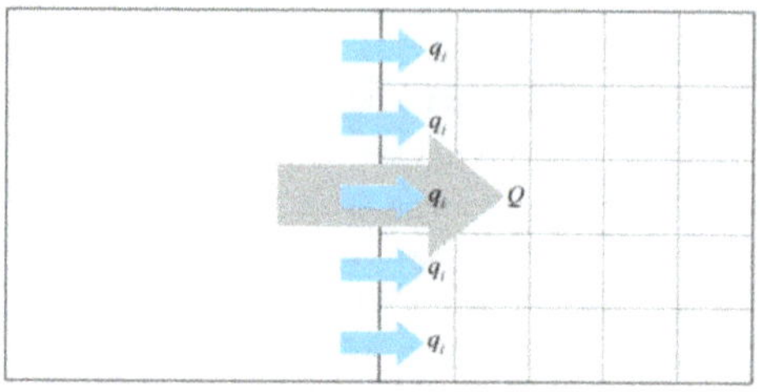

Figure 4: Boundary flow arrangement.

where q_i is the flow rate through the boundary of a fine grid [m^3/s]; K_i is the weighting factor; $\sum K_i = 1$; p is the number of fine grids divided by a coarse grid in the interface.

In this paper, uniform velocity distribution on the interface is assumed such that the flux redistributed to the fine grid based on the cross-section area of a fine grid. Since all the fine cells have the same cell size, the discharge is in proportion to the water depth of each fine cell. Therefore, the weighting factor can be written as

$$K_i = d_i \Big/ \sum_{i=1}^{p} d_i \tag{6}$$

where d_i is the water depth of the fine grid [m].

3 MODEL APPLICATION

3.1 Synthetic case study

The topography of the synthetic area was a 10 km x 10 km symmetric area. The local-refined case adopted two grid resolutions, 40 m and 200 m, for model validation. The elevation of 200 m grids was generated by means of the average elevations of 40 m grids. Two local-refined regions with 2 km x 2 km area each using 40 m grid resolution were used for evaluating the model performance (Fig. 5). The benchmark case adopts original 40 m fine grid in whole

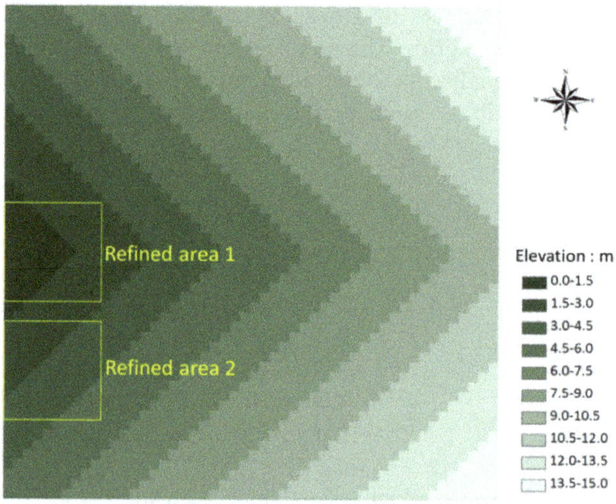

Figure 5: Topography of synthetic case study.

global domain. The building was represented as BCR setting in two refined areas including the coarse grids and the fine grids in both local-refined case and the benchmark case. The settings of BCR in refined area include 0.04, 0.09, 0.25, 0.36, 0.49 and 0.64, respectively. In all cases, a 24-hr uniform rainfall (10 mm/hr) and a constant Manning's roughness coefficient of 0.20, which was widely adopted for urban inundation modelling, were applied in simulations. Two indicators were chosen for evaluating the model performance. The root mean square error (RMSE) shows the errors of water depths between local-refined case and the benchmark case. The coefficient of determination (R square, R^2) indicates the proportion of the variance between the local-refined case and the benchmark case.

The simulation result of BCR 0.64 of case study at 24 hr is shown in Figure 6. The left part is the water depth of benchmark case. The results of the refined area, which contain the local-refined case and benchmark, are illustrated in right side respectively. Table 1 shows the RMSE of water depths between the local-refined case and the benchmark in refined area 1.

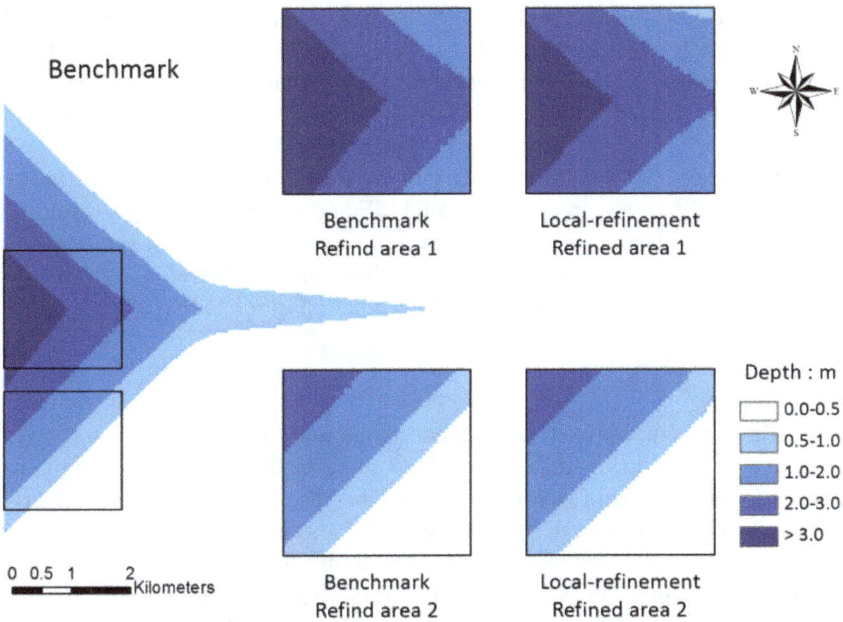

Figure 6: Simulation result of BCR 0.64 of synthetic case study at 24hr.

Table 1: RMSE of refined area 1 for synthetic case study (m).

BCR	6 hr	12 hr	18 hr	24 hr
0.04	0.006	0.008	0.006	0.025
0.09	0.007	0.008	0.008	0.028
0.16	0.007	0.009	0.010	0.030
0.25	0.010	0.014	0.014	0.038
0.36	0.022	0.028	0.027	0.066
0.49	0.052	0.053	0.053	0.152
0.64	0.090	0.098	0.089	0.233

Table 2: RMSE of refined area 2 for synthetic case study (m).

BCR	6 hr	12 hr	18 hr	24 hr
0.04	0.004	0.004	0.004	0.018
0.09	0.004	0.004	0.004	0.020
0.16	0.004	0.004	0.007	0.019
0.25	0.004	0.004	0.020	0.018
0.36	0.006	0.005	0.045	0.011
0.49	0.009	0.009	0.064	0.013
0.64	0.020	0.032	0.058	0.060

As can be seen, the RMSE increases slightly with the simulation time and the BCR, i.e. the lower BCR produces the smaller error. The RMSE is smaller than 0.1 m even the BCR arise to 0.36, whereas, the RMSE larger than 0.1 only appears at 24 hr while the BCR is bigger than 0.49. In refined area 2, the simulation result of the local-refined case is very consistent with the benchmark; the largest RMSE is only 0.06 m (Table 2). These results show that local-refined case has high correlation with benchmark.

From the result, the R^2 of the refined area 1 and area 2 are larger than 0.96 of each BCR at any simulation time. From the analysis of the RMSE and R^2, the simulation results of local-refined cases may have some differences in water depths, but they also indicate high correlation between the local-refined cases and benchmark. Therefore, the local-refined approach can adequately illustrate the detail flow situation in refined area.

3.2 Real case

The Zenwen River originates from the Alishan range and has the fourth longest length of 138.5 km in Taiwan. The downstream of the Zenwen River flow into the Jianan plain, which is the largest alluvial floodplain in southwest Taiwan. According to the land use data (Fig. 7), the simulation area is mainly covered by the agricultural land and most population is concentrated in urban settlement located in different area. These areas that tend to be more vulnerable to flood were selected as local-refined areas for inundation modelling. There are six refined-areas which has the higher BCR comparing with the BCR of whole district (Table 3).

Two cases using 40 m and 200 m grid were simulated to evaluate the performance of coarse grid in whole basin. Table 4 shows the inundation areas of two different grids of the typhoon Morakot event. This result shows that the coarse grid can maintain the accuracy and reveal the good representation of flood extent and water depth. Therefore, the simulation result of the coarsen grid can provide the adequate boundary condition to the local-refined area.

Table 5 and Table 6 are the RMSE and R^2 of the refined areas. The largest RMSE of each refined area is 0.233 m, 0.290 m, 0.331 m, 0.307 m and 0.342 m, which occurs at 42 hr, 48 hr, 48 hr, 42 hr, 42 hr and 18 hr, respectively. The largest R^2 of each refined area is 0.86, 0.92, 0.96, 0.91, 0.98 and 0.99, which occurs at 48 hr, 48 hr, 36 hr, 42 hr, 42 hr and 48 hr, respectively. Figure 8 is the simulation depth of different grid in area 6 at 48 hr; the simulation results of fine grid and local-refinement grid show the good consistence in distribution of water depth.

In this study, the operating system using for simulation is Microsoft Windows 7 Enterprise 64 bit, the processor is Intel i7-920 Quad core CPU @2.67GHz with 6.0 GB memory. Total simulation time using 40 m fine grid in whole basin is 76,380 sec (21 hr and 13 min)

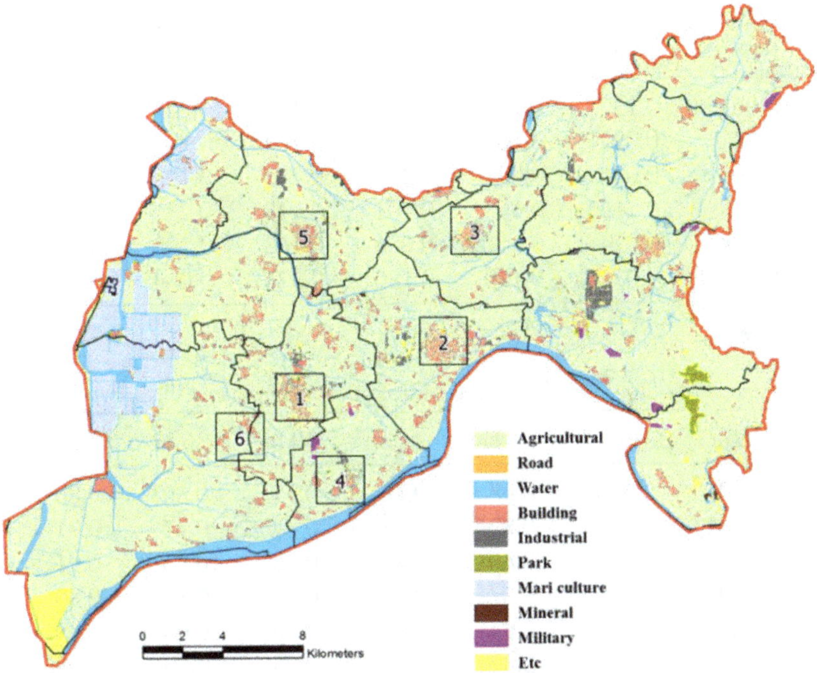

Figure 7: Land use of the Zenwen River Basin.

Table 3: BCR of refined areas in the Zenwen River Basin.

Refined-area	BCR of refined area	BCR of whole district
Area 1	0.25	0.11
Area 2	0.29	0.10
Area 3	0.17	0.07
Area 4	0.13	0.08
Area 5	0.18	0.07
Area 6	0.14	0.04

Table 4: Inundation area of different grid size of the Morakot typhoon event (ha).

Depth(m)	Fine Grid (40 m)	Coarse Grid (200 m)
0.5–1.0	10046	9972
1.0–1.5	15529	14944
1.5–2.0	5772	5576
Above 2.0	2278	1836
Total Area	33626	32328

Table 5: RMSE of the Morakot typhoon event (m).

	6 hr	12 hr	18 hr	24 hr	30 hr	36 hr	42 hr	48 hr
Area 1	0.061	0.132	0.162	0.130	0.185	0.222	0.233	0.173
Area 2	0.021	0.092	0.102	0.087	0.104	0.201	0.227	0.290
Area 3	0.006	0.025	0.081	0.234	0.283	0.105	0.261	0.331
Area 4	0.033	0.112	0.168	0.101	0.148	0.231	0.307	0.164
Area 5	0.018	0.074	0.104	0.086	0.163	0.298	0.342	0.257
Area 6	0.007	0.071	0.190	0.186	0.121	0.064	0.060	0.094

Table 6: R^2 of the Morakot typhoon event.

	6 hr	12 hr	18 hr	24 hr	30 hr	36 hr	42 hr	48 hr
Area 1	0.05	0.46	0.68	0.78	0.77	0.80	0.80	0.86
Area 2	0.82	0.70	0.82	0.83	0.87	0.77	0.91	0.92
Area 3	0.96	0.94	0.86	0.82	0.88	0.96	0.91	0.88
Area 4	0.07	0.28	0.62	0.89	0.91	0.87	0.80	0.88
Area 5	0.80	0.87	0.94	0.97	0.96	0.95	0.98	0.97
Area 6	0.96	0.87	0.82	0.88	0.98	0.99	0.98	0.99

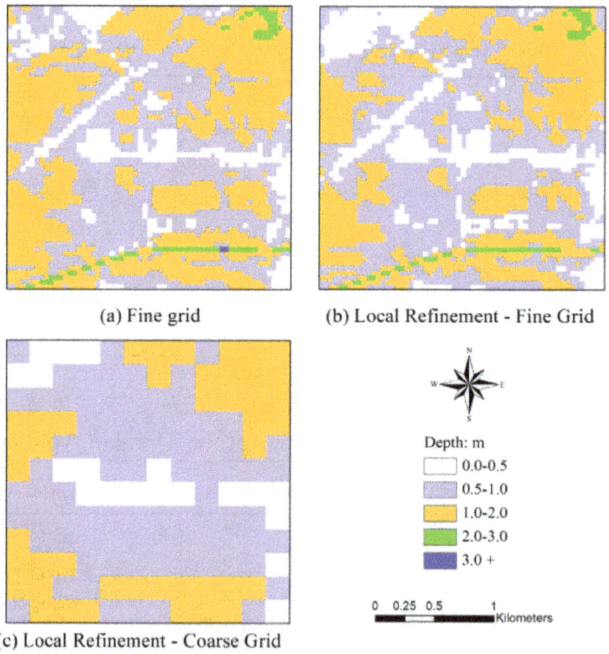

Figure 8: The simulation result of different grid in area 6 at 48 hr.

and it can be obviously reduced to 6,653 sec (1 hr and 51 min) by the local-refinement model. Consequently, the local grid refinement approach can be applied in real-time flood forecasting.

4 CONCLUSION

A local-refined inundation model was developed to represent the effect of building blockage in urban area. In this paper, we proposed a local grid refinement approach to simulate the detail surface runoff of local regions in a large basin. From the results, it can be concluded that the multi-scale model can maintain the accuracy in local-refined area and has significant improve the model efficiency.

ACKNOWLEDGEMENTS

The authors appreciate the Ministry of Science and Technology for the grant support (MOST 105-2625-M-002-023).

REFERENCES

[1] Blumberg, A.F., Georgas, N., Yin, L., Herrington, T.O. & Orton, P.M., Street-scale modeling of storm surge inundation along the New Jersey Hudson river waterfront. *Journal of Atmospheric and Oceanic Technology*, **32**(8), pp. 1486–1497, 2015. https://doi.org/10.1175/jtech-d-14-00213.1

[2] Meesuk, V., Vojinovic, Z., Mynett, A.E. & Abdullah, A.F., Urban flood modelling combining top-view LiDAR data with ground-view SfM observations. *Advances in Water Resources*, **75**, pp. 105–117, 2015. https://doi.org/10.1016/j.advwatres.2014.11.008

[3] Sampson, C.C., Fewtrell, T.J., Duncan, A., Shaad, K., Horritt, M.S. & Bates, P.D., Use of terrestrial laser scanning data to drive decimetric resolution urban inundation models. *Advances in Water Resources*, **41**, pp. 1–17, 2012. https://doi.org/10.1016/j.advwatres.2012.02.010

[4] Tsubaki, R. & Fujita, I., Unstructured grid generation using LiDAR data for urban flood inundation modelling. *Hydrological Processes*, **24**(11), pp. 1404–1420, 2010. https://doi.org/10.1002/hyp.7608

[5] Aronica, G., Tucciarelli, T. & Nasello, C., Multilevel model for flood wave propagation in flood-affected areas. *Journal of Water Resources Planning and Management*, **124**(4), pp. 210–217, 1998. https://doi.org/10.1061/(asce)0733-9496(1998)124:4(210)

[6] Begnudelli, L., Sanders, B.F. & Bradford, S.F., Adaptive Godunov-based model for flood simulation. *Journal of Hydraulic Engineering*, **134**(6), pp. 714–725, 2008. https://doi.org/10.1061/(asce)0733-9429(2008)134:6(714)

[7] Hu, K., Mingham, C.G. & Causon, D.M., A mesh patching method for finite volume modelling of shallow water flow. *International Journal for Numerical Methods in Fluids*, **50**(12), pp. 1381–1404, 2006. https://doi.org/10.1002/fld.1100

[8] Ruge, J.W., McCormick, S.F. & Yee, S.Y.K., Multilevel adaptive methods for semi-implicit solution of shallow-water equations on a sphere. *Monthly Weather Review*, **123**(7), pp. 2197–2205, 1995. https://doi.org/10.1175/1520-0493(1995)123<2197:mamfsi>2.0.co;2

[9] Lamby, P., Muller, S. & Stiriba, Y., Solution of shallow water equations using fully adaptive multiscale schemes. *International Journal for Numerical Methods in Fluids*, **49**(4), pp. 417–437, 2005.
https://doi.org/10.1002/fld.1004

[10] Henonin, J., Ma, H. T., Yang, Z. Y., Hartnack, J., Havno, K., Gourbesville, P. & Mark, O., Citywide multi-grid urban flood modelling: the July 2012 flood in Beijing. *Urban Water Journal*, **12**(1), pp. 52–66, 2015.
https://doi.org/10.1080/1573062x.2013.851710

[11] Huang, C.J., Hsu, M.H., Chen, A.S. & Chiu, C.H. Simulating the storage and the blockage effects of buildings in urban flood modeling. *Terrestrial Atmospheric and Oceanic Sciences*, **25**(4), pp. 591–604, 2014.
https://doi.org/10.3319/tao.2014.02.11.01(hy)

THE POTENTIAL OF TWO TYPES OF URBAN FLOODING TO CAUSE MATERIAL DAMAGES IN LISBON, PORTUGAL

MIGUEL LEAL & CATARINA RAMOS
Centro de Estudos Geográficos, Instituto de Geografia e Ordenamento do Território,
Universidade de Lisboa, Portugal

ABSTRACT

The incapacity to cope with heavy rainfall events leads to flooding in urban areas. Urban flooding is usually divided into pluvial flooding, sewer flooding, among others. Determining which of these types was responsible to cause a flooding occurrence is, in most cases, an error-prone task, because it can be triggered by more than one of them and there is frequently an overlap in their extent.

Given these problems, a classification based on the hydro-geomorphological features was applied to Lisbon, in which the relief and the ancient floodplains are still crucial factors in the city's current overland flow behavior. Two types of flooding are proposed: flooding related to natural drainage network (FREN) and flooding unrelated to natural drainage network (FUNN). In order to determine the material damages associated to FREN and FUNN and their spatial distribution, an insurance flooding database (2000–2011) was used. Through the accurate location of the APS (the Portuguese Association of Insurers) database flooding records triggered by rainfall and the reconstruction of the Lisbon's natural drainage network, it was possible to define which type of flooding caused each reported claim.

There are different spatial patterns in Lisbon concerning FREN and FUNN. FREN occurs along the valley bottoms, while FUNN has a more scattered spatial pattern, meaning that FUNN can occur where there are overland flow difficulties. 53% of the claims were caused by FUNN; however, higher payouts are associated with FREN (58%). Resorting only to the claims recorded during an extreme rainfall event (in 18 February 2008), the weight of FREN is even more pronounced with 58% of the claims and 71% of the payouts. This proves the higher FREN's potential to cause material damages when compared to FUNN's. This knowledge can be applied to the flooding mitigation or adaptation measures to be included in urban planning.
Keywords: FREN; FUNN; insurance database; Lisbon; material damages; Urban flooding.

1 INTRODUCTION

Material damages caused by flooding in developed countries are increasing due to the built-up areas expansion, population growth and wealth and, possibly, climate change [1–8]. In the European Union, the average annual costs related to flooding reach 4.2 billion € [4].

Flooding is one of the most important, dangerous and destructive natural hazards in the Mediterranean region [8–13]. In Portugal, flooding was the most deadly natural disaster during the 20th century [14], in which the Lisbon Metropolitan Area (LMA) is a particularly affected region. More than half of the total fatalities registered in the country between 1865 and 2010 due to flooding were recorded here [15]. However, most of the mentioned facts are due to (fluvial) floods, mostly flash floods, and not to urban flooding, whose potential to cause damages is lower [16].

There are several terms applied to the flooding that occurs in urban environment, there being no consensus in the scientific community [17, 18]. Urban flooding is probably the most used, but its definition is not clear. It can be understood as all flooding that occur inside a city [5], or as the flooding that occurs in urban areas as a result of heavy rainfall events, causing overland flow and the surcharging of the drainage systems [19, 20]. Surface water flooding (SWF) is another concept broadly used, which can be defined as the combined flooding during a heavy rainfall event, in which are included pluvial flooding, sewer flooding, flooding from small open-channel and culverted urban watercourses and overland flows from groundwater springs [17]. Therefore,

DOI: 10.2495/SAFE-V7-N2-190-200

pluvial, sewer or culverted watercourses flooding are more restricted terms/types of flooding, which can be seen as subcategories of urban flooding or SWF [21]. Despite that, some studies do not make any distinction between urban flooding/SWF and pluvial flooding [20, 22].

As was pointed out, there is no consensus and there is an overlap in the extent of these concepts. In these conditions, determining which of these types of flooding was responsible to cause a given occurrence is, in most cases, an error prone task, once it can be triggered by more than one. On the other hand, they should not be included in the same category because they often have different triggering conditions, resulting in different consequences [23]. Considering that, it is proposed a different classification based on hydro-geomorphological criteria: FREN means: flooding related to ancient natural drainage network. FUNN means: flooding unrelated to present or ancient natural drainage network.

The material damages in Lisbon are estimated resorting to an insurance database provided by the APS (the Portuguese Association of Insurers), which is available between January 2000 and October 2011. Through this database, it was possible: (a) to define which records have been caused by each proposed type of flooding; (b) to determine the material damages associated with each type of flooding in Lisbon; (c) to understand if an extreme rainfall event, like the one that occurred in 18 February 2008, is able to produce different consequences and to modify the standard pattern concerning the prevailing type of flooding; and (d) to represent the spatial distribution of material damages in Lisbon.

2 STUDY AREA

Lisbon is the Portuguese capital and comprises an area of 85 km^2, which represents 3% of the LMA's area. It is the most built-up municipality of the 18 that compose this region. More than 70% of its surface corresponds to built-up areas, where 506.892 people lived in 2015.

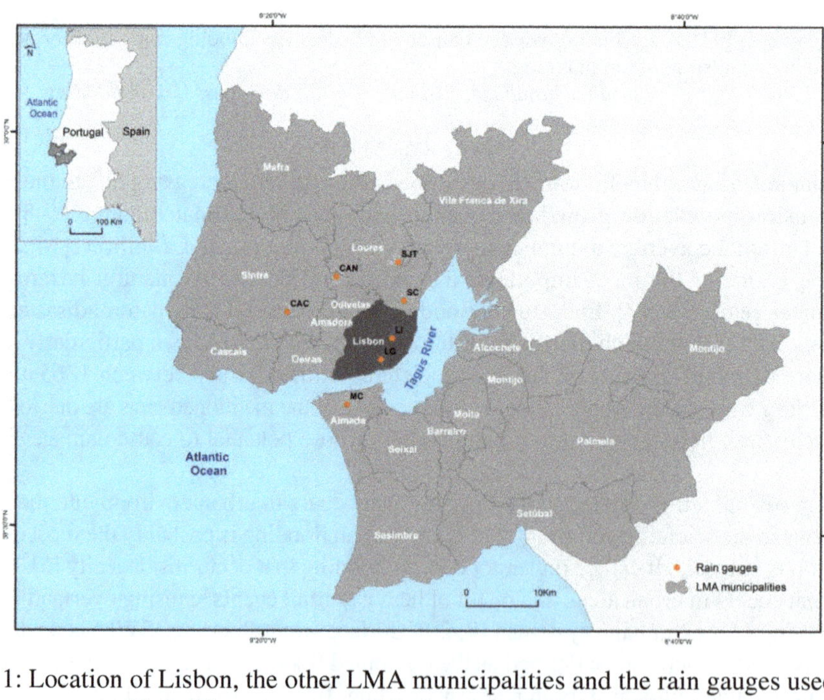

Figure 1: Location of Lisbon, the other LMA municipalities and the rain gauges used.
Note: The rain gauges' labels can be found in Table 1.

This city is located near the Atlantic coast and in the North bank of the Tagus River (Fig. 1). In Lisbon, elevation ranges between 0 (by the Tagus River) and 227 meters (in the Monsanto Mountain), being known as 'the city of the seven hills' essentially due to the natural hydrographic network action. This municipality is drained by a set of small drainage basins, whose sizes vary between 42 km^2 and less than 1 km^2. With the exception of some upstream sections, all watercourses were progressively being culverted or buried mainly in the 20th century. As a result, Lisbon is almost exclusively affected by urban flooding.

3 DATA AND METHODS

3.1 The APS database and the support rainfall data

The APS database stores the material damages caused by flooding for the period between January 2000 and October 2011 in Portugal, comprising about 60% of the national universe of insurance policies. The APS database records are designated as APS claims, containing data on: (a) communication date (when the APS claim was reported to the insurance company); (b) location based on the postal code (only accurate and localizable with seven digits); (c) capital sum affected; and (d) payouts. Capital sum affected corresponds to the capital sum insured in the insurance policy when an APS claim is reported. Payouts are the amount of money (€) paid by an insurance company to a policyholder when an APS claim is reported (the maximum value is equal to the capital sum affected). The ratio between payouts and capital sum affected is designated by loss. It was calculated for each APS claim and is expressed as a percentage; however, as the overall losses are low, the total results are expressed as per 1000.

An APS event corresponds to a flooding situation triggered by rainfall, causing one or more APS claims. Nevertheless, the APS database has no information about the flooding cause. Thus, there are APS claims caused by rainfall, coastal flooding or when pipes burst in the street or inside home. To fulfil the purpose of this research, it was necessary to define the APS events and to determine which APS claims were triggered by rainfall. To achieve this, the newspapers' information was used. Five hundred and fifty localizable APS claims were identified, corresponding to 57 flooding events related to rainfall.

In order to maximize the number of usable APS claims, daily rainfall data of seven rain gauges inside or near Lisbon were collected (Fig. 1 and Table 1). The Lisboa/Geofísico rain gauge belongs to IPMA (Portuguese Institute for Sea and Atmosphere), whereas the other six

Table 1: Characteristics of the rain gauges collected.

Code	Rain gauges	Altitude (m)	Start date	End date	Number of records (daily data)	Days with records (%)
CAC	Cacém	173	01-10-1979	Active	10,763	81.2
CAN	Caneças	277	01-10-1980	Active	11,731	89.9
LG	Lisboa/Geofísico	77	01-10-1941	Active	25,536	100
LI	Lisboa/INAG	52	13-01-2005	20-05-2009	1,155	72.7
MC	Monte da Caparica	96	01-05-1985	Active	8,456	74.3
SC	Sacavém de Cima	46	01-10-1980	30-09-2002	8,035	100
SJT	São Julião do Tojal	6	10-01-1955	Active	21,018	98.7

rain gauges belong to the SNIRH (National Water Resources Information System) network. This option was taken because there are APS claims not reported in newspapers occurring in dates with high daily rainfall values. So, if at least one of the rain gauges has recorded a value greater than or equal to 20 mm/day, and there were one or more APS claims in the same date, an APS event is considered and those claims are related to rainfall. The value of 20 mm/day is slightly higher than the 90th percentile (19.8 mm/day) of the rain gauges (4) with more than 30 years of daily rainfall data. When high rainfall values dates (higher than the 90th percentile) and the APS claims dates are coincident, it was assumed that these claims were triggered by rainfall. One hundred and two more localizable APS claims and 52 more flooding events related to rainfall were detected resorting to this methodology. These correspond to minor events in the amount of APS claims and material damages; otherwise, it would have been reported in newspapers. In total, 652 localizable APS claims and 109 flooding events related to rainfall were accounted for the considered period in Lisbon. Incorrect records (misplaced, duplicated and with missing data) were eliminated.

In the APS database, the claim date may not correspond to the date of the flooding occurrence, but to the date the situation was communicated to the insurance company. The APS estimates that the lag time between the flooding date and the communication date is approximately 5 business days. Therefore, an APS claim related to rainfall was considered when it has occurred up to 5 business days after the last day of a flooding event. When two APS events are less than 5 business days apart, there are some APS claims that may belong to both events, corresponding to 4% of the total APS claims. In those circumstances, it was decided not to associate those APS claims to any APS event. It must be recognized that the average lag time of 5 business days between the communication date and the flooding date may not allow determining with absolute safety if some APS claims are related to rainfall or not.

The total number of localizable APS claims for the 2000–2011 period (652) corresponds to 56% of the total of claims reported in the APS database for Lisbon, which is a very acceptable sample to achieve the goals of this paper.

3.2 Classification and determination of types of flooding

The APS claims were classified as FREN and FUNN. All the situations occurred in places where there are (through pipes) or there were watercourses are FREN. This includes: flooding from culverted watercourses, flooding from buried watercourses and flooding from old watercourse routes. To a certain extent, FREN corresponds nowadays to the floods that occurred in small streams before the land artificialization. This type of flooding occurs along the Lisbon's ancient floodplains, proving that the relief is still a crucial factor in the city's current overland flow behavior [24]. All the situations not related to natural drainage network are FUNN, which means that their occurrence is not influenced by watercourses. FUNN occurs in flat areas, low-lying areas and natural depressions. It can also occurs when grated inlets are obstructed by leaves or debris and whenever there are situations of overland flow blocking, whether they are natural or human induced. This type of flooding is highly related to the urban grid features.

Through the location of the APS claim, its hydro-geomorphological context (floodplains, valley bottoms or topographic natural depressions) and the reconstruction of the natural drainage network (using a set of the 19th and the 20th centuries topographic maps with 1:10,000 or 1:25,000 scales), it was possible to determine the type of flooding for each APS claim.

4 RESULTS

4.1 The APS database global results for Lisbon

Through the APS database, 109 flooding events with material damages were identified in Lisbon between January 2000 and October 2011, causing 652 localizable APS claims (55 per year) and 2,266,083 € in payouts (191,500 € per year). On average, 6 APS claims and 20,790 € of payouts per event were registered.

The 18 February 2008 event had a major influence in the APS database: 161 claims (25% of the Lisbon's total) and 1,020,521 € of payouts (45% of the Lisbon's total). The second most important event (29 October 2010) totalled 41 APS claims and 134,725 € in payouts. The 2008 event can be considered an extreme rainfall event, fitting in what Fragoso [25] defined as a torrential rainfall episode for the LMA (greater than or equal to 100 mm/day). In the São Julião do Tojal rain gauge (SJT - Fig. 1 and Table 1), which is often used as the reference rain gauge for the LMA due to the duration and reliability of the daily rainfall series [26], 140.9 mm in 24 h were registered. This value exceeded by far the average monthly value of February in this rain gauge (89.4 mm), being only surpassed by the value recorded in 18 November 1983 (163.7 mm/day). The other record above 100 mm/day in the SJT rain gauge was registered in 25/26 November 1967 (137 mm), when a very destructive flooding event caused about 700 fatalities [14]. Considering the Lisboa/Geofísco rain gauge (located inside the Lisbon municipality – Figure 1 and Table 1), the highest value in its daily rainfall series was recorded during the 2008 event (118.4 mm).

Lisbon is the LMA's municipality with the highest number of APS claims (33%) and payouts (36%). Payouts per claim reached 3471 € in Lisbon, being above the LMA's average (3229 €). This is partly explained by the fact of being the most built-up municipality (73%). Impervious surfaces lead to increases in volume and speed of overland flow [27–29], and, consequently, to increases in flooding frequency and magnitude [17, 30–32]. Lisbon has also the highest number of insurance policies (22%), dwellings (22%) and inhabitants (19%), which means a higher flooding exposure. On the other hand, factors like the hydro-geomorphological and the urban grid features, the undersized urban drainage network and the tidal influence contribute to the high amount of APS claims in Lisbon.

The Portuguese capital suffers the effects of urban flooding with some frequency, being a heavy rainfall episode that lasts a couple of hours enough to cause damages. Resorting to the APS database, it is possible to confirm that a high number of APS claims are not necessarily demanding of big rainfall events. In fact, the same amount of rainfall can cause few or many APS claims, or may not even trigger any APS claim. This is expressed by the low correlation coefficient (r = 0.51) between the number of APS claims per event in Lisbon and the daily rainfall values recorded in the Lisboa/Geofísico rain gauge. Several factors can explain this: (a) the quality and availability of rainfall and insurance data [33, 34]; (b) the spatial and temporal distribution of the triggering rainfall; (c) the geomorphological, lithological and hydrological features; (d) the extent of built-up areas; (e) the amount of exposed (and insured) dwellings; (f) the capacity of the stormwater drainage systems to cope with heavy rainfall events; (g) the obstruction by leaves or debris of the grated inlets [35]; (h) the urban grid features; and (i) the tidal and/or storm surge influence.

4.2 Types of flooding and their spatial distribution

All the previously mentioned factors make Lisbon particularly prone to urban flooding. However, the water depth and volume reached used to be low [24]. This partially justifies the absence of fatalities in this municipality since 1966 [15, 16]. The progressive culverting of Lisbon's watercourses in the 1940s, 1950s and 1960s led to the disappearance of flash floods, which is the most dangerous type of flooding in the Mediterranean region [13]. Despite the human damages caused by flooding are currently low, the same does not happen with material damages.

During the analyzed period, Lisbon achieved 36% of the APS claims caused by urban flooding (FREN and FUNN) recorded in the LMA. It should not be forgotten that this municipality occupies only 3% of the LMA's area, showing the weight of the Lisbon's urban flooding in this region.

As expected, FREN and FUNN have distinct behaviors regarding the amount of claims and the consequences both are capable of generate. This is predictable due to their proper characteristics and the places where these types of flooding occur. FREN only takes place along the ancient floodplains, presenting high densities of APS claims in some sectors located in the most important Lisbon's drainage basins (Fig. 2). On the contrary, FUNN has a scattered pattern, which is more dependent on the urban grid (Fig. 2). Generally, rainwater is ponded where there are overland flow difficulties due to blocking situations or in flat areas, having no relation with drainage basins.

Considering the total records, FUNN presented a higher number of APS claims (53%) (Fig. 3a), which is understandable taking into account the larger area that may be affected by this type of flooding. Nevertheless, FREN was able to cause higher payouts (58%) (Fig. 3b). In order to eliminate the effect of the absolute numbers, payouts per claim were obtained (Fig. 3c).

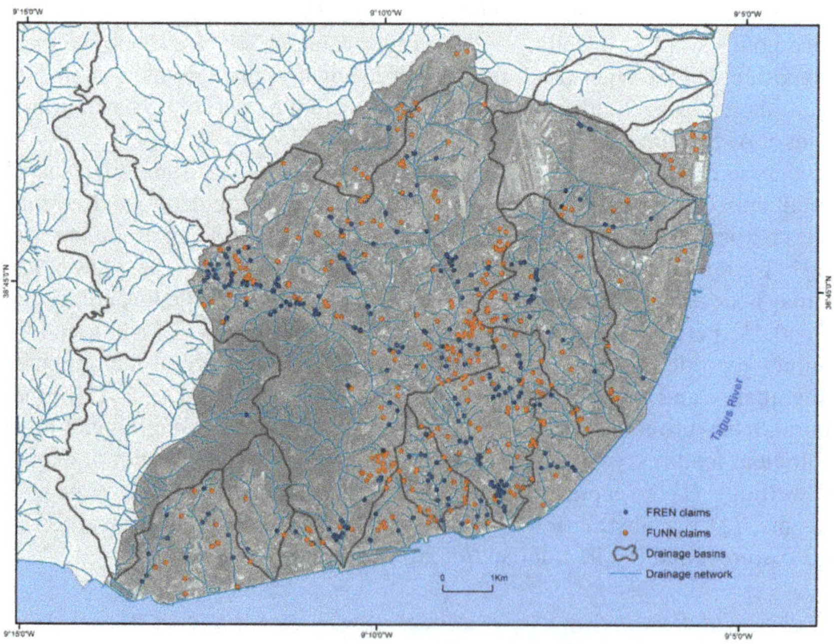

Figure 2: Spatial distribution of FREN and FUNN claims in Lisbon.

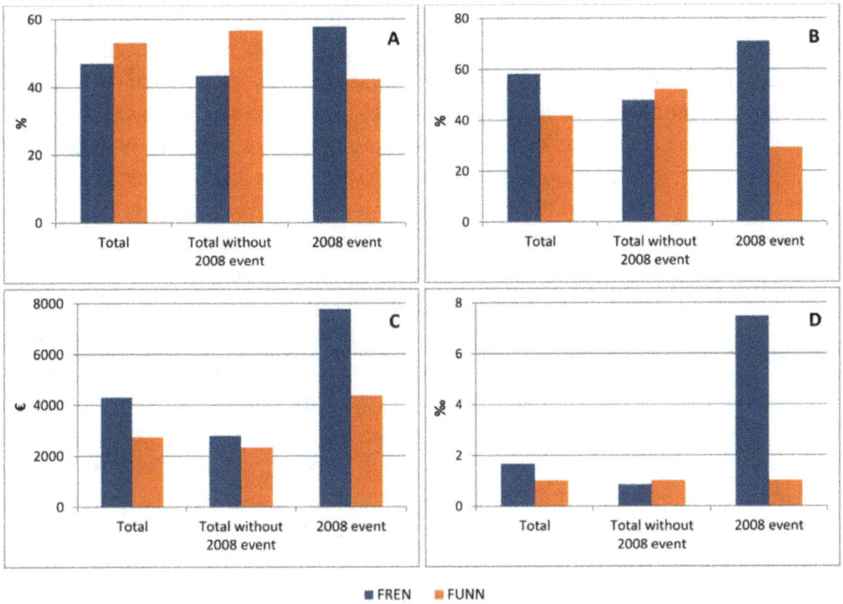

Figure 3: Material damages per type of flooding (2000–2011) in Lisbon: (a) APS claims, (b) Payouts, (c) Payouts per APS claim, (d) Losses.

This relation pointed in the same direction. FREN had 4313 € per claim and FUNN had 2735 € per claim. Losses in FREN were also higher than in FUNN (1.66‰ against 1.01‰) (Fig. 3d).

As it was previously said, the 2008 event has a major relevance in the APS database. This event highlighted the weight of FREN in the material damages recorded for Lisbon. Isolating the 2008 event, this type of flooding was responsible for a higher number of claims (58%), payouts (71%), payouts per claim (7779 € against 4370 €) and losses (7.48‰ against 1.01‰) when compared to FUNN (Fig. 3).

The weight of this event in the total records can be seen if it is excluded (total without the 2008 event – Figure 3). In fact, if the 2008 event had never happened, the results would be substantially different, once payouts per claim would drop 1513 € in FREN and 400 € in FUNN. On the other hand, the prevalence of FREN would no longer exist because APS claims, payouts and losses would be higher for FUNN (Fig. 3). The only exception would be payouts per claim, in which FREN would continue to present a higher value.

5 DISCUSSION AND CONCLUSIONS

The APS database was used in this research to determine the location of the flooding claims triggered by rainfall and to account the material damages caused by this natural hazard in Lisbon. The spatial planning decisions mainly taken in the 1940s, 1950s and 1960s, by culverting the majority of the watercourses, led to the flash floods disappearance. However, the expansion of built-up areas has increased the urban flooding problems, since increases in volume and speed of the overland flow are generated, and the stormwater drainage systems design capacity is frequently exceeded during heavy rainfall events.

Six hundred and fifty-two APS claims with an accurate location were registered in Lisbon between January 2000 and October 2011. However, it is necessary to take into account that the material damages reported here are an estimation of the total material damages occurred in Lisbon in the studied period because: (a) the 652 localizable APS claims only correspond to 56% of the total APS claims recorded by the APS database in the study area; (b) not all residential, commercial and industrial spaces have insurance policies; (c) insurance companies may refuse to make or to renew an insurance policy or they may require a very high premiums due to past flooding occurrences; (d) payouts generally do not cover all damages, because there are limits stablished in the insurance policies or some types of damages may not be included [19]; and (e) the paid amounts depend on the evaluation of the insurance companies' experts.

Lisbon presented the highest number of APS claims and payouts of the 18 municipalities that compose the LMA. Beyond the hydro-geomorphological and urban grid features, also the largest percentage of built-up areas, the highest capital sum insured and the highest number of insurance policies, dwellings and inhabitants can influence this result. It must be kept in mind that the APS database only comprises buildings insurance policies. Thus, the higher the building densities, the higher the flooding exposure.

In Lisbon, FUNN was the most common type of flooding, presenting a scattered spatial pattern. If valley bottoms are excluded, FUNN can practically occur wherever there is a (exposed) building. Nevertheless, FREN reached higher payouts, payouts per claim and losses, being essentially concentrated in some critical sectors of the main drainage basins. These facts confirm the FREN's larger destructive potential. This is even more notorious when an extreme rainfall event, like the 2008 event, occurs. The presented numbers and values support the idea that, when a high magnitude rainfall event occurs, the relief is the key factor in Lisbon and, in practice, the ancient watercourses respond in a similar way as before they were culverted. This happens because the overland flow is directed to the built-up valley bottoms, not being able to enter into the drainage system. In these cases, watercourses are formed along the streets, in the same places they existed before. In this sense, high magnitude rainfall events more easily cause FREN and the more frequent events (with lower magnitude) tend to trigger FUNN, there being less likely to cause FREN.

Based on the precise location of APS claims and the Lisbon's hydro-geomorphological features, two types of flooding are proposed. Given the difficulties in defining which of the normally used types of flooding (pluvial flooding, sewer flooding or other) causes a given APS claim, the separation in FREN and FUNN can be an important contribution in this matter. Resorting to this classification, it is possible to determine the material damages caused by two different types of flooding that occur in urban areas. This knowledge can be applied to the flooding mitigation or adaptation measures to be included in urban planning.

ACKNOWLEDGEMENTS

The authors thank the Portuguese Association of Insurers (APS) and the CIRAC project (Floods and Risk in Climate Change Scenarios) for their support and for providing the insurance data used in this research. Miguel Leal was funded by the Portuguese Science and Technology Foundation (FCT) through grant SFRH/BD/96632/2013.

REFERENCES

[1] Mills, E., Insurance in a climate of change. *Science,* **309**, pp. 1040–1044, 2005. https://doi.org/10.1126/science.1112121

[2] Botzen, W. J. W., van den Bergh, J. C. J. M. & Bouwer, L. M., Climate change and increased risk for the insurance sector: A global perspective and an assessment for the Netherlands. *Natural Hazards,* **52**, pp. 577–598, 2010. https://doi.org/10.1007/s11069-009-9404-1

[3] Aerts, J. C. J. H. & Botzen, W. J. W., Climate change impacts on pricing long-term flood insurance: A comprehensive study for the Netherlands. *Global Environmental Change,* **21**, pp. 1045–1060, 2011. https://doi.org/10.1016/j.gloenvcha.2011.04.005

[4] Jongman, B., Hochrainer-Stigler, S., Feyen, L., Aerts, J.C.J.H., Mechler, R., Botzen, W.J.W., Bouwer, L.M., Pflug, G., Rojas, R. & Ward, P.J., Increasing stress on disaster-risk finance due to large floods. *Nature Climate Change,* **4**, pp. 264–268, 2014. https://doi.org/10.1038/nclimate2124

[5] Hammond, M.J., Chen, A.S., Djordjević, S., Butler, D. & Mark, O., Urban flood impact assessment: a state-of-the-art review. *Urban Water Journal,* **12**, pp. 14–29, 2015. https://doi.org/10.1080/1573062x.2013.857421

[6] Paudel, Y., Botzen, W. J. W. & Aerts, J. C. J. H., Influence of climate change and socio-economic development on catastrophe insurance: a case study of flood risk scenarios in the Netherlands. *Regional Environmental Change,* **15**, pp. 1717–1729, 2015. https://doi.org/10.1007/s10113-014-0736-3

[7] Surminski, S., Aerts, J.C.J.H., Botzen, W.J.W., Hudson, P., Mysiak, J. & PPrez-Blanco, C.D., Reflections on the current debate on how to link flood insurance and disaster risk reduction in the European Union. *Natural Hazards,* **79**, pp. 1451–1479, 2015. https://doi.org/10.1007/s11069-015-1832-5

[8] Diakakis, M., Deligiannakis, G., Pallikarakis, A. & Skordoulis, M., Factors controlling the spatial distribution of flash flooding in the complex environment of a metropolitan urban area. The case of Athens 2013 flash flood event. *International Journal of Disaster Risk Reduction,* **18**, pp. 171–180, 2016. https://doi.org/10.1016/j.ijdrr.2016.06.010

[9] Llasat, M.C., Llasat-Botija, M., Barnolas, M., López, L. & Altava-Ortiz, V., An analysis of the evolution of hydrometeorological extremes in newspapers: the case of Catalonia, 1982–2006. *Natural Hazards and Earth System Science,* **9**, pp. 1201–1212, 2009. https://doi.org/10.5194/nhess-9-1201-2009

[10] Llasat, M.C., Llasat-Botija, M., Prat, M.A., Porcú, F., Price, C., Mugnai, A., Lagouvardos, K., Kotroni, V., Katsanos, D., Michaelides, S., Yair, Y., Savvidou, K. & Nicolaides, K., High-impact floods and flash floods in Mediterranean countries: the FLASH preliminary database. *Advances in Geosciences,* **23**, pp. 47–55, 2010. https://doi.org/10.5194/adgeo-23-47-2010

[11] Camarasa-Belmonte, A.M. & Soriano-García, J., Flood risk assessment and mapping in peri-urban Mediterranean environments using hydrogeomorphology. Application to ephemeral streams in the Valencia region (eastern Spain). *Landscape and Urban Planning,* **104**, pp. 189–200, 2012. https://doi.org/10.1016/j.landurbplan.2011.10.009

[12] Ruiz-Bellet, J.L., Balasch, J.C., Tuset, J., Barriendos, M., Mazon, J. & Pino, D., Historical, hydraulic, hydrological and meteorological reconstruction of 1874 Santa Tecla flash floods in Catalonia (NE Iberian Peninsula). *Journal of Hydrology,* **524**, pp, 279–295, 2015. https://doi.org/10.1016/j.jhydrol.2015.02.023

[13] Pereira, S., Diakakis, M., Deligiannakis, G. & Zêzere, J. L., Comparing flood mortality in Portugal and Greece (Western and Eastern Mediterranean). *International Journal of Disaster Risk Reduction,* **22**, pp. 147–157, 2017. https://doi.org/10.1016/j.ijdrr.2017.03.007

[14] Ramos, C. & Reis, E., Floods in Southern Portugal: their physical and human causes. *Mitigation and Adaptation Strategies for Global Change,* **7**, pp. 267–284, 2002. https://doi.org/10.1023/a:1024475529524

[15] Zêzere, J.L., Pereira, S., Tavares, A.O., Bateira, C., Trigo, R.M., Quaresma, I., Santos, P.P., Santos, M. & Verde, J., DISASTER: A GIS database on hydro-geomorphologic disasters in Portugal. *Natural Hazards,* **72**, pp. 503–532, 2014. https://doi.org/10.1007/s11069-013-1018-y

[16] Pereira, S., Zêzere, J.L. & Quaresma, I., *Mortality associated to hydro-geomorphologic disasters in the Great Lisbon area in the last 150 years.* International Conference on Urban Risks, Lisbon, 2016.

[17] Falconer, R.H., Cobby, D., Smyth, P., Astle, G., Dent, J. & Golding, B., Pluvial flooding: New approaches in flood warning, mapping and risk management. *Journal of Flood Risk Management,* **2**, pp. 198–208, 2009. https://doi.org/10.1111/j.1753-318x.2009.01034.x

[18] Priest, S.J., Parker, D.J., Hurford, A.P., Walker, J. & Evans, K., Assessing options for the development of surface water flood warning in england and wales. *Journal of Environmental Management,* **92**, pp. 3038–3048, 2011. https://doi.org/10.1016/j.jenvman.2011.06.041

[19] CNT. *The prevalence and cost of urban flooding: a case study of cook county, IL.* Center for Neighborhood Technology, 2014.

[20] Yu, D. & Coulthard, T.J., Evaluating the importance of catchment hydrological parameters for urban surface water flood modelling using a simple hydro-inundation model. *Journal of Hydrology,* **524**, pp. 385–400, 2015. https://doi.org/10.1016/j.jhydrol.2015.02.040

[21] Parker, D.J., Priest, S.J. & McCarthy, S.S., Surface water flood warnings requirements and potential in England and Wales. *Applied Geography,* **31**, pp. 891–900, 2011. https://doi.org/10.1016/j.apgeog.2011.01.002

[22] Pitt, M., *Learning lessons from the 2007 floods. floods review,* 2008. https://doi.org/10.1007/s13398-014-0173-7.2.

[23] Jha, A.K., Bloch, R. & Lamond, J., Cities and flooding: a guide to integrated urban flood risk management for the 21st century. *Journal of Regional Science*, **52**(5), pp. 885–887, 2012. https://doi.org/10.1596/978-0-8213-8866-2

[24] Oliveira, P.E. & Ramos, C., Inundações na cidade de Lisboa durante o século XX e seus factores agravantes. *Finisterra*, **74**, pp. 33–54, 2002.

[25] Fragoso, M., *O contexto atmosférico dos episódios de precipitações intensas no Sul de Portugal.* Relatório 39, Área de Investigação de Geografia Física e Ambiente, Centro de Estudos Geográficos, Lisbon, 2004. ISBN: 972-636-148-6.

[26] Trigo, R.M., Ramos, C., Pereira, S.S., Ramos, A.M., Zêzere, J.L. & Liberato, M.L.R., The deadliest storm of the 20th century striking Portugal: Flood impacts and atmospheric circulation. *Journal of Hydrology,* **541**, pp. 597–610, 2016. https://doi.org/10.1016/j.jhydrol.2015.10.036

[27] Patton, P., Drainage basin morphometry and floods. In: V. Baker, R. Kochel, & P. Patton, (eds) *Flood geomorphology*. Wiley, New York, pp. 51–65, 1988. ISBN: 978-0471625582.
[28] Smith, L. & Ward, R., *Floods: physical processes and human impacts*. Wiley, Chichester, 1998. ISBN: 978-0-471-95248-0
[29] Butler, D. & Davies, J.W., *Urban drainage*. 2nd edition. Spon Press, London, 2004. ISBN: 0-203-34190-2.
[30] Suriya, S., Mudgal, B.V. & Nelliyat, P., Flood damage assessment of an urban area in Chennai, India, part I: Methodology. *Natural Hazards,* **62**, pp. 149–167, 2012. https://doi.org/10.1007/s11069-011-9985-3
[31] Huong, H.T.L. & Pathirana, A., Urbanization and climate change impacts on future urban flooding in Can Tho city, Vietnam. *Hydrology and Earth System Sciences,* **17**, pp. 379–394, 2013.
https://doi.org/10.5194/hess-17-379-2013
[32] Chen, C.F. & Liu, C.M., The definition of urban stormwater tolerance threshold and its conceptual estimation: An example from Taiwan. *Natural Hazards,* **73**, pp. 173–190, 2014.
https://doi.org/10.1007/s11069-013-0645-7
[33] Merz, B., Kreibich, H., Thieken, A. & Schmidtke, R., Estimation uncertainty of direct monetary flood damage to buildings. *Natural Hazards and Earth System Science,* **4**, pp. 153–163, 2004.
https://doi.org/10.5194/nhess-4-153-2004
[34] Spekkers, M.H., Kok, M., Clemens, F.H.L.R. & Ten Veldhuis, J.A.E., A statistical analysis of insurance damage claims related to rainfall extremes. *Hydrology and Earth System Sciences,* **17**, pp. 913–922, 2013.
https://doi.org/10.5194/hess-17-913-2013
[35] Cherqui, F., Belmeziti, A., Granger, D., Sourdril, A. & Le Gauffre, P., Assessing urban potential flooding risk and identifying effective risk-reduction measures. *Science of The Total Environment,* **514**, pp. 418–425, 2015.
https://doi.org/10.1016/j.scitotenv.2015.02.027

MANAGING THE SOCIOECONOMIC IMPACTS OF EXTREME WEATHER EVENTS IN THE SOUTHWEST PACIFIC BASIN

VIVIENNE SAVERIMUTTU & MARIA E. VARUA
School of Business, Western Sydney University, Sydney.

ABSTRACT
Cyclones are a recurrent feature in the Southwest Pacific basin. Australia, Papua New Guinea, Fiji and New Zealand are responsible for monitoring and forecasting cyclone activity in the region. In dealing with extreme events of this nature, the small islands and atolls have had to depend heavily on their more developed neighbours for assistance. Under the FRANZ and Pacific Island Forum (PIF) arrangements Australia, New Zealand and France have an agreement in place to assist partner countries when such disasters strike. The last category five cyclone (Australian scale), to make landfall in this region was Cyclone Winston, which devastated the Island of Fiji in 2016. Fiji is one of the 16 independent member States of the PIF, which includes Australia and New Zealand. Following the Hyogo Framework for Action 2005–2015, an outcome of the World Conference on Disaster Reduction held in January 2005, the PIF convened a meeting, in Madang, Papua New Guinea, to develop a regional Framework for implementation. Foremost in mind were the social and economic structures of these countries where entire communities could lose their livelihoods or face severe disruption as a result of one disaster. This paper explores disaster management, of the impact of severe cyclones, by these PIF countries (excluding Australia and New Zealand), specifically in terms of disaster prevention, preparedness, relief and recovery. Specific focus will be on the islands of Fiji, Tonga and Vanuatu, where category five cyclones made landfall within the last three years and Samoa, where a category four cyclone made landfall in 2012.
Keywords: cyclones, Pacific islands, regional cooperation, socioeconomic impacts and management

1 INTRODUCTION

Natural disasters are a persistent characteristic of the Southwest Pacific (SWP) basin. Fourteen member States of the Pacific Island Forum (PIF) located within this region are Small Island Developing States (SIDS). The social and economic structures of these SIDS are incapable of coping effectively with severe natural disasters. Thus, when natural disasters strike, more often than not, these SIDS have had to seek assistance from Australia and New Zealand, their more developed neighbours. Australia and New Zealand are also members of the 16 nation PIF regional partnership.

A study by the World Bank [1], assessing the period between 1950 and 2004, reported that more than 3.4 million people have been affected by natural disasters in the SWP region. Reported fatalities during this period, excluding the more than 2,000 deaths resulting from the 1998 Tsunami that struck the Aitape Province of Papua New Guinea [2], were 1747. There is no doubt that a tsunami would have the most devastating impact in terms of fatalities and infrastructure damage on these fragile nations let alone Australia or New Zealand. Further, as reported by the World Bank [1] drought has affected the most number of people in this region. However, the overwhelming majority of natural hazards reportedly affecting the SWP basin are tropical cyclones (TCs), which had the highest frequency of occurrence during this 50 year period. The most recent category five TC (Australian cyclone category scale), to make landfall in this region at the time of writing this article, was cyclone Winston, which struck the island of Fiji, on the 20th of February 2016 [3].

DOI: 10.2495/SAFE-V7-N2-201-212

A category 5 TC based on the Australian scale is a severe cyclone accompanied by extremely destructive wind gusts of speeds greater than 280 km/h, which equates to Beaufort 12. On the Beaufort scale this is the highest category and is classified as a Hurricane [4]. The island economies located in the SWP include a high dependence on tourism, agriculture and fisheries in general. Severe cyclones not only disrupt economic activity, but also result in, damage to essential infrastructure and social sectors, coastal erosion as a result of storm surges, disruption of access to basic needs and because of budgetary and human resource problems a slow recovery. This paper will explore the management of the impact of TCs in the PIF SIDS located in the SWP region. Specific focus will be on the islands of Fiji, Tonga and Vanuatu, countries that have experienced category five cyclones making landfall within the last three years and Samoa, where a category four cyclone was experienced in 2012. Section 2 will assess the impact of cyclones on these SIDS, with particular focus on the productive sector of tourism, followed by an exploration of pre-disaster prevention and preparedness strategies and a post disaster assessment of relief and recovery.

2 IMPACT, EXPOSURE AND VULNERABILITY

Although Vanuatu and Fiji, SIDS that are a part of the Melanesian ethno-geographic grouping, experienced higher incidences of cyclones during the period 1950 to 2004, the impact of such events on social and economic structures was considered greater in Samoa and Tonga, SIDS that belong to the Polynesian grouping [1]. The extent of the destruction caused by a cyclone increases as its strength increases. Some authors [5, 6] contend that the intensity of cyclones has increased in recent decades as a result of rising sea surface temperature (SST), generating more category 4 and 5 cyclonic events since the 1970s. The events of the last 5 years within the SWP appear to justify this viewpoint.

Samoa experienced a category four cyclone, Evan, in December 2012, which while killing 14 people and displacing 7,500 people or more, damaged and destroyed homes and key infrastructure and disrupted the livelihoods of many [7]. A category five cyclone, Ian, destroyed nearly 11% of Tonga's GDP in 2014, while affecting approximately 5500 people and damaging or destroying almost 75% of the Tongan island of Ha'apai's housing stock and other buildings [8]. When a category five cyclone, Pam, struck approximately a third of Vanuatu's islands in 2015, it caused 11 fatalities and affected nearly 188,000 people with 75,000 needing emergency shelter. The total impact was estimated at 64% of Vanuatu's GDP [9]. In 2016, a category five cyclone, Winston, affected 15% of Fiji's population, and the damage was estimated at approximately 32% of GDP with 44 fatalities [10]. The following sub-sections examine the impact of cyclones on the productive, social and infrastructure sectors of these four countries, especially growth rates in the tourism industry, which is of growing importance within the SWP region, followed by their exposure to risk.

2.1 Impact of cyclones on the tourism industry

Quarterly time series data on cyclone activity [11] and tourist arrivals by air to Fiji [12], Samoa [13], Tonga [14] and Vanuatu [15] for the years 2002–2016 was gathered for the purpose of analysing the impact of cyclones on tourism, an important industry in terms of employment and economic growth. The augmented Dicky-Fuller tests indicate that the data for both growth rates of tourist arrivals and cyclone activity are stationary, a necessary condition to run simple regression analysis of time series data (Table 1).

Table 2 displays negative correlation coefficients for each country with p values significant at 1% for Fiji, Tonga and Vanuatu and 10% for Samoa confirming that cyclonic activity

Table 1: ADF (unit root test) Tau statistic.

Variable Name	Fiji	Samoa	Tonga	Vanuatu
Arrivals- Quarterly growth rate	-7.533***	-9.508***	-10.265***	-7.334***
Quarterly data cyclone activity	-7.628***	-7.843***	-8.236***	-8.665***

***significant at 1% level of significance

Table 2: Impact of cyclones on growth rates of tourist arrivals from 2002 to 2016.

	Cyclone	
Variable Quarterly Growth Rate Arrivals	Correlation coefficient	P-value
Fiji	-0.37787	0.003171***
Samoa	-0.23050653	0.090436*
Tonga	-0.47413	0.000149***
Vanuatu	-0.60806054	8.48289E-07***

***,**,* significant at 1%, 5% and 10% respectively

during the 15 years affected the tourism industry. The correlation coefficient values indicate that Vanuatu experienced the greatest impact. This result is not surprising considering that Vanuatu experiences a higher frequency of cyclones being located at the centre of the Pacific cyclone belt [16].

2.2 Impact of cyclones on other sectors

Cyclones have also affected other productive sectors such as agriculture, livestock and fisheries, manufacturing and commerce in Samoa, Tonga and Vanuatu and additionally the mining sector in the case of Fiji. Table 3 below details the total of the estimated damage and loss from all sectors in US dollars. The exchange rate used is the rate prevailing at the time the severe cyclone was experienced by each of the four countries during the years 2012 to 2016. Although three of the countries experienced a category five cyclone, the extent of the damage depended on the cyclone track and the level of development along its path. Thus, comparisons of the estimated monetary value of the impact based on the strength of the cyclone are futile.

Fiji, the most developed and wealthiest of the four countries, sustained the highest level of damage, estimated at US$ 1.3 billion, when cyclone Winston made landfall on the largest island, Viti Levu, followed by storm surges affecting the second largest island, Vanua Levu, with the entire island losing power [17]. In Vanuatu, the worst impacts of cyclone Pam were felt on the larger islands Efate, Tanna and Erromango. However, estimates of the total of damage and loss are based on the best available information and therefore incomplete [16]. The impact of cyclone Evan on Samoa was significant considering the relatively small size of the Samoan economy. Major damage was sustained in Apia and the central agricultural areas of Upolu [18]. The damage caused by cyclone Ian occurred as it passed directly over the northeast islands of Ha'apai in Tonga. The information on damage and loss resulting from cyclone Ian is also incomplete [19, 20].

Table 3: Damage and loss caused by cyclones in the SWP 2012–2016.

Sectors	Fiji 2016 US$ Millions [17]	Samoa 2012 US$ Millions [18]	Tonga 2014 US$ millions [19–20]	Vanuatu 2015 US$ millions [16]
Productive Sectors				
Agriculture, livestock, fisheries	254.46	32.80	20.8	57.74
Manufacture & commerce	67.04	17.12	N/A	31.89
Tourism	56.34	21.89	0.86	90.65
Mining	14.79	-	-	-
Social Sectors				
Education, health, housing	407.46	24.89	27.42	142.57
Infrastructure Sectors				
Electricity, transport, water & sanitation,	116.71	75.43	8.92	88.86
Cross-sector				
Environment	422.39	31.86	N/A	50.75
Total	1,339.19	203.99	58.0	462.46

2.3 Hazard risk, exposure and vulnerability

Despite increasing research on the frequency of TCs, no trend has been identified other than increasing destructiveness relating to greater intensities of the TCs [5]. The risks associated with TCs are based on clear-cut aspects of hazard, exposure and vulnerability with consequences relating to different timeframes. The immediate problems to be dealt with are damage, mortality, morbidity, displacement and asset damage. The hazard arising from TCs relate to storm surges and floods resulting from strong winds and torrential rainfall. In the SWP, with increasing development in coastal areas and subsequent increases in population densities, the exposure to risk from TCs is growing [21]. Further, the vulnerability of coastal populations is higher due to storm surges. Populations living in remote rural areas are also more vulnerable to risk, though in urban areas slum dwellers appear to be more at risk in comparison with urban residents living in better constructed dwellings [22].

2.4 Socioeconomic profiles of the SIDS

Vanuatu heads the World Risk Index [23], for both exposure and vulnerability to natural hazards, followed by Tonga. Fiji is 16th on the list and Samoa is not listed in 2015 (51st in 2012). In Vanuatu, although it is difficult to identify instances of extreme poverty, the remoteness of some rural areas and lack of access to basic infrastructure, exacerbated by budgetary problems of the government, reduces connectivity [16] creating problems for effective pre-disaster management. Figures 1 and 2 below [24] display access to electricity by the rural and urban populations of the four SIDS. Vanuatu has only 12.7% of the population living below the poverty line. Yet, access to electricity is lowest in Vanuatu reflecting the remoteness of the rural community and a GDP per capita, which is below the Pacific average.

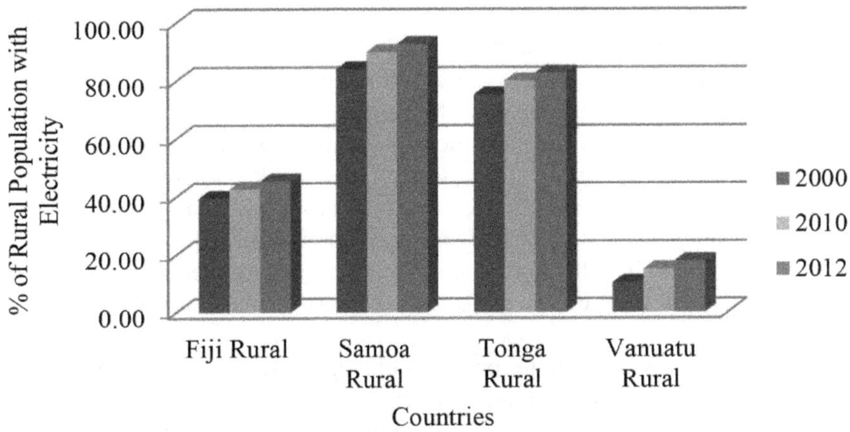

Figure 1: Access to Electricity – Rural Population

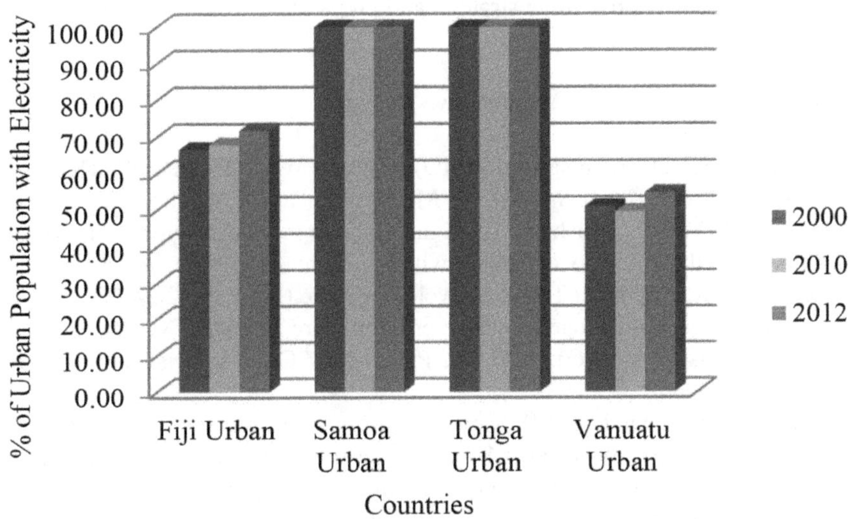

Figure 2: Access to Electricity – Urban Population

Despite being a relatively wealthier country, the data for Fiji too reflects the remoteness of some islands and also the relatively high percentage (35%) of the total (44% rural and 26% urban) population living below the poverty line. Samoa and Tonga have 26.9% and 22.5%, respectively, of the population existing below the poverty line [25]. However, they have policies in place that have ensured high access to electricity by the rural and urban populations of both countries. In Samoa, nearly 77% of residential electricity usage is accessed on a prepayments basis. In Tonga, the government has concentrated on energy efficiency and least cost access [26].

3 PREVENTION, PREPAREDNESS, RELIEF AND RECOVERY

A cyclone cannot be prevented, but its impact can be managed such that the affected communities can cope with its worst effects when it strikes, minimise the losses in the medium term and recover in the long term without a permanent loss in wellbeing. Therefore, prevention is discussed in terms of pre-disaster monitoring and forecasting activities resulting in timely warnings, regarding the cyclone's category and track, facilitating the possibility of better preparedness. The SWP region has dealt with cyclones over centuries. To some extent, the oral transmission of traditions by the indigenous population has resulted in some knowledge of coping strategies surviving through to present times. Such knowledge though useful may be inadequate because of changes in climate due to global warming and changes in frequency of occurrence and intensity of the TCs along with increased levels of development. Also much knowledge has been lost due to the introduction of modern methods and further traditional knowledge has not been enhanced following improvements and advances in science and technology.

The emphasis here is on being better prepared utilising both indigenous coping strategies and modern methods ensuring that the impact of the worst cyclone does not exceed the capacity of the affected community to cope. Currently though, this is an aspiration for countries in the SWP, including the SIDS that are the focus of this study, due to their, in some cases, excessive, dependence on the international community for assistance with pre-disaster monitoring and forecasting as a preventive strategy and preparedness as well as for post-disaster relief and recovery.

3.1 Monitoring and forecasting as a preventive strategy

The World Meteorological Organisation (WMO), under its World Weather Watch Programme [27], has established a Regional Specialised Meteorological Centre (RSMC) in Nadi, Fiji. This is one of six RSMCs established around the globe to monitor cyclone activity and provide warnings and advisories to countries within its area of responsibility. The area of responsibility of the RSMC-Nadi Tropical Cyclone Centre (160E to 120W, Equator to 25S) includes, Samoa, Tonga and Vanuatu. In close proximity are Tropical Cyclone Warning Centres (TCWCs) established in, Port Moresby, Papua New Guinea, Brisbane, Australia and Wellington, New Zealand.

Australia and New Zealand, as PIF members, are responsible for regional security and therefore monitor cyclone activity in the SWP area ensuring that other member countries are able to maintain the stability of their economies. An event categorised as a disaster, signifies that the impact of the event exceeds the capability of the affected SIDS to cope.

Additionally, there are National Meteorological Services in Samoa and Vanuatu. In the event of a cyclone, timely information fed effectively into the Early Warning Systems (EWS) is essential so that preparation to meet the threat could get underway speedily. However, there are problems with the EWSs in place. In Samoa, it was reported that only the Joint Typhoon Warning Centre (JTWC) located in the US, which also issues cyclone warnings for the SWP, correctly predicted the track for cyclone Evan in 2012. Communication to the public was also not timely due to procedural delays at the RSMC in naming the cyclone pointing to human resource constraints [18]. The post-disaster assessment on cyclone Pam raised concerns on the lack of a reliable scientific basis underlying the prediction and forecasting of cyclones [16] in 2014. In communicating the information to the public in Vanuatu, the use of social media, radio and mobile phones was critical as in Fiji when preparing for cyclone Winston in 2016 [17]. Figure 3 below details the increase in mobile phone usage in the four SIDS since 1998.

However, if a cyclone damages the communications systems, as it did in Vanuatu, then remote rural communities as well as urban communities can experience reduced connectivity in the post disaster relief period.

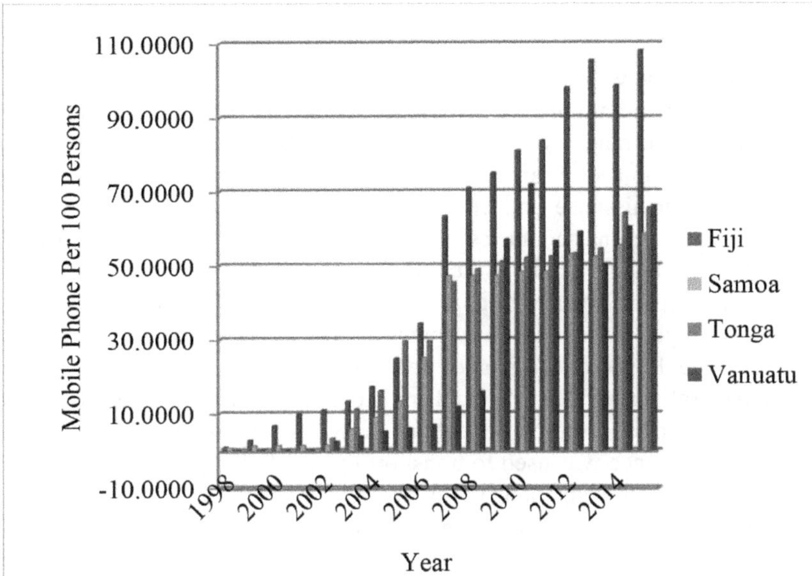

Figure 3: Increase in mobile phone usage per 100 persons 1998–2015.

3.2 Preparedness

Samoa, having experienced a tsunami in 2009, had disaster response plans in place. However, it was reported that at community level, preparedness was generally low even though most people felt that they had a good understanding of cyclones having experienced it before. This community sentiment was probably due to the fact that the cyclone warning was not timely for reasons stated previously. Further, officially designated evacuation centres were operational for the first time but only in urban areas. Overall, a need was expressed for a multi-hazard, community level, disaster management plan to address the preparedness gaps that were experienced [18]. In Tonga, the Meteorological Service was praised by communities for timely alerts on the impending cyclone Ian. Warnings continued even after the cyclone had moved on to Fijian waters. However, a need for improving community response was expressed, as feedback indicated that many residents accessed the evacuation centres only after they had been exposed to danger [28].

The Vanuatu Meteorological and Geo-hazards Department (VMGD) was effective in alerting the public of the approaching hazard, cyclone Pam, in a timely and accurate manner. This fact and the response of the public to the warnings to seek appropriate shelter, is credited with having limited the loss of life to 11 persons. There were no cyclone proof emergency shelters in Vanuatu. So public buildings, which had not been purpose built to withstand category 5 cyclones, were used as emergency shelters. A cyclone with devastation to such an extent occurred nearly 30 years prior to cyclone Pam. Therefore, the residents were inexperienced in taking steps to minimise damage to property. The needs of vulnerable members of society too were not taken into consideration [16]. In Fiji, disaster risk management was considered proactive having evolved significantly over the past three decades. There was effective transmission of monitoring data to accurately predict the cyclone's path and issue timely and accurate warnings on what needed to be done in preparation. However, here too, inexperience of storms of such intensity resulted in poorly located evacuation centres [17].

3.3 Relief

In the event of an extreme weather event or any other disaster the four countries are dependent on the international community for assistance, not only to fund relief operations but also for logistical support and human resource skills. Under the FRANZ arrangement signed in December 1992, three countries, France, Australia and New Zealand, stand ready, if requested, to provide aerial reconnaissance and co-ordinate relief assistance working with national counterparts in their 12 partner countries within the SWP. This arrangement recognises the vulnerability of these Pacific island SIDS when a disastrous event occurs, which often exceeds the capacity of the country to cope with on its own. All four SIDS, which are the focus of this study, requested assistance during the period 2012 to 2016.

Following cyclone Evan, an assistance team with essential equipment was sent by France, while Australia and New Zealand assisted with relief supplies. In response to cyclone Ian, relief supplies from New Zealand were flown to Tongatapu. A smaller plane (CASA) supplied by France was used to transport these relief supplies, as well as supplies stockpiled by Australia in Tongatapu (held by the Tongan Red Cross), to Ha'apai. The CASA is a useful smaller plane, which can easily land and take off on shorter tracks [29]. Relief assistance following cyclone Pam included aerial reconnaissance to assess the damage as well as cargo flights transporting humanitarian assistance teams and supplies under the FRANZ partnership [30].

Similar assistance was offered to Fiji following cyclone Winston. France supplied a CASA to transport supplies and humanitarian work teams, and a helicopter to Fijian authorities for aerial reconnaissance and transport of medical equipment [31]. Australia despatched HMAS Canberra carrying defence force personnel, emergency medical supplies, food and water [3]. Both Fiji and Vanuatu have remote areas that were affected and required additional support. New Zealand too assisted with aerial reconnaissance.

Assistance in terms of emergency supplies and support personnel, for all four SIDS, were also provided by the PIF SIDS that were not affected, International Organisations and various NGOs. Most importantly, financial assistance was supplied in the form of grants from various countries and international organisations, including the World Bank. The governments of the affected countries too approved tax concessions and drew from their own funds set aside for disaster relief efforts. Such funding was required not only for short term relief efforts but also the medium term and longer term recovery plans.

3.4 Recovery

The first priority, in terms of relief efforts, was to restore communications and electricity, clean up debris to secure access to airports and affected areas and ensure food supplies, water, sanitation and hygiene products (WASH), shelter and protection for those affected. Social structures, relating to education and health, that had sustained damage, also required attention with temporary structures set in place with as little delay as possible. In each case a state of emergency was declared by the Prime Minister of the affected country. While processes to attend to these immediate term needs were underway, post disaster needs assessments (PDNA) were being prepared to seek much needed funds and plan the management of the medium term and long term recovery. The damage and loss sustained for which funding was required has been illustrated in Table 3 under section 2.2.

In Samoa long term recovery projects focussed on strengthening resilience by constructing weather resistant roads and relocating people away from the coast line (Enhanced road access project), improving response to disasters through the Agriculture and fisheries cyclone response project, and, improving financial management of future shocks through the Samoa development policy operations project [7]. In Tonga's Ha'apai islands, the Tonga cyclone reconstruction and climate resilient project focussed on improving housing resilience through safe home construction and providing new technology for housing damage assessments [8]. Projects are also under way to make electricity and transport infrastructure more resilient to cyclone damage. In Vanuatu, extensive infrastructure rehabilitation work, including an improved EWS has received attention for long-term funding [9]. In Fiji [32], the rural areas were the worst hit. As such, long-term recovery needs include financial support for vulnerable communities through its social protection system as well as those needs identified in the PDNA (Table 3, Section 2.2).

4 CONCLUSION AND RECOMMENDATIONS

The management of the impact of severe cyclones during the past 5 years, in four SIDS located in the SWP, was examined in terms of pre-disaster monitoring and forecasting (prevention) and preparedness followed by post disaster relief and recovery assessments. Each of these SIDS experienced a severe cyclone within the last five years. In each case, managing the socioeconomic impacts, of these extreme weather events, required the assistance of Australia and New Zealand in particular, as well as other countries and international organisations. Some lessons learned and recommendations based on good practice in managing these events are elucidated below.

As a preventive measure effective monitoring and forecasting is essential. In Vanuatu, the EWS was damaged and a recommendation was made to protect the EWS infrastructure and ensure sound monitoring and forecasting based on scientific methods [16]. Delays experienced in Samoa emphasise the need for timeliness of warnings, which will ensure better preparation in securing assets that could sustain damage. In the pre-disaster period, it is important to ensure that a mix of indigenous and modern methods prevail to achieve the best results. For example the cultivation of cyclone resistant foods based on indigenous traditions while utilising modern cyclone resistant storage facilities for food and water. Considering that loss of communication and power was an issue in all cases, it is essential to ensure that alternatives are in place, when preparing for the disaster, to fall back on in the post disaster period. Although mobile, phones and high frequency (HF) radios proved to be useful communication tools, both the HF antennae and cellular towers were damaged in Vanuatu and communication had to be conducted via internet. For power shortages and disruptions, solar lanterns are a cheap and durable alternative as a lighting source, especially for the rural poor, when power lines are damaged. As cyclones are a seasonal occurrence in this region, clearing and securing potential sources of debris and damage to essential infrastructure prior to the cyclone season is an option. Also as highlighted by Samoa in its PDNA, a community level disaster management plan would ensure timely evacuation, which was not the case for many in Tonga in 2013.

Budgetary and human resource constraints and the extent of generally unplanned development in these SIDS, impose severe restrictions in terms of coping with disastrous events without international assistance for relief and recovery. Additionally, Fiji and Vanuatu face the problem of remoteness and difficulty of access to some of their rural areas. Budgetary and human resource constraints faced by these SIDS are unlikely to change in the near future and recovery will be slow. Most of the projects that commenced as a response to a severe cyclone, as early as in 2013, have had their original end dates extended to 2017. What can be done is to ensure that funds released for recovery be utilised to address the human resource constraints as well as the financial management skills of local disaster management offices in addition to reconstruction and recovery. Further, the problem of coastal development could also be addressed. Relocation of some communities from coastal areas has already commenced in some SIDS. Traditionally, the indigenous population would relocate to higher ground following a severe cyclone.

Finally, the projects that focus on cyclone resilient durable infrastructure, implicitly acknowledge that the impact of climate change is serious. Funding will most certainly be inadequate to ensure that future damage will be minimal. Perhaps, an alternative would be to consider the use of Pacific architecture [33] like the *fale (house)* in Samoa, a fabrication that will be damaged by cyclones but replaceable easily and cheaply and could be managed to fall within the budgetary capabilities of the SIDS.

REFERENCES

[1] World Bank. *Not if but when. Adapting to natural hazards in the Pacific Islands region: A policy note,* 2006. Available at: http://documents.worldbank.org/curated/en/2006/01/6589273/not-adapting-natural-hazards-pacific-islands-region-policy-note. (Accessed 15 April 2017).

[2] Geoscience Australia. *The Aitape Papua New Guinea Earthquake and Tsunami*, 1998. Available at: https://data.gov.au/dataset/the-aitape-papua-new-guinea-earthquake-and-tsunami-1998. (Accessed 15 April 2017).

[3] DFAT Australia. *Tropical cyclone Winston*, Available at: http://dfat.gov.au/aid/topics/investment-priorities/building-resilience/humanitarian-preparedness-and-response/tc-winston/Pages/default.aspx. (accessed 16 April 2017).

[4] Bureau of Meteorology Australia. *Tropical cyclone intensity impacts*. Available at: http://www.bom.gov.au/cyclone/about/intensity.shtml. (Accessed 15 April 2017).

[5] Emanuel, K.A., Increasing destructiveness of tropical cyclones over the past 30 years, *Nature*, **436**(7051), pp. 686–688, 2005.
https://doi.org/10.1038/nature03906

[6] Webster, P.J., Holland, G.J., Curry, J.A. & Chang, H.–R. Changes in tropical cyclone number, duration and intensity in a warming environment. *Science, New Series,* **309**(5742), pp. 1844–1846, 2005.
https://doi.org/10.1126/science.1116448

[7] World Bank. *Resilient recovery in Samoa after cyclone Evan*, 2014. Available at: http://www.worldbank.org/en/results/2014/09/08/resilient-recovery-samoa-after-cyclone-evan. (Accessed 18 April 2017).

[8] World Bank. *Building back better in Tonga after cyclone Ian*, 2014. Available at: http://www.worldbank.org/en/results/2014/10/01/building-back-better-tonga-cyclone-ian. (Accessed 18 April 2017).

[9] World Bank. *Supporting resilient recovery in Vanuatu after cyclone Pam*, 2016. Available at: http://documents.worldbank.org/curated/en/906741468196132305/pdf/103269-BRI-Vanuatu-Box394863B-PUBLIC-ADD-SERIES-Stories-of-impact.pdf. (Accessed 8 April 2017).

[10] World bank. *World Bank commits $50 million to support Fiji's long-term cyclone Winston recovery*, 2016. Available at: http://www.worldbank.org/en/news/press-release/2016/06/30/world-bank-commits-50m-to-support-fijis-long-term-cyclone-winston-recovery. (Accessed 18 April 2017).

[11] Australian Severe Weather. Available at: http://www.australiasevereweather.com/cyclones/summary_jtwc.htm. (Accessed 7 April 2017).

[12] Reserve Bank of Fiji. Available at: http://www.rbf.gov.fj/. (Accessed 4 April 2017).

[13] Central Bank of Samoa. Available at: http://www.cbs.gov.ws/. (Accessed 4 April 2017).

[14] Tonga department of Statistics. Available at: http://tonga.prism.spc.int/social/migration-statistics. (Accessed 7 April 2017).

[15] Vanuatu National Statistical Office. Available at: http://www.vnso.gov.vu/. (Accessed 20 March 2017).

[16] World Bank. *Vanuatu: Post-disaster needs assessment, tropical cyclone Pam*, March 2015. Available at: http://dfat.gov.au/about-us/publications/Documents/post-disaster-needs-assessment-cyclone-pam.pdf. (Accessed 16 April 2017).

[17] World Bank. *Fiji: Post-disaster needs assessment, tropical cyclone Winston*, 2016, Available at: https://www.gfdrr.org/sites/default/files/publication/PostDisasterNeedsAssessmentsCYCLONEWINSTONFiji2016(Online%20Version).pdf. (Accessed 16 April 2017).

[18] World Bank. *Samoa: Post-disaster needs assessment, tropical cyclone Evan*, 2012. Available at: http://www.gfdrr.org/sites/gfdrr/files/SAMOA_PDNA_Cyclone_Evan_2012.pdf. (Accessed 16 April 2017).

[19] FAO, Tonga. *Cyclone Ian in Ha'apai: Rapid damage assessment to the agriculture and fisheries sectors report*, 2014. Available at: http://reliefweb.int/report/tonga/cyclone-ian-ha-apai-rapid-damage-assessment-agriculture-and-fisheries-sectors-report. (Accessed 16 April 2017).

[20] National Emergency Operations Committee. *Tonga, Tropical cyclone Ian response plan* 30 January 2014. Available at: http://reliefweb.int/report/tonga/tropical-cyclone-ian-response-plan. (Accessed 16 April 2017).

[21] Noy, I., The socio-economics of cyclones. *Nature Climate Change,* **6**, pp. 343–348, 2016. Available at: www.nature.com/natureclimatechange. (Accessed 17 April 2017). https://doi.org/10.1038/nclimate2975

[22] Peduzzi, P., Chatenoux, B., Dao, H., De Bono, A., Herold, C., Kossin, J., Mouton, F. & Nordbeck, J., Global trends in tropical cyclone risk. *Nature Climate Change,* **2**, pp. 289–294, 2012. Available at: www.nature.com/natureclimatechange. (Accessed 17 April 2017).
https://doi.org/10.1038/nclimate1410

[23] UNU-EHS. *World risk report,* 2015, Available at: www.WorldRiskReport.org. (Accessed 20 April 2017).

[24] World Development Indicators. Available at: http://databank.worldbank.org/data/reports.aspx?source=world-development-indicators. (Accessed 23 March 2017).

[25] ADB. *Poverty in Samoa, share of population below the poverty line*. Available at: https://www.adb.org/countries/samoa/poverty. (Accessed 20 April 2017).

[26] Promoting energy efficiency in the Pacific – phase 2. Available at: http://www.ee-pacific.net/index.php/database/country-information/. (Accessed 20 April 2017).

[27] WMO. Available at: http://www.wmo.int/pages/prog/www/tcp/Advisories-RSMCs.html. (Accessed 20 April 2017).

[28] SPREP. *Lessons learnt by Tonga Met Service after cyclone Ian*. Available at: http://www.sprep.org/climate-change/lessons-learnt-by-tonga-met-service-after-cyclone-ian. (Accessed 23 April 2017).

[29] The FRANZ arrangement. Available at: https://www.mfat.govt.nz/assets/_securedfiles/Aid-Prog-docs/NZDRP-docs/Franz-Arrangement-Brochure.pdf. (Accessed 23 April 2017).
[30] UNICEF. *Cyclone Pam situation report No: 3*. Available at: https://www.unicef.org/pacificislands/Cyclone_Pam_SitRep_3.pdf. (Accessed 25 April 2017).
[31] French help after the passage of cyclone Winston. Available at: https://fj.ambafrance.org/French-Help-after-the-passage-of-cyclone-Winston-February-24-2016. (Accessed 25th April 2017).
[32] World Bank. *World Bank commits $50 million to support Fiji's long-term cyclone Winston recovery*. Available at: http://www.worldbank.org/en/news/press-release/2016/06/30/world-bank-commits-50m-to-support-fijis-long-term-cyclone-winston-recovery. (Accessed 25 April 2017).
[33] Mckay, B., Maori architecture: transforming western notions of architecture. *Fabrications,* **14**(1&2), pp. 1–12, 2004.
https://doi.org/10.1080/10331867.2004.10525189

FLOOD FORECASTING SYSTEM OF URBAN AREAS IN SOUTH KOREA

YOUNG-IL MOON, JI-HYEOK CHOI, MIN-SEOK KIM & JUNG-HWAN LEE
Department of Civil Engineering, University of Seoul, Seoul, South Korea

ABSTRACT
The needs of safety and the security of the homeland from unexpected natural disasters have been growing among people recently. Flood damages have been recorded every year and those damages are greater than the annual average of 2 billion US dollars since 2000 in South Korea. Due to the rapid growth of urbanization and climate change, the frequency of concentrated heavy rainfall has increased causing urban flood that results in casualties and property damage. Despite the emerging importance of a flooding situation, the studies related to the development of an integrated management system for reducing floods are insufficient in South Korea. In addition, it is difficult to reduce floods effectively without developing integrated operation system taking into account of sewage pipe network configuration with the river level. Since the floods result in increasing damages to infrastructure, as well as life and property, structural and non-structural measures should be urgently established in order to reduce the flood effectively. Therefore, in this study, we developed an integrated flood analysis system that systematized technology to quantify flood risk and flood forecasting in urban areas. The purpose of this study is to introduce integrated Landside-River Combined System by reducing the risk of floods in urban areas using radar and satellite information. Therefore, we developed the flood forecasting system so people can have enough time to evacuate in case of emergency.
Keywords: flood risk, inland-river combined flood system, urban flood

1 INTRODUCTION

Severe natural disasters, including flooding and landslides, occur repeatedly in the Korean Peninsula due to the heavy rainfall induced by typhoons and localized torrential downpours during the summer season, resulting in considerable casualties as well as severe social and economic loss [1, 2]. Throughout the Korean Peninsula, two thirds of the annual precipitation (1283 mm) is concentrated in the rainy season from June to September, so the risk of flood damage is present every year. Moreover, it has been observed that the frequency of localized heavy rain has greatly increased and the scale of damage tends to be larger in the city of Seoul [3]. Additionally, in Seoul, due to its urbanization, when flood damage occurs in lowland, it will cause enormous human casualties and property damage [4]. Thus, it is needed to make a systematic flood control plan. Therefore, this article is going to introduce the causes of flood damage and flood forecasting system.

2 MAJOR CAUSES OF FLOOD IN SEOUL

Flooding that occurs in the city basin causes progressive damage on social infrastructure as well as on human life and property. It is urgent to build measures for both structural and non-structural inland drainage flood prevention that can be effective in defence. The following summarizes the major causes of floods in Seoul and the types of damage.

- Due to recent climate change, localized heavy rain has resulted in flooding.
- The increased of water level in the stream caused poor river drainage of inland basin.
- The flood occurred by the overflow from the river.

DOI: 10.2495/SAFE-V7-N2-213-220

Figure 1: Flood-damaged areas in Seoul (1984–2011).

- Inland damage occurred because of the lack of sewerage capacity as well as capacity of a rain pumping station.
- Various factors including landslides worked complexly.
- The greatest damage was done to river facilities including sewerage and small rivers.

3 FLOOD FORECASTING SYSTEM OF URBAN AREA

Through the analysis of the major causes of inundation in Seoul, a system which can alert and forecast urban flooding must be developed to prevent property damage along with casualties. The following figure shows a system framework of flood forecasting system.

The system provide forecast for urban flood through real-time operation. This is a system for analysing river and inland inundation through a short-term rainfall prediction.

The critical technology of system is divided into three parts: (A) forecasting rainfall; (B) constructing sewer networks; and (C) setting up inundation data base.

In the first step, the short-term rainfall forecasting, using the radar data or radar and satellite data were implemented. It can be estimated for 120-min before rainfall prediction with 10-min lead time using the TREC (Tracking Radar Echoes by Correlation) technique.

In the second step, rainfall-runoff model is needed to apply forecasted rainfall. In this study, the runoff analysis of urban basin has used in SWMM 5.1 model. In result of the constructing sewer networks (all basin of Dorim area), the number of sewers is 31,480. The more number of sewers increase, the more analysis run-time of the model increase. If the model is applied to the real-time urban flood forecast using the total sewer network, there is a problem of missing the golden time due to the time spent on analysing the model. Thus, to reduce simulation time of model, sewer network simplification by automatic simplification technology must be developed.

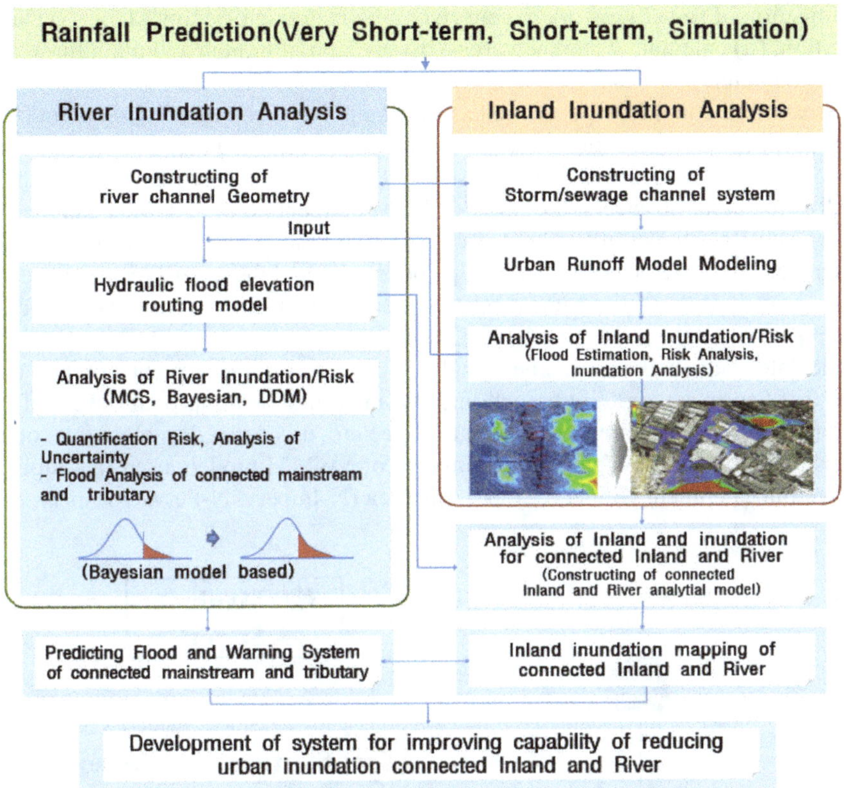

Figure 2: Framework of flood forecasting system.

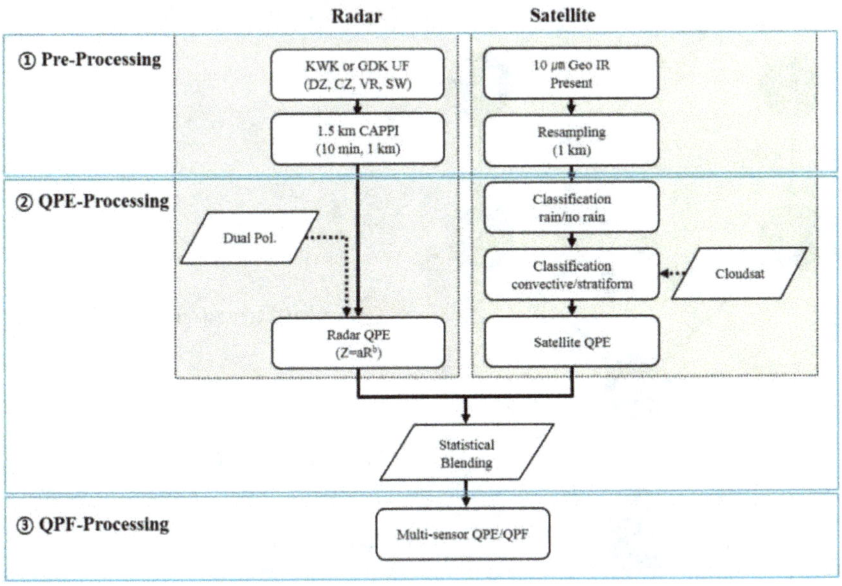

Figure 3: Short-term rainfall forecasting.

Simplification of sewer network except river area is divided by four steps: (i) check initial condition, (ii) calculate watershed area, (iii) division of branch and main lines, and (iv) calculate parameter.

(i) Check initial condition is classified as starting point of sewer, dividing point of sewer, confluent point of sewer and outlet of sewer. Calculate watershed area can be calculated by adding starting point of sewer to outlet of sewer. The sewer 1 is starting of sewer, the sewer 2, 3 are dividing point of sewers, the sewer 6, 7 are confluent point of sewers and the sewer 8 is outlet of sewer as shown in Figure 4. In the case of branch line, a sewer pipe having the large discharge area is selected as a main pipe of branch, and the other pipes were set as partial pipe of branch. If they have same discharge area, a pipe located at front side is defined as the main pipe.

(ii) Calculate watershed area can be calculated by adding starting point of sewer to outlet of sewer. (iii) Sewer network are branch line and main line. If the product of runoff coefficient and planned drainage area is more than 12 hectare, it is defined as main line, otherwise it is set as branch line. (iv) Calculate parameters of SWMM models used in simplification process are Subcatchment (Imperv, Slope), Subarea (N-Imperv, N-Perv, S-Imperv, S-Perv),

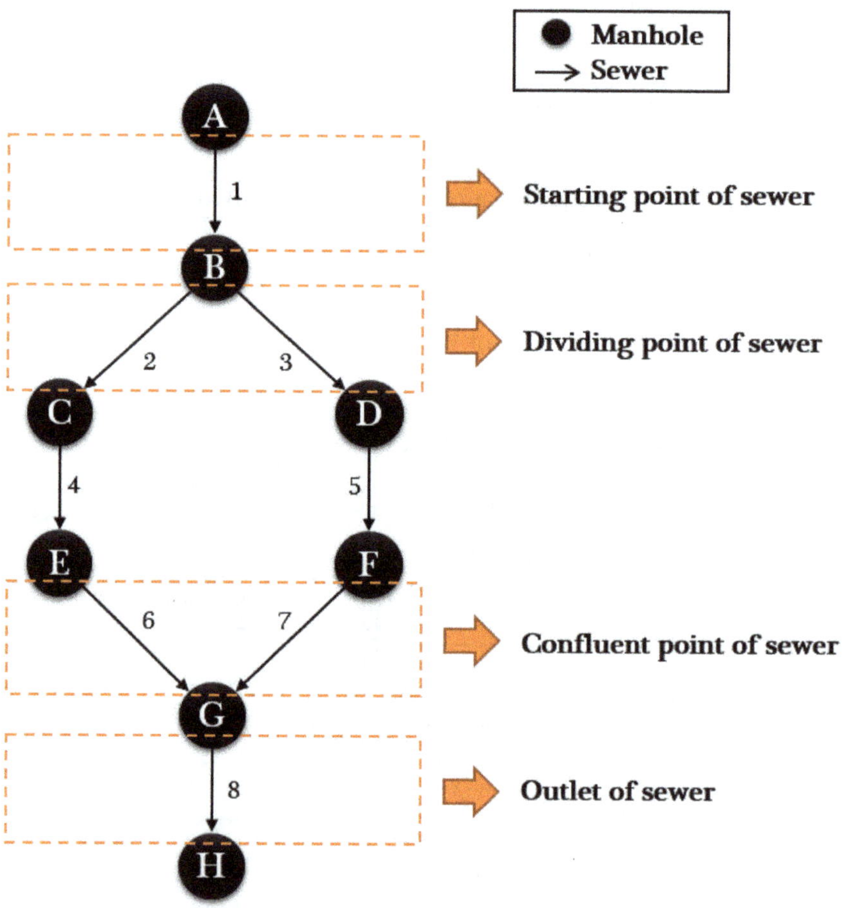

Figure 4: Check initial condition.

Infiltration (Max Rate, Min Rate, Decay, Drytime) and Conduits (Roughness). At last, Sewer network simplification consist of sewer networks by calculate reasonable watershed area. A node or link less than 12 hectare of cumulative watershed area as a basis for dividing branch and main line are deleted in the simplification process, so the parameters of the node and link to be deleted are calculated by applying area weighting. Through this process, a simplified sewer network can be constructed.

Figure 5a presents the sewer network of test basin before simplification. Figure 5b presents the result of sewer network simplification based on cumulative basin area.

The sewer network has simplified by 85% and the simulation time of rainfall-runoff model is reduced by 90% as shown in Table 1. This means when the runoff analysis is carried out

(a)

(b)

Figure 5: Simplification technology of sewer network.

Table 1: Comparative simulation time according to the simplified.

Duration of rainfall	Simulation time (a) Total sewer network	(b) Simplified sewer network	(c) Reduction rate of time (%)
60 min	00:11:00	00:01:12	89.1
180 min	00:40:00	00:03:37	91.0
360 min	01:20:21	00:07:24	90.8

using the simplified sewer network for real-time urban flood forecasting, it calculates 90% of the golden time compared to the entire network due to short simulation time.

There is a classification in accordance with utilization between total sewer network and simplified sewer network. Simplified sewer network (Fig. 5b) used real-time urban flood forecast. Total sewer network (Fig. 5a) is used as a rainfall-runoff model for inundation analysis in inland. It used 2D Analysis which performed by interlocking SWMM with TUFLOW. In the final step, different scenarios (Fig. 6a–d) using total sewer network, constructed the maximum inundation area database according to various rainfall and durations. Various 2D database used real-time flood forecasting due to the reason that real-time 2D takes a longer time to process.

4 CONCLUSION

As part of the measures to deal with the increase of inland flood damage, it is necessary to build a systematic city flood prevention system that systematizes technology to quantify flood risk as well as flood forecast considering both inland and river water. This Combined

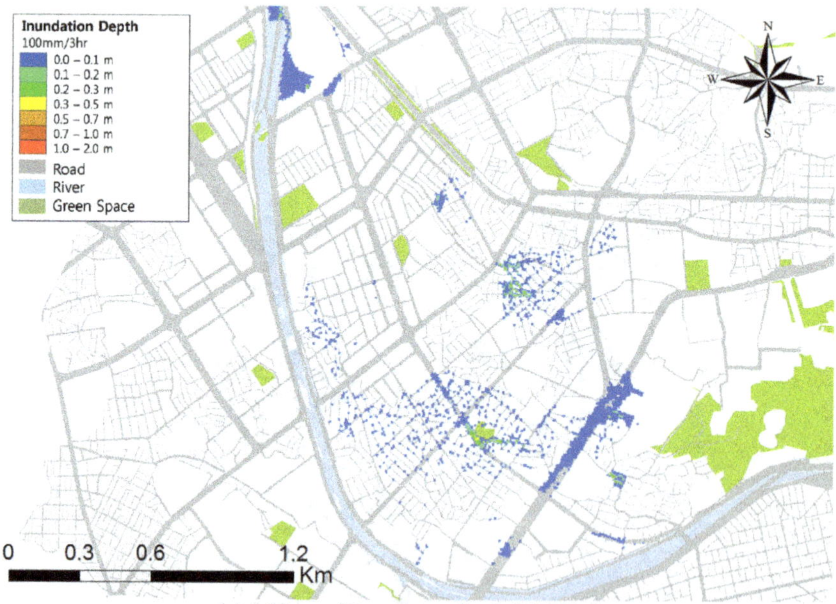

(a) 100mm (Duration of rainfall: 3hr)

Figure 6: Maximum inundation area database.

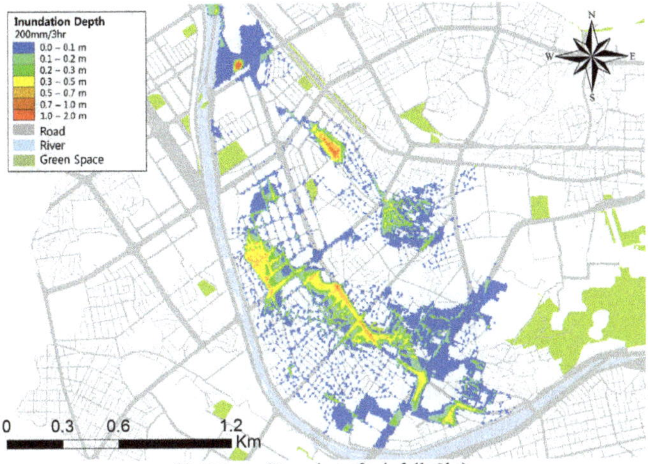

(b) 200mm (Duration of rainfall: 3hr)

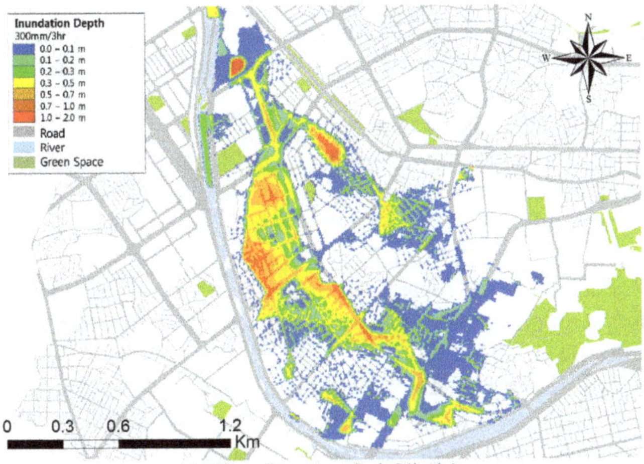

(c) 300mm (Duration of rainfall: 3hr)

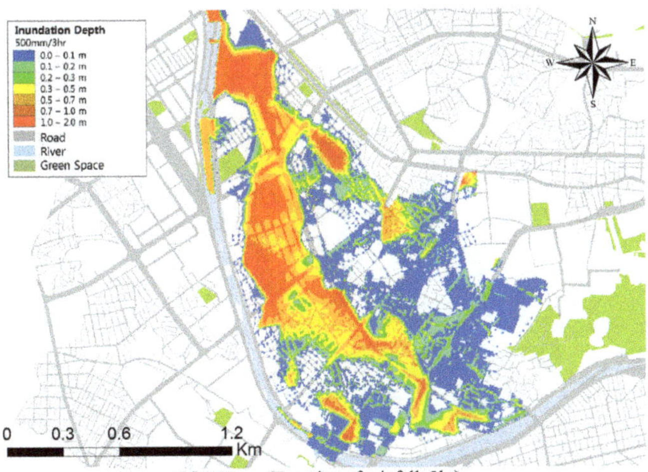

(d) 500mm (Duration of rainfall: 3hr)

Figure 6: (*Continued*)

Inland-River Flood Analysis System conducts prediction on flash rain or short-term rainfall by using radar and satellite information and performs prompt and accurate prediction on the inland flooded area. It also supports synthetic decision-making for prevention through real-time monitoring. Climatic conditions in the 21st century which can cause disasters may happen to us anytime in the future. The Korean Peninsula has no exception. Although we cannot prevent damage from typhoons or localized heavy rain, we can minimize that damage with accurate and prompt forecast and a prevention system. To realize this, it is demanded to integrated decision-making system to forecast super typhoons or sudden rainfalls and monitor them in real time, which can generate forecast or alarms.

ACKNOWLEDGEMENTS

This research was supported by a grant (16AWMP-B066744-04) from Advanced Water Management Research Program (AWMP) funded by Ministry of Land, Infrastructure and Transport of Korean government.

REFERENCES

[1] Kim, J.S. & Jain, S., Precipitation trends over the Korean peninsula: typhoon-induced changes and a typology for characterizing climate-related risk. *Environmental Research Letters*, **6**, p. 034033, 2011. https://doi.org/10.1088/1748-9326/6/3/034033

[2] Kim, J.S., Li, R.C.Y. & Zhou,W., Effects of the Pacific-Japan teleconnection pattern on tropical cyclone activity and extreme events over the Korean peninsula. *Journal of Geophysical Research*, **117**(D18), p. 109, 2012. https://doi.org/10.1029/2012jd017677

[3] Kim, S.M., Tachikawa, Y. & Takara, K., Recent flood disasters and progress of disaster management system in Korea. *Annuals of Disaster Prevention Research Institute Kyoto University*, **50B**, pp. 15–31, 2007.

[4] Park, B.C., Oh, K. S. & Lee, D. W., Flood damage prediction for provincial governments in Korea based on climate change scenarios. *International Journal of Environmental Science and Development*, **8**(6), pp. 435–441, 2017. https://doi.org/10.18178/ijesd.2017.8.6.993

USING A COASTAL STORM HAZARD INDEX TO ASSESS STORM IMPACTS IN LISBON

BOGDAN JARANOVIC, JORGE TRINDADE, JOÃO RIBEIRO & ADÉLIO SILVA
Centre of Geographical Studies - University of Lisbon, Universidade Aberta and Hidromod - Modelling & IT Specialists

ABSTRACT

Coastal areas are among the most dynamic earth systems as they are exposed to powerful agents. Near-shore wave energy is one of the most important triggering factors for erosion and flooding and is often neglected for severe infrastructure damaging, property losses and loss of life. These consequences are amplified with high population density and heavy infrastructure implantation as it happens in Lisbon (Portugal). In this context, it is of great importance for coastal stakeholders, decision-makers and civil protection entities to estimate precisely the spatial distribution of storm hazard for prevention and mitigation purposes, as well as to design adjusted answers for calamity responses. We apply a coastal storm hazard index (CSHI) considering triggering and conditioning variables involved in the effects of an extreme storm, namely: 100-year return period of SWAN modelled H_s, and its spatial distribution across the study area, land use, number of buildings, height, slope, geology, geomorphology, erosion/accretion rates, width of the systems, exposure of the coastline, bathymetry and legally protected areas. The variables were weighted according to a hierarchical analysis process and classified into five classes of exposure. A validation process was then implemented by comparing the occurrences identified in the last two decades newspapers and the storm hazard classification, showing a satisfactory validation results. The results show a classified storm hazard map that identifies the most and the less exposed areas. High values of CSHI occur in areas with excessive human pressure, low heights sandy systems with significant costal erosion rates. The main type of consequences identified are associated with inland flooding and erosion, resulting in the destruction of coastal protection infrastructures, and population displacement leading to great economic and social impacts and loss of life.

Keywords: Coast, hazard index, numerical modelling, return period, waves

1 INTRODUCTION

Coastal areas have great economic advantages, due to navigation, coastal fisheries, tourism and leisure. Inevitably, these areas are also exposed to a significant number of hazards related to its specific hydrodynamic characteristics. Coastal storms are among the most powerful and dynamic hazards, frequently leading to severe structural damages, coastal erosion, affecting people and leading to loss of life.

In the Portuguese coast, the impact of storms is frequent. The last research about the quantification of coastal storms in the country was done by (Sousa [1]) who, through the modelled data from the *Sines* buoy (*micore* project – Dodet *et al.* [2]), found a total of 409 storms in the period from 1953 to 2009. 94.8% of the storms occurred between September and April, and the investigation concluded: (i) that the average significant wave height (H_s) of these events was 5.08 m and (ii) the storm direction was from the NW in 65.18% of the cases. The 60' was the decade with the higher number of storms, totalling 86, but no increase/decrease tendency in storminess was noted.

To assess the coastal storm hazard, several methodologies have been developed. A commonly used index for defining the coastal vulnerability is the CVI (coastal vulnerability index). This index allows the classification of different variables, relating them to one another and quantifying their relations [3–7]. The analysis of the coastal areas that may be affected by these extreme events is frequently done with geographic information systems, which

DOI: 10.2495/SAFE-V7-N2-221-233

simplifies the spatial analysis of the results. Other methodologies use numerical models for the characterization of storm wave patterns in the nearshore [8]. It is also possible to analyse the coastal vulnerability with a statistical approach. Logistical regression has been used to relate coastal vulnerability to several explanatory variables [9]. The triggering and conditioning factors considered in this kind of model implementation do not always coincide in the scientific literature, although it is common to find clusters of factors (i) that characterize the coastal morphology and its dynamics (heights, slope, rates of erosion, exposure), (ii) that correspond to land use and other anthropogenic variables and (iii) that define wave energy and other hydrodynamic parameters (wave parameters, tides and currents) and, finally, (iv) that predict the sea level near future evolution. Wave energy is one of the most important independent variable for the research of storms impact on the coast, but it is also lacking in measured data from wave buoy. When wave measured data time series is available, usually it does not translate wave parameters between the breaker zone and the beach face, where storm incidence is more likely to cause damage. This includes the disregard of several processes that change the wave characteristics (white-capping, wave-wave interactions, bottom friction, shoaling, depth-induced braking, refraction and diffraction) [10, 11]. The progress of technologies has enabled the development of models that simulate this process and provide correct wave data nearshore. One of the most common models is SWAN (Simulating Waves Near-shore), which has been accepted and used extensively for this purpose [12–15].

Independently of the chosen approach to assess the coastal storm hazard, the accuracy of a coastal storm hazard map must be demonstrated through a validation process. The most frequent method consists in the comparison of the classified coastal storm hazard map results and the areas that have been affected previously by storms.

The main goal of this investigation is to assess the coastal storm hazard of the coastline between *Cabo Raso* and *Cabo Espichel* (Lisbon, Portugal). This would be carried through the analysis of the hydrodynamic distribution of a 100-year return-period scenario, and by establishing the spatial distribution of the morphologic, hydrodynamic and anthropic parameters related to the storm events, and combining the different variables into a single validated measure of coastal storm hazard mapping.

2 STUDY AREA

The Portuguese West coast is characterized by high energetic wave regime during maritime winters. Most of the NW storms are defined by the general Atlantic atmospheric circulation and the mean significant wave height (H_s) is 2.5 m, reaching frequently 5 m. In this mesotidal regime, the maximum tide amplitude reaches 4 m [16].

The Lisbon coastal area is located in the mid-west coast of Portugal (Fig. 1), between the *Cabo Raso* and the *Cabo Espichel*, and has almost 60 km of diversified sandy and rocky coastal systems. Because of the position of the *Cabo Raso* headland, predominant storm wave direction is deflected from its original direction (NW) at this point and high amount of wave energy is spent here. Along with *Cabo Raso*, *Cabo Espichel* headland is also characterized by high incident wave energy concentration, while the coastline near the Tagus estuary is less exposed to NW storms and less prone to wave action.

The study area is composed by the Lisbon and the Setubal peninsulas. The first one, between *Cabo Raso* and Lisbon, is dominated by a rocky coast, although with low altitudes and scattered pocket beaches, usually surrounded by cliffs. Besides the high level of human occupation and building density, the erosion rates present in this coast are very low, due to the predominantly hard limestone and sandstone cliffs [12, 17]. The Setubal peninsula is a low

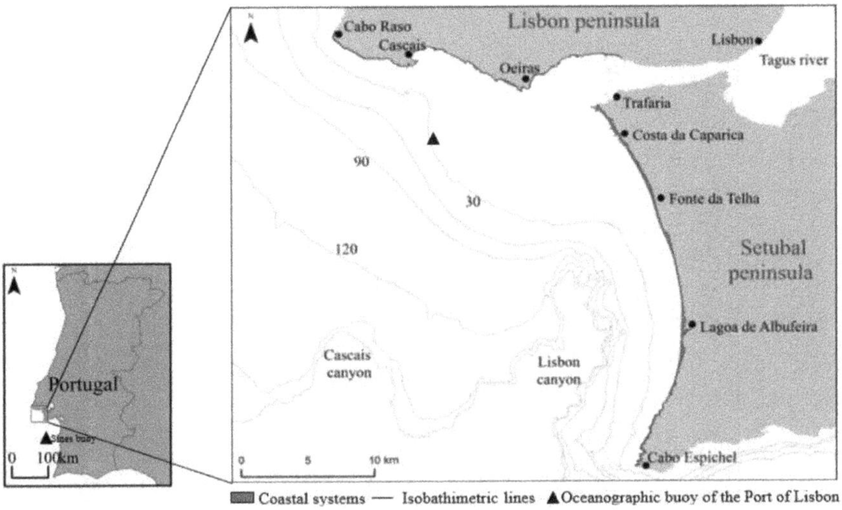

Figure 1: The Lisbon and Setubal peninsulas.

altitude sandy coast from the extreme north (*Trafaria*) until the southern area of *Lagoa de Albufeira*, where highest erosion rates dominate [1, 18] essentially where urbanization occupies large areas – *Costa da Caparica* and *Fonte da Telha*. The northern coastal stretch of the Setubal peninsula is the most affected area by coastline recession over the last century [19]. For that reason, hard protection mitigation measures were taken in the 70's and in the 80's of the past century to solve erosion problems. The construction of nine groins, a seawall, and sand nourishment were only some of the drastic measures taken. Southwards from this erosion hotspot, it is also possible to observe some sectors with accretion rates [1].

The nearshore bathymetric configuration is very important for wave propagation and, thus, for wave energy dissipation. Exception made for the Tagus estuary (Fig. 1), the inner continental shelf platform of the study area is relatively narrow, reaching less than 20 km wide with high slope values. The *Cascais* and Lisbon canyons (Fig. 1) have direct influence on the hydrodynamics of the coast, leading to the accumulation of wave energy.

3 MATERIALS AND METHODS

To implement, validate and analyse the coastal storm hazard index (CSHI), the first step was to delimitate the study area. Sandy coastal systems were delimited including the beach, the seaward and the landward sides of the foredune. The limits of cliff systems were established between the coastline and the cliff top, including the shore platform and the cliff toe and face. In highly urbanized areas, the delimitation of the study area followed land use criteria, by using the first frontal urban blocks as the inner limit. Generally, these coastal areas also included heavy protection structures, such as groins and sea walls.

The second stage was to carefully choose the most significant triggering and conditioning factors for coastline storm hazard exposure. A set of 12 variables was selected based on the literature review, the study area specificity and on the available spatial data (Table 1) [3, 12, 20]. After the system was delimited, this was parcelled out in 50 m equidistance cells.

The main triggering factor for storm hazard exposure is the high energetic waves that hit the coast. Considering a high wave energy event scenario, the 100-year return period of H_s was calculated from a 50-year wave database, available at (Dodet *et al.* [2]) for the coastal

Table 1: Classification of the variables used in the CSHI, with the respective sources and resolution.

CSHI IHP	Independent variables	Classification of the variables trough the storm hazard favourability					Source	Resolution or scale
		Very low (1)	Low (2)	Medium (3)	Hight (4)	Very high (5)		
a	100-year return period H_s (m)	0–2.7	2.7–4.8	4.8–6.3	6.3–7.7	7.7–12.4	WAVEWATCH III / SWAN modelled data	50 m
b	Land use	Forest	Shrubby and herbaceous vegetation	Low density vegetated areas	Agricultural pond	Industrial or commercial area; Urban area	COS 2007	1: 25 000
c	Numb. of buildings	0	0.28–2.85	2.28–12	12–33	33–44	INE	1: 25 000
d	Altitude (m)	30–131	15–30	10–15	5–10	0–5	LIDAR 2011	2 m
e	Slope (°)	50–84	30–50	15–30	6–15	0–6	LIDAR 2011	2 m
f	Geology	Volcanic and metamorphic	Limestone	Sedimentary rocks	Dunes and other unconsolidated deposits	Beach sand	LNEG	1: 50 000
g	Geomorphology	Artificial shorelines	High altitude rocky system	Low altitude rocky system	Beach-cliff system	Sandy coast	-	-
h	Erosion / Accretion rates	Accretion	Null	Low erosion	Medium erosion	High erosion	-	-
i	Width of the system (m)	363–680	254–363	179–254	102–179	1–102	-	-
j	Exposition	SE	S	WSW	W	WNW	-	-
k	Bathymetry (m)	-2.2–0	-5.3—2.2	-8.5 – -5.3	-12.5—8.5	-20.6—12.5	LIDAR 2011	2 m
l	Protected areas	No protection areas	-	-	-	Protection areas	ICNF	-

nearby area (Sines nearshore coastal sector, *a* in eqn (1)). The method described by Gilleland and Katz [21] is frequently used for wave ocean modelling.

The Generalized Extreme Value distribution is usually used to explain the patterns of the maximum values distribution, and this analysis used the r-largest annual maxima method. The final 100-year H_s value obtained was 12.9 m. The representation of the 95%-confidence intervals is shown in the Figure 2. The value of μ, that represents the centre of a distribution, is 6.75, σ, the scale, is 1.08 and ξ, the shape, is 0.009.

This H_s 12,9 m value is used for the simulation of the worst-case scenario, using the SWAN model. The SWAN model computes short-crested wind-generated waves in coastal regions and inland waters, and enables the implementation of several physical processes that influence wave progression [10, 11], as it takes into consideration the wind generation of waves, the interactions between waves (*triads*), and whitecapping. In this simulation, the mean sea level is held at 2.08 m, according to the *Cascais* tide gauge data. For the dissipation of waves, due to the sea floor interaction, the JONSWAP mode is used, and the growth of the waves was considered linear (JANSSEN). The dissipation and the diffraction processes were considered constant. The parameter AGROW is established as 0.0030, since Ribeiro *et al.* [22] states that this modification introduces a quality improvement of the simulated data. Two different bathymetric data are used, one for the western Iberian margin (EMODNET – 250 m resolution) and another for the studied area (Hydrographic Institute of Portugal – 100 m resolution). A validation process was implemented by comparing the SWAN simulated wave data and the measured wave data from the Port of Lisbon buoy (latitude 38º37´33.6´´N e longitude 09º23´16.8´´W – WGS84 coordinate system). For the validation process, the interaction between the WAVEWATCH III and SWAN models was used. The WAVEWATCH III model was developed by NOAA/NCEP [23] and is used for wind generated waves on a large-scale area that may output boundary conditions for SWAN nesting. Nesting consist of running, in the first place, the WAVEWATCH III model on a larger scale and using the outputs that this model provided to start a new run in SWAN, using boundary conditions on a finer and more specific scale. The results of the validation for this wave climate models are presented in Figure 3. The result is considered satisfactory due to the general coincidence of the simulated and measured wave patterns, and reasonable statistical results – correlation value of 0.91; RMSE of 0.07 and BIAS value of -0.07.

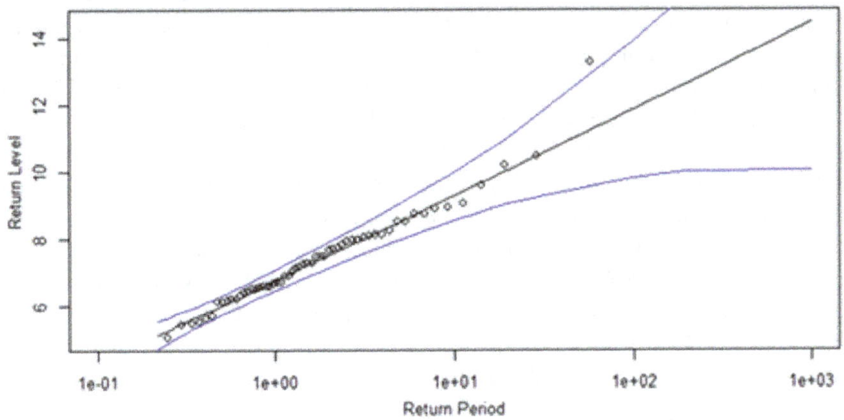

Figure 2: Distribution of the maximum annual data in 95% confidence intervals.

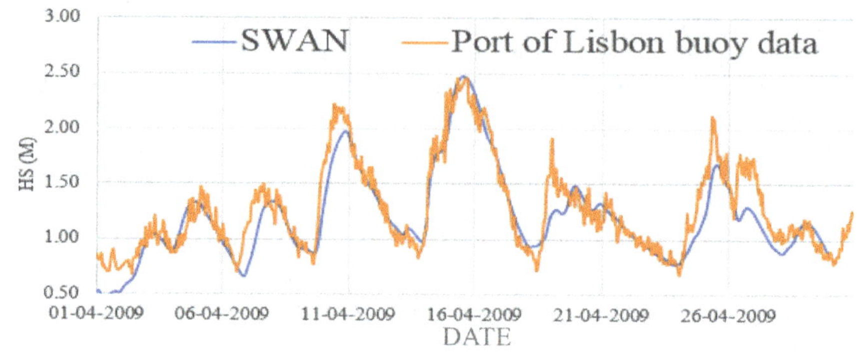

Figure 3: Comparison of the SWAN data and the data in the Port of Lisbon buoy.

The worst-case scenario simulation was only carried out when the validation process was over, obtaining a spatial distribution of the H_s in the study area. In contrast to the validation method, carried out in the non-stationary mode, the simulation of the worst-case scenario was carried out in the stationary mode. Finally, the spatial distribution of the 100-year H_s was assumed to be the values estimated at a distance of 500m seaward from the coastline, thus including the breaker zone extreme wave H_s.

Since the strongly occupied areas are the ones where the largest number of accidents occur, the land use classification (*b* in eqn (1)) and the number of buildings (*c* in eqn (1)) are used to represent the human pressures over the coastal systems. Altitude (*d* in eqn (1)) and slope (*e* in eqn (1)) are also used to define the influence that an extreme storm may have on the coast. Geology (*f* in eqn (1)) is considered in this study in terms of materials consolidation and propensity to mechanical erosion due to wave action. Geomorphology (*g* in eqn (1)) defines the type of coastal systems in presence. Local erosion and accretion rates (*h* in eqn (1)) were also considered in this study, as well as the width of the system (*i* in eqn (1)), for it represents a buffer zone to the storm impacts. Coastline orientation is an important variable to consider, especially if the dominant storm direction is known. In Portugal, the most frequent storms are generated in the NW Atlantic Ocean and that is the more prone orientation to hazard incidence. The bathymetric data (*k* in eqn (1)) was an important independent variable for WAVEWATCH III/SWAN wave modelling, because it drives the water masses to the coast proximity under the influence of waves and currents. The legally protected areas (*l* in eqn (1)) where considered because the consequences of an extreme storm on the coast cannot be resumed to human and infrastructure loss and should consider also the loss of biodiversity of the legally protected areas in this coastal stretch. Table 1 shows all the factors considered and its classification.

An analytic hierarchy process was used to order the relative importance of the twelve variables for the CSHI [24]. The variables were classified in accordance with the groups represented in the Table 1. This method defines a way of measurement through pairwise comparisons and is based on the judgment of experts to derive the priority scale [25]. In this investigation, it was used the nine-level scale, where the biggest number represents a higher importance of the factor [25]. The result shows 66 comparisons between the variables with a consistency index of 0.06. After the classification of the 50m equidistant parcels with the representative value of each variable, the CSHI was calculated, using the formula below [3, 26]. The letters represent the different variables in the same order as in Table 1.

$$CSHI = \sqrt{\frac{25{,}6a*11{,}9(b*c*d*e)*4{,}2(f*g*h*i*j*k)*1{,}6l}{12}} \tag{1}$$

A storm consequences geodatabase was created with the analysis of 3150 newspaper editions, during maritime winters, from September to April. The analysis on the 2000 to 2015 editions of the *Diário de Notícias* newspaper was essential for the validation of the final CSHI map.

4 RESULTS

The spatial distribution of the eight independent variables with the highest IHP score values are shown in Figure 4 and classified according to Table 1.

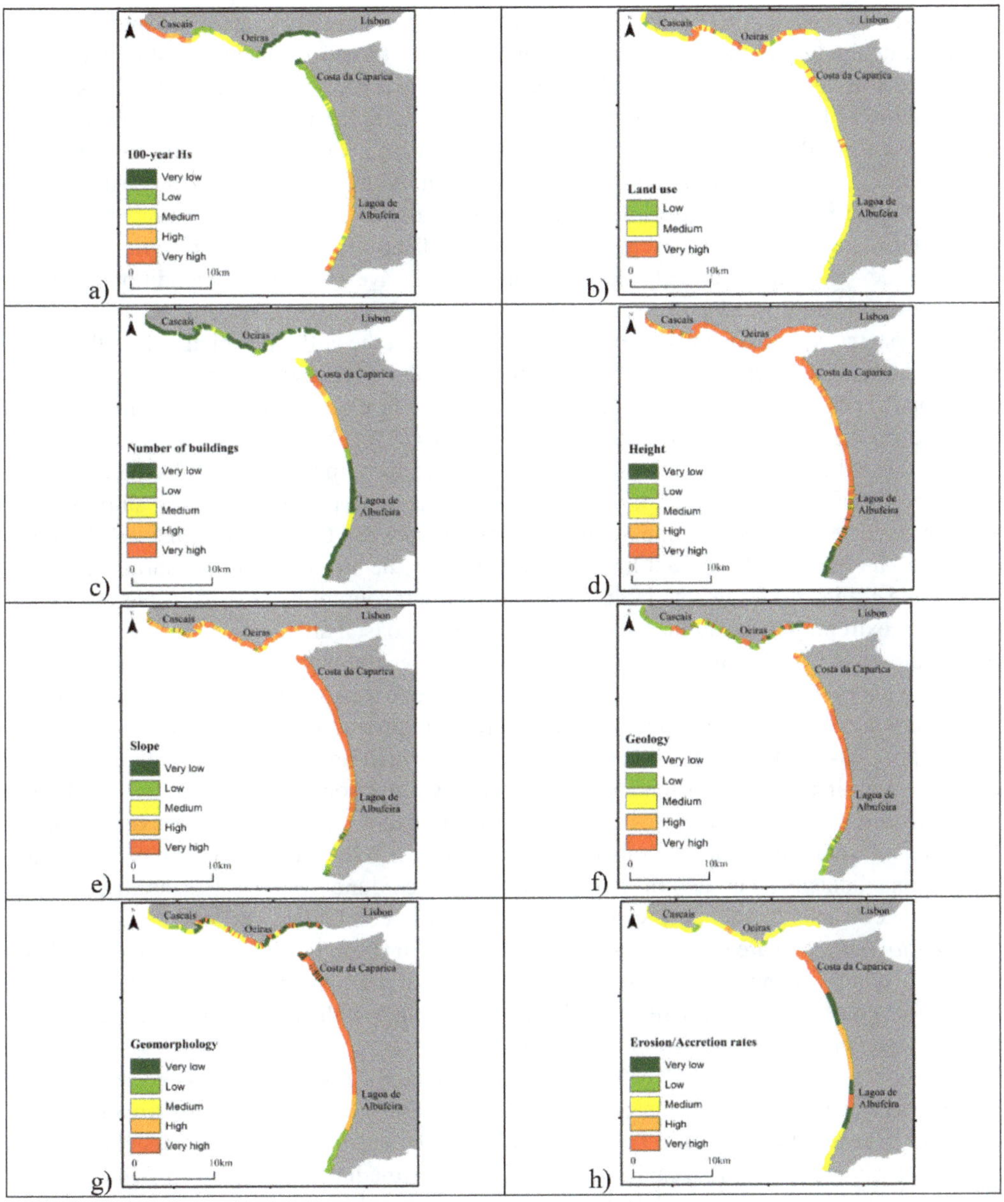

Figure 4: Classified variables according to its favourability to storms occurrence: (a) 100-year return period H_s; (b) land use; (c) number of buildings; (d) height; (e) slope; (f) geology; (g) accretion/erosion rates; (h) geomorphology.

(a) 100-year H_s – represents a probability of an extreme storm event and its spatial distribution throughout the Lisbon coast. The reference value for the SWAN numerical wave modelling was 12.9 m. Spatial distribution of maximum values of 100-year H_s can be observed near the *Cabo Raso*, on the west side of *Cascais*, and *Cabo Espichel*, in the south side of *Lagoa de Albufeira*. Gradually, the 100-year H_s decreases towards the Tagus estuary area, between Lisbon and *Costa da Caparica*.

(b) Land use – depicts the natural and the types of human presence and pressure at the Lisbon coast. High to low vegetated areas include forests to low herbaceous plant strata, and from high to low vegetation density with little anthropic pressure. These areas are spatially concentrated in the southern coastal sector, while the northern Lisbon coastal stretch is more densely urbanized. Heavy coastal protection structures are found near the Tagus estuary.

(c) Number of buildings – The spatial distribution reflects the presence of anthropic structures within the limits of the coastal systems. Higher values are present in *Costa da Caparica*, where building densities are very high. Although the northern coastal sector has higher population densities, in the nearby coastline areas the cliff dominant coastal systems are narrower and exclude land use classifications related to anthropic presence.

(d) Altitude – influences directly the landward storm wave progression, conditioning flood extent and duration. Lisbon coast is essentially a low lying coastal system, despite the cliff subsystems, very frequently associated with a beach at the cliff toe, which reduce the overall altitude. High system altitude values can be found at the extreme south of the study area.

(e) Slope – can be an important independent variable to explain, not only storm wave inland progression, but also the anthropic presence [27]. Namely, coastal systems near densely urbanized areas and with low slope values have more favourable conditions for recreational use and building, than those with higher slope values. Lower slope values are consistently present at the Setubal peninsula, between *Costa da Caparica* and *Lagoa de Albufeira*. This leads to a high classification in the storm hazard favourability scale (Table 1, Fig. 4*e*) for this variable. Low to medium favourability classifications are found at the extreme south and inconsistently at *Cascais* coastal area.

(f) Geology and (g) Geomorphology – Geology defines the physical processes that lead to wave energy dissipation, and erosion and coastline recession [26]. Thus, unconsolidated materials, like beach and dune sediments, are more prone to coastline recession and run-up progression, than consolidated rocky coasts, which are more favourable to a stable behaviour in terms of erosion and to wave energy reflection back to the nearshore. In the study area, these two variables are highly linked together. The type of the coastal system is conditioned by geology: consolidate materials are associated with cliff systems, and unconsolidated materials are associated with beach/dune systems. Essentially, unconsolidated sandy beach/dune systems are found between *Costa da Caparica* and *Lagoa de Albufeira*, and are classified as high and very high storm hazard favourable areas. Due to the composite nature of the beach-cliff systems of the Lisbon peninsula, the *Cascais – Oeiras* coastal scratch has the more diffuse distribution of the favourability classes. Here, low and very low classes are predominant.

(h) Erosion/accretion rates – Coastline recession can be the result of successive storm wave impacts over decades on the coast. Thus, the spatial distribution of erosion/accretion rates over the study area is a helpful evidence of the last decease storm impact hotspots. The Lisbon peninsula coastal scratch reaches intermediate storm hazard favourability. Higher values are observed at *Costa da Caparica*, where erosion rates, over the last century, lead to heavy coastal protection interventions, including groins, sea-walls and beach artificial nourishment.

5 RESULTS

The outcome of the CSHI appears as the spatial classification of the Lisbon coastal areas prone to 100-year storm hazard impact. It is a result of the weighted favourability triggering and conditioning factors for extreme wave impacts on the coast. The 100-year H_s triggering factor was assumed to have NW to SE direction, similar to the most frequent storms in the west coast of Portugal. Therefore, CSHI results are highly related to this hydrodynamic extreme storm context, which is particularly relevant in the Lisbon coastal area, given the different general coastal exposure and coastal systems geology and geomorphology.

Although a set of 12 variables are included in the CSHI eqn (1), several independent variables were not considered, tidal range, storm surge and sea level rise are considered to have no spatial variability at this analysis scale. Therefore, they act as spatially neutral variables for the model implementation.

5.1 CSHI Model validation

CSHI map validation was assessed through the comparison between the CSHI classification and the 67 occurrences from the newspaper geodatabase. The distribution of these occurrences does not shows the total area that might be affected by storms, essentially because these registers report only the events that have impacts on people and infrastructures. Areas with low atrophic pressures tend to have less or no newspaper geodatabase entries.

The validation process is based on prediction-rate curves, and the area accuracy of the map can be directly measured with the area under the curve (Fig. 5).

50.7% of the newspaper occurrences are coincident with the highest class of CSHI (Fig. 6). The two highest classes explain 74.6% of the occurrences. This is considered a satisfactory result mainly because 'occurrences' variable was not taken into account in the CSHI model, and the model explains ¾ of the total 'occurrences' in the most hazardous classes.

5.2 CSHI classification and spatial distribution

As stated before, the CSHI mapping quantifies classified areas more or less prone to the effects of coastal storm driven consequences at the Lisbon coast. This is a major tool to

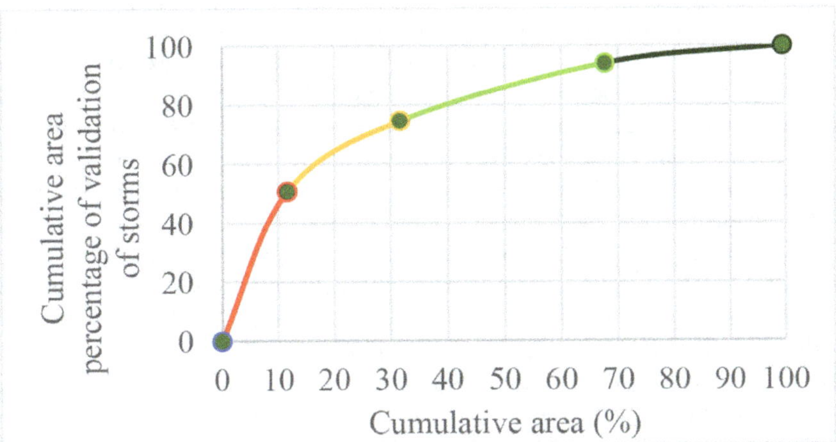

Figure 5: Predicted curve for validation of the model. Colours are related to the classes of CSHI (Fig. 6).

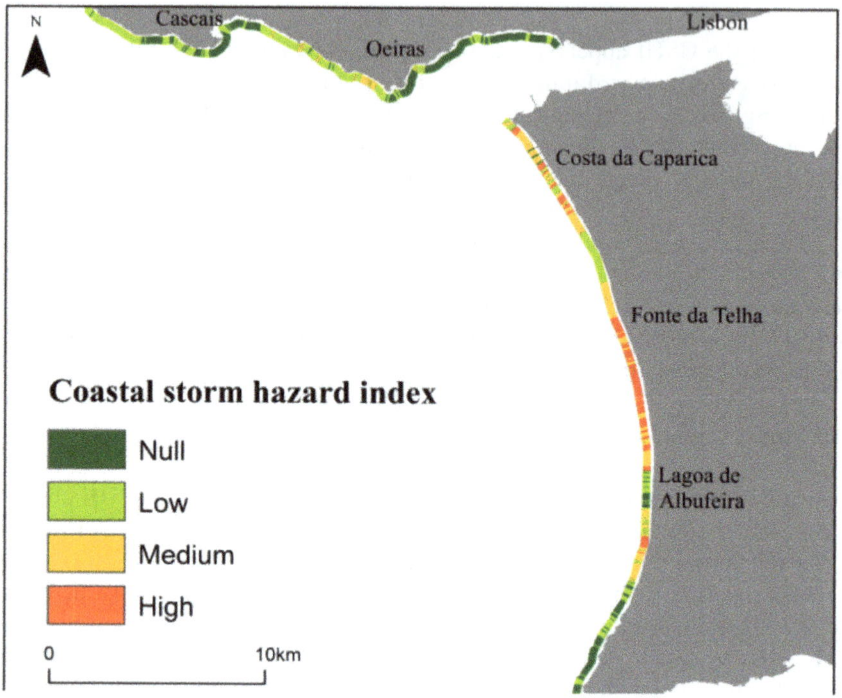

Figure 6: Coastal storm hazard index.

support decision making, stakeholder's action, population preparedness and hazard mitigation by the authorities.

In the last 15 years, 67 occurrences related to storm wave action on the coast of Lisbon can be found in the newspaper reports. This means that this coast is, at least, exposed to this kind of hazard.

The CSHI is classified into four classes, from null to high storm hazard impacts (Fig. 6).

Extreme storm wave height is considered in this work as the main triggering factor for storm induced consequences in the coast of Lisbon. As seen in Figure 6, the most prone coastal areas to the effects of an extreme storm (CSHI, Fig. 6) are not exactly the same areas where to have the higher incident wave energy is expected (100-year H_s in Figure 4a). In fact, coastal sectors to the west of *Cascais* and to the south of *Lagoa de Albufeira* with 'Low' and 'Null' CSHI classifications have the higher favourability for wave 100-year H_s. On the other hand, extreme storm wave sheltered coastal areas of *Costa da Caparica* have less favourable 100-year H_s and are classified with higher values of CSHI.

'Low" CSHI values (35% of the study area) dominate the area to the west of *Cascais*, which is characterized by rocky cliff systems, with medium to low erosion rates, and high incident wave energy. They are also present where accretion rates dominate the historical coastline dynamics between *Costa da Caparica* and *Fonte da Telha*.

The 'Null' CSHI classification (31.8% of the study area) is dominant in the southern sector of *Lagoa de Albufeira* (Fig. 6) where coastal systems correspond to high limestone cliffs. This class is also present from *Oeiras* to Lisbon and in the *Cascais* bay. These are the sheltered coastal stretches for the assumed storm scenario of 100-year H_s. The 'Medium' and

'High' CSHI classes appear in 33.2% of the study area. These classes are present only in low height and slope areas and unconsolidated sedimentary sectors that are indeed the most vulnerable areas to storms [1]. In the Lisbon peninsula, the most exposed areas to storm incidence are the widest beach systems in the *Oeiras* coast (Fig. 6). Higher classes of CSHI can be found also between *Costa da Caparica* and *Lagoa de Albufeira*, where anthropic pressure over the beach/dune systems is very high with observed historical high erosion rates [1, 19] (*Costa da Caparica* and *Fonte da Telha*, Fig. 6) or in coastal areas with low anthropic presence (south of *Fonte da Telha*) but with higher incident storm wave action (Fig. 4a)

6 CONCLUSION

As a powerful and harmful event, capable of causing significant damage and life loss, storms are a frequent coastal phenomenon in the Portuguese coast.

Using twelve independent variables this research explored the spatial patterns of the coastal storm vulnerability index (CSHI) validated classification over the 60 km coastal area.

CSHI showed that 33.2% of the study area is medium to highly prone to the effects of an extreme storm scenario (100-year H_s), while 66.8% of the coast of Lisbon have low exposure.

It is to be noted that the most favourable areas for wave action were not found to be the most exposed to an expected extreme storm event. During the occurrence of an extreme storm event, high level of wave energy dissipation is expected at the extreme west and south of the study area, but CSHI classification points out the relative low exposures to the effects of such an event. On the other hand, between *Costa da Caparica* and *Fonte da Telha* wave modelling sowed low 100-year H_s values but these are medium to high vulnerable areas.

Coastal systems height, slope, type, coastline rate of change (erosion/accretion rates) and anthropic pressure are the conditioning factors controlling the spatial distribution of the CSHI. Low lying, sandy, eroded and highly dense urbanized coastal stretches are more vulnerable to extreme storms, even though they have not high 100-year H_s favourability.

The improvement of the CSHI is possible, whether it is by the introduction of a less frequent storm event from the SW direction, which would affect the coastal area with a different spatial distribution, or by the improvement of the validation process with the contribution of more newspaper reports of the storm consequences. Nevertheless, the authors believe that the CSHI show results with an acceptable level of confidence and can be used to support coastal decision makers, stakeholders, and civil protection entities.

BIBLIOGRAPHY

[1] Sousa, N., *Dinâmica da linha de costa e vulnerabilidade à erosão no setor não artificializado do arco Caparica-Espichel.* Masters dissertation, Institute of Geography and territorial planning, Lisbon, 2015.

[2] Dodet, G., Bertin, X. & Taborda, R., Wave climate variability in the North-East Atlantic Ocean over the last six decades. *Ocean Modelling*, **31**, pp. 120–131, 2010. https://doi.org/10.1016/j.ocemod.2009.10.010

[3] Gornitz, V.M., Daniels, R.C., White, T.W. & Birdwell, K.R., The development of a coastal risk assessment database: vulnerability to sea-level rise in the U.S. Southeast. *Journal of Coast Research*, **12**, pp. 327–338, 1994.

[4] Shaw, J., Taylor, R.B., Forbes, D.L., Ruz, M.H., & Solomon, S., Sensitivity of the coasts of Canada to sea-level rise. *Bulletin – Geological Survey Canada*, **505**, pp. 1–79, 1998. https://doi.org/10.4095/210075

[5] Silva, S.F., Martinho, M., Capitão, R., Reis, T., Fortes, C.J. & Ferreira, J.C., An index-based method for coastal-flood risk assessment in low-lying areas (Costa de Caparica, Portugal). *Ocean & Coastal Management*, **144**, pp. 90–104, 2017. https://doi.org/10.1016/j.ocecoaman.2017.04.010

[6] Kumar, T.S., Mahendra, R.S., Nayak, S., Radhakrishnan, K. & Sahu, K., Coastal vulnerability assessment for Orissa state, East Coast of India. *Journal of Coastal Research,* 3, pp. 523–534, 2010. https://doi.org/10.2112/09-1186.1

[7] Islam, A., Mitra, D., Dewan, A., & Akhter S.H., Coastal multi-hazard vulnerability assessment along the Ganges deltaic coast of Bangladesh:a geospatial approach. *Ocean & Coastal Management*, **127**, pp. 1–15, 2016. https://doi.org/10.1016/j.ocecoaman.2016.03.012

[8] Seenath, A., Wilson, M. & Miller, K., Hydrodynamic versus GIS modelling for coastal flood vulnerability assessment: Which is better for guiding coastal management? *Ocean & Coastal Management*, **120**, pp. 99–109, 2016. https://doi.org/10.1016/j.ocecoaman.2015.11.019

[9] Dickson, M. & Perry, G., Identifying the controls on coastal cliff landslides using machine-learning approaches. *Environmental Modelling & Software*, **76**, pp. 117–127, 2016. https://doi.org/10.1016/j.envsoft.2015.10.029

[10] Booij, N., Ris R.C. & Holthuijsen, L.H., A third-generation wave model for coastal regions, Part I, Model description and validation. *Journal of Geophysical Research*, **104**, pp. 7649–7666, 1999. https://doi.org/10.1029/98jc02622

[11] Holthuijsen, L.H., *Waves in Oceanic and Coastal Waters*, Delft University of Technology and UNESCO-IHE, pp. 387, 2007.

[12] Rangel-Buitrago, N. & Anfuso G., *Risk assessment of storms in coastal zones: case studies from Cartagena (Colombia) and Cadiz (Spain),* Springer Briefs in Earth Sciences, pp. 63, 2015.

[13] Huang, Y., Weisberg, R., Zheng, L. & Zijlema, M., Gulf of Mexico hurricane wave simulations using SWAN: Bulk formula-based drag coefficient sensitivity for Hurricane Ike. *Journal of Geophysical Research: Oceans*, **118**, pp. 3916–3938, 2012. https://doi.org/10.1002/jgrc.20283

[14] Gorrell. L., Raubenheimer, B., Elgar, S. & Guza, R., SWAN predictions of waves observed in shallow water onshore of complex bathymetry. *Coastal Engineering*, **58**, pp. 510–516, 2011. https://doi.org/10.1016/j.coastaleng.2011.01.013

[15] Rusu, E., Pilar, P. & Soares, G.S., Evaluation of the wave conditions in Madeira Archipelago with spectral models. *Ocean Engineering*, **35**, pp. 1357–1371, 2008. https://doi.org/10.1016/j.oceaneng.2008.05.007

[16] Pires, H.O. & Pessanha, V.E., Wave power climate of Portugal. Proceedings *IUTAM Symposium, Hidrodynamics of Ocean Wave-Energy Utilization*, pp.157–167, 1986. https://doi.org/10.1007/978-3-642-82666-5_12

[17] Davidson-Arnott R., *An introduction to coastal processes and geomorphology.* Cambridge University press, pp. 396–400, 2010.

[18] Veloso-Gomes, F., Pinto, F.P., Barbosa, J.P., Costa, J. & Rodrigues, A., The Defensive Works at Costa da Caparica. 2. *as Jornadas de Hidráulica, Recursos Hídricos e Ambiente*, pp. 23–32, 2007.

[19] Rocha, J., Ferreira, J.C., Simões, J. & Tenedório, J.A., Modelling coastal and land use evolution patterns through neural network and cellular automata integration. *Journal of Coastal Research*, **50**, pp. 827–831, 2007.

[20] Coelho, C.R., Silva, F., Veloso-Gomes, F. & Taveira-Pinto, F., A vulnerability analysis approach for the Portuguese west coast. *Risk Analysis V: Simulation and Hazard Mitigation*, **91**, pp. 251–262, 2009. https://doi.org/10.2495/risk060241

[21] Gilleland, E., & Katz, R., extRemes 2.0: an extreme value analysis package inR. *Journal of Statistical Software*, **72**(8), pp. 1–39, 2016. https://doi.org/10.18637/jss.v072.i08

[22] Ribeiro, J., Silva, A.J.R. & Leitão, J.C., Modelos Operacionais de previsão da agitação para suporte à navegação e gestão de riscos, *6as Jornadas Portuguesas de Engenharia Costeira e Portuária*, pp. 22–38, 2009.

[23] Tolman, H. – *User manual and system documentation of WAVEWATCH-III version 2.22*, U.S. Department of Commerce, National Oceanic and Atmospheric Administration, pp. 133, 2002.

[24] Denner, K., Phillips, M.R., Jenkins, R.E. & Thomas, T., A coastal vulnerability and environmental risk assessment of Loughor Estuary, South Wales. *Ocean & Coastal Management*, 116, pp. 478–490, 2015. https://doi.org/10.1016/j.ocecoaman.2015.09.002

[25] Saaty, R.W., The analytic hierarchy process – What it is and how it is used. *Math Modelling*, **9**, pp. 161–176, 1987. https://doi.org/10.1016/0270-0255(87)90473-8

[26] Thieler E.R. & Hammar-Klose E.S., National Assessment of Coastal Vulnerability to Sea Level Rise: preliminary results for the US. *Atlanta coast USGS*, pp. 99 – 593, 1999.

[27] Nageswara, K., Subraelu, P., Venkateswara, T., Hema, B., Ratheesh, R., Bhattacharya, S. & Ajai, A.S.R., Sea-level rise and coastal vulnerability: an assessment of Andhra Pradesh coast, India through remote sensing and GIS. *Journal of Coastal Conservation*. **12**, pp. 195–207, 2008. https://doi.org/10.1007/s11852-009-0042-2

SAFETY AND SECURITY OF HOSPITALS DURING NATURAL DISASTERS: CHALLENGES OF DISASTER MANAGERS

SEYED PAYAM SALAMATI NIA & UDAYANGANI KULATUNGA
Centre for Disaster Resilience at university of Salford, Greater Manchester, United Kingdom.

ABSTRACT

The purpose of this research is to explore some challenges that hospital disaster managers face in dealing with natural disaster events. In such events, it is crucial that hospitals remain safe and functional during and after the disaster; thus, hospitals at all levels need to plan for natural disasters and be aware of the requirement to remain in an operational condition both during and after any event. Evidence from around the world suggests that the malfunctioning of hospitals during a disaster have extensive impacts on both inbound and outbound patients; as such, disaster preparedness is a significant concern for hospital disaster managers. Hospital disaster management is important because of the critical services that healthcare facilities provide for injured people and existing patients. Therefore, developing a good management system for natural disaster events can help to ensure better efficiency and economy in the use of facilities and human resources within hospitals. Although appropriate disaster management can mitigate the impact of natural disasters in hospitals, there are some barriers that can prevent the effective management of these facilities in such events. For this study, secondary information was retrieved from the Internet and via academic database on sudden-onset natural disasters, and it was found that the: awareness, knowledge, disaster preparedness of hospital staff; allocation of building codes, and the relocation of buildings to higher levels, need be improved. Also, equipping health care facilities at the time of natural disaster events is important. To manage the challenges facing hospital disaster managers, a national strategy for the disaster management planning for hospitals is required.

Keywords: Challenges, disaster manager, hospitals, natural disasters

1 INTRODUCTION

The impact of a disaster can seriously disrupt or destroy the functioning of a community or a society, and involve widespread human, material, economic or environmental losses, which exceed the ability of the affected community or society to cope using its own resources [1]. Although often caused by nature, disasters can have human origins [2]; they can happen abruptly and be classified as dangerous and calamitous, which can thus overwhelm and disrupt essential infrastructures. From a different perspective, a disaster can also be described, for instance, on a household scale, where such an event can cause major sickness and social calamity, or an essential economic catastrophe. At the level of community impact, a disaster can manifest as a fire, flood, earthquake, the demolition of buildings, or as displacement through conflict. When it happens, particularly across a local district or city, a large number of the population can be affected [3]. Biswas and Choudhuri [4] define a natural disaster as the outcome when people are affected by natural hazards. In fact, they believe that, with a lack of proper emergency management, human vulnerability can be increased at such times, and that these events can also have an impact on the environment and on financial resources. Catastrophic and calamitous incidents that result from natural resources form a part of the natural disasters group, which includes tornadoes, eruptions, earthquakes, and floods [4]. Indeed, Shaluf [5] states that, due to their severity and unpredictability, natural disasters are often called, 'Acts of God'. As communities worldwide face an increasing frequency and variety of disaster, there is an urgent need to mitigate the risk from these events [6]. Therefore, in this context,

DOI: 10.2495/SAFE-V7-N2-234-246

disaster management is significant as it can reduce the negative impacts; such efforts aim to avoid or mitigate potential losses, appropriate assistance for the victims of a disaster, and deliver an effective and quick recovery. However, disaster management becomes even more important when considering the mitigation of risk in relation to natural disasters. Sabharwal and Swarup [7] define disaster management as, 'activities taken by the organisation in regarding unexpected incidents, which are adversely affecting people or resources and threatening the continued operation of the organization'. As such, the focus of most researchers in this field has been on the analysis of disaster management and relatively little attention has been paid to disaster prevention [8]. To better appreciate the significance of disaster management, it is important to understand the procedures required; Orlando, Danna, Giarratano, Prepas, and Johnson [9] state that the disaster management process includes four main phases, which are mitigation, preparedness, response, and recovery, and these should all be considered in an event. Indeed, disaster management processes may present several challenges that are overwhelming and complex, where additional actions in each step are required to mitigate the impact of the disaster [10]. Nevertheless, they argue that this four-stage process should be followed, and such an experience also used to assist in preparing for future events. Disaster management becomes even more critical for hospitals as the health sector has historically been particularly vulnerable to disaster. Health facilities at all levels deserve special attention in the case of natural disasters, as they are expected to continue patient treatment as well as care for those injured by the event [11]. For example, during the Bam earthquake (2003), almost all of the hospital buildings were damaged and there was chaos in relation to its operational capacities [12]. Furthermore, in order to continue providing a health service during the disaster, a large number of patients were relocated from the hospital to shelters during the earthquake, and some medicines and medical equipment also had to be moved [13]; this also caused severe disruption to the organisation. At any given time, hospitals have a population of patients, staff, visitors, and transient patients, but in a disaster event, the number may rapidly and substantially increase. The security and safety of all patients and occupants must be secured whilst staff continue to provide ongoing treatments and support services [14]. Hospitals are expected to be ready to play an essential role in reducing death and injury, and hospital readiness has been defined as the ability to effectively maintain hospital operations, sustain a medically safe environment, and adequately address the increased and potentially unexpected medical needs of the affected population [15]. Nevertheless, managing hospitals during a disaster presents particular challenges when there is a lack of proper disaster management planning before an event. Thus, this study focuses on the importance of hospitals in natural disaster events, and explores some challenges facing Iranian hospital disaster managers in natural disaster events. This paper is structured as follows: firstly, the concept of disaster, and the importance of disaster management in hospitals are reviewed. Secondly, some key challenges for hospital disaster management will be discussed, and finally, the paper provides a discussion that focuses on those challenges of disaster managers in hospitals, and identifies some improvement that can be considered in the light of some past disasters and their impacts on hospitals.

2 LITERATURE REVIEW

2.1 Importance of hospitals in natural disaster events

A hospital has been defined by PAHO [16] as a hotel, laboratory, office building and warehouse, and their importance is apparent in their function in providing critical services to patients at any time. In disaster events it is crucial that hospitals remain safe and functional during and after incidents. When facing disaster, health facilities at all levels deserve special attention, in that they must continue to provide patient treatment for those staying in their facilities whilst also treat persons injured by the event [17]. It is also important that their promotion and prevention programs are not suspended, such as prenatal care and haemodialysis, which places further strain on the organisation. Thus, to ensure the continuity of services in the face of natural disaster, a hospital must implement formal plans to deal with such eventualities. Furthermore, the construction plan of the building and its equipment must also remain in serviceable conditions. Many hospital authorities already recognise these requirements, which is why they have drawn up plans for dealing with disasters, and developed strategies to understand and mitigate these risk by using specially designed construction plans [18]. However, hospitals still need to incorporate prevention in such plans, in order to strengthen the impact of the committee's hospital disaster risk management system. Hence, it is crucial to formalise these considerations in the design and construction of buildings in order to provide security and better preserve certain critical areas of the hospital, such as the emergency department, intensive care units, diagnostic facilities, operating rooms, pharmacy, food and medicine stores, and services for registration and booking. Therefore, when planning a hospital, the architectural and structural designs should address safety specifications, not only in relation to aspects of the natural physical phenomena that can affect hospitals but also with regard to the social, economic and human needs [19]. Medical facilities are one of the most critical services in any region; however, the size and the priority of health equipment usage in a disaster event can be challenging to manage. The typical hospital depends on different factors, such as the: continuity of its utility supplies; sufficiency and availability of normal and skilled staff; accessibility and availability of medical and equipment supplies; and its accessibility for its daily operations. In fact, the failure of any of these factors can affect the continuity of medical services.

2.2 Impact of natural disasters on hospitals

Generally, hospitals are required to undertake special consideration in how they will manage the impact of a disaster. This is due to the need to continuously provide healthcare facilities for patients currently in the hospitals as well as deal with the unanticipated injuries of the people during the event. Hospitals and health facilities are the world's largest industry and such organisations (especially public hospitals) need a substantial bureaucratic structure to provide full services to patients [20]. According to World Health Organization [21] (also cited by Noralfishah, Thayaparan, Haidaliza, and Kulatunga [22] hospitals play critical roles in disaster events by providing communities with essential medical care. However, due to the nature of disasters the demand for health care services can rapidly and unexpectedly increase.

Thus, without sufficient planning, a disaster can overwhelm the functional capacity and safety of a hospital.

There are different impacts in disaster events, such as the ability of the hospital to function, the direct impact to patients and health equipment, and the physical damage to hospital buildings [23]. In disaster events, access to patient documentation and medicine can be affected and it is important to keep these in the safe. Also, supplying food and essential medicines, especially in the early hours of a disaster are crucial, and must be considered by disaster managers [24]. In terms of the time period, disasters can have different impacts on hospitals. The most typical impact was identified by the Disease Control Priorities Project [25], which noted a range of damage and loss, such as the loss of health equipment, a failure in the hospital's energy sources, a lack of staff, difficulty in accessing patients' documents (for example, the medical records), and issues in accessing the hospital buildings. For example, the earthquake that struck Mexico City in 1985 resulted in the collapse of 13 hospitals. In just three of the hospitals' buildings, 866 people died, 100 of whom were health personnel, and nearly 6000 hospital beds were lost across the city's metropolitan facilities. This illustrates the severity and extent of the challenges that can face hospital disaster managers when planning for such events. The 2003 Bam earthquake, with a magnitude of 6.7 on the Richter scale, destroyed the city of Bam in the South-East of Iran. The Bam earthquake is clasified as one of the major earthquakes in the 21st Century, when around 35,000 people were injured and more than 12,000 people had to evacuate their houses. This placed enormous, urgent demand on the disaster response agencies; however, the damage to health care infrustructures not a consideration at that time [13]. Table 1, provides an illustration of some of the damage sustained by health care infrastructures due to the Bam earthquake.

Further examples of the extent of the impact of natural disasters on hospital and health care facilities include; the earthquake that struck Mexico City in 1985, which resulted in the collapse of 13 hospitals. In just three of those buildings, 866 people died, 100 of whom were health personnel, and nearly 6000 hospital beds were lost across the metropolitan facilities. As a result of Hurricane Mitch in 1998, the water supply systems of 23 hospitals in Honduras were damaged or destroyed, and 123 health centres were affected. Peru reported that nearly 10% of the country's health facilities suffered damage as a result of the El Niño events in 1997 and 1998 [26]. Disasters can have two different impacts on hospitals in terms of the time period. According to the Disease Control Priorities Project's [26] report, some typical impacts can include damage and loss, which include: the loss of health equipment, a failure in the hospital's energy resources, a lack of staff, problems in

Table 1: Damage to health care infrastructures due to the Bam earthquake [13].

Health facility	Damage percentage%	Number of health facilities
Dormitory of the facility of nursing	100%	1500 Sq.m
District health care management care	100%	1
Facility of nursing and paramedics	100%	2000Sq.m
Emergancy station (115)	100%	1
Behvarz training centre	100%	1
Emam district hospital (public)	50%	136
Aflatoonyan hospital (private)	100%	65
District health network expansion centre	100%	1
Rural health centre (RHC)	100%	14
Health posts (Urban)	100%	5
Health house	100%	95
Madiheh Maternity Hospital (public)	40%	54 beds
Urban health centre (UHC)	100%	10

accessing patients' documents, and issues with access to buildings. In addition to this, patient documents and medicines can be affected and it is important to protect these. Also, supplying food and necessary medicine especially in the early hours of a disaster are crucial, which must be considered by disaster managers [24]. Therefore, it is important to manage hospitals appropriately so that they are able to better able to withstand the impacts of disasters. For instance, it is important to provide security and preserve certain critical areas of the hospital, such as the emergency department, intensive care units, diagnostic facilities, operating rooms, pharmacy, food and medicine stores, and services registration and booking. These all need to be a focus for hospital disaster management. The role, significance and specific needs of a hospital means that their design needs to carefully considered, arguably more so than for other more conventional buildings with less critical purposes, such as homes and offices [27]. Furthermore, during a disaster, it may become necessary to evacuate non-ambulant and ambulant patients; therefore, the response to disaster including evacuation procedures should also be well established and considered in a building design. These issues also demonstrate that proper disaster management is highly important for hospitals and health centres.

2.3 Disaster management in Iranian hospitals

Communities worldwide have been facing an increasing frequency and variety of disasters, and such incidents can have direct and indirect impacts, thus there is an urgent need to reduce disaster risk [28]. This is also linked with the development of resilient communities capable of recovering from disasters, which is of increasing concern in many countries [29]. For example, Iran is located in Central Eurasia, and in the Middle East; its population is about 75 million and as a country it is exposed to a wide range of man-made and natural hazards. According to EM-DAT, more than 180 disasters were recorded in Iran between 1900 and 2013, which were connected with 160,000 deaths and impacted 45,000 people. The most important natural hazards in this region are earthquakes, floods, and drought [23]. In Iran, the Emergency Management Centre (EMC) is under the control of the Deputy of Health, and the Centre's most important duties are the planning, supervision, and coordination in emergencies. Due to the potential impact of disasters on human life, health facilities have since been moved to the top of the agenda in terms of disaster management [32]. The 2003 Bam earthquake particularly influenced this, when the functions of the city's health system collapsed. Indeed, due to this impact, the event impact influenced a rapid reshaping of health provision in different regions; furthermore, training was provided and regional emergency operational services were planned at a different level to the district level. These steps aimed to ensure that health services were more resilient in the face of disaster, and thus able to respond appropriately.

The experience of the Bam earthquake also influenced the development of guidelines regarding disease surveillance in disasters, which have since been applied to subsequent disaster events [23]. Therefore, after the Bam earthquake, the progress of disaster risk management was significantly accelerated, and many improvements were made. Despite the valuable achievements and learning from the Bam earthquake and other disaster events, there is a room available for improvement in the disaster management systems in Iran, and in particularly in the Iranian health system. This challenge has changed in the last decade; since the disaster, the health system has been improved and upgraded, moving from the Centre for Environmental Health to the Emergency Management Centre, and now falls under the remit of the Deputy of Health Minister. Furthermore, the health system has since been successfully dealing with health-related problems, which range from the lowest to the highest level of the country's disaster management strategies. However, to secure better results it has been necessary to ensure effective collaboration between different sectors [33]. As briefly discussed, hospital disaster management in Iran is very important because of the critical services that such healthcare facilities provide in a country that has experienced a number disaster events. Not only do healthcare facilities need to continue to provide services to patients already undergoing treatment, but also to extend their services to those who are affected by a disaster. As such, disaster managers face additional and unpredictable challenges when managing hospitals in disaster events. In the sections 4 and 5, some challenges for hospital disaster managers will be reviewed.

3 RESEARCH METHODOLOGY

This research is based on secondary data analysis that defines and explains disaster impacts on hospitals, and reviews some challenges facing hospital disaster managers in the light of natural disaster events. The data have been gathered from peer reviewed journals, and some books and conference papers. Databases were used to gather information regarding: the context of disaster management in hospitals, the local pre-disaster availability of hospitals and health facilities, and the sudden impact of natural disasters on hospitals. For this phase of the study, a qualitative content analysis will form the major research instrument to analyses the secondary data. At this point, the primary data collection and analysis has not yet been completed, which will collect by interview relevant expert views regarding hospital disaster management. This current paper will thus purely adopt a qualitative data collection strategy, and consider a variety of secondary sources accessed through the Internet and academic databases.

4 CHALLENGES FOR HOSPITAL DISASTER MANAGERS

How hospitals respond to the demand for medical attention during and after a disaster event can be extremely challenging. Natural disasters impact on healthcare organisations, and remind disaster managers to pay attention to the safety of their staff, particularly their first responders and physicians [34]. Usually disaster managers in hospitals are faced with a range of challenges in such events, which include unexpected changes to newly injured or ongoing patient management, communication challenges, a lack of coordination, a lack of preparedness, and logistical deficiencies. For example, in August 2012, an earthquake occurred close to the cities of Ahar and Varzeghan in the East-Azerbaijan province in North-West Iran. This earthquake caused suffering for over 2 million people and forced many people sleep outside for two nights.

The earthquake killed 327 people, claimed more than 3000 injuries, and left more than 30,000 people homeless. Buildings in the stricken area experienced different levels of damage; however, most of the adobe buildings in the villages had collapsed and several masonry and framed buildings were damaged. Furthermore, some of the roads were damaged due to the surface faulting and geotechnical instabilities [35]. Although some of the bridges suffered damage, they remained serviceable after the event. Nevertheless, many of the essential facilities, such as hospitals, in the stricken areas were damaged and some industrial plants experienced economic losses due to the unsafe shutdown process after the earthquake. The performance of the hospitals was not acceptable in that they were not operable after the disaster event. One particular example of this was the Bagher-aloom Hospital in Ahar city, which suffered remarkable non-structural damages and was put out of commission after the earthquake. In particular, the collapse of the infill walls and of the false ceilings, and the overturning of medical equipment were major failures in the non-structural components of

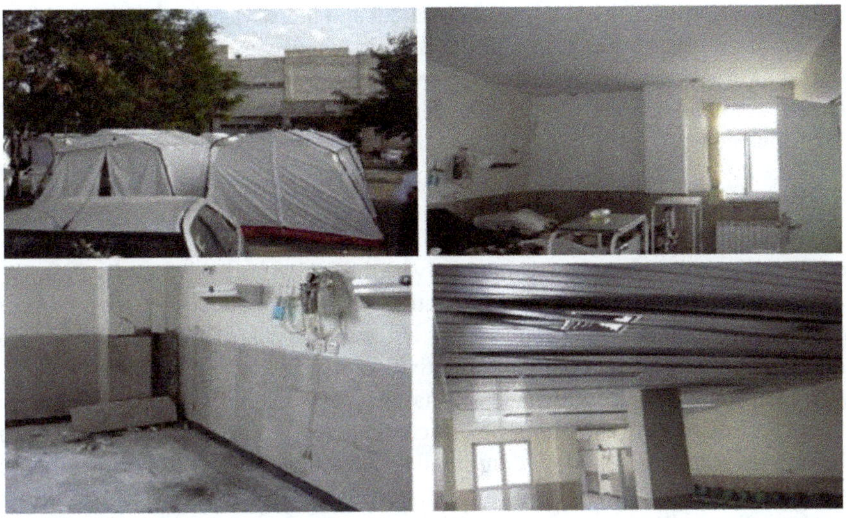

Figure 1: Herris Hospital and the Temporary Tents Outside Of Hospital Building [35].

the hospital. Due to such damage, treatment was carried out in temporary hospitals set up in tents, as it can be seen in Figure 1, and rescue teams had to transfer injured people to the hospitals in nearby cities. Heavy traffic jams in Tabriz-Varzaghan, Tabriz-Ahar and along other primary roads of the affected area caused significant trouble for the rescue teams, and the main reason for these was the large number of volunteers going to the affected areas to help the rescue teams. Thus, effective disaster planning could have mitigated some of the potential damage of this natural disaster, and provide a better health service during and after disaster to patients in the worst affected hospital. In the following sections, some specific challenges of disaster managers will be explained [35].

4.1 Injured management

The first issue for the majority of hospitals when facing a disaster can be in managing the numbers and requirements of injured people. Examples of this include the Herris hospital (in North–West Iran), which was destroyed in the 2012 earthquake, and the Ahar hospital, which saw its first and ground floors destroyed. Although both hospitals had to be evacuated, there were no special instructions prepared to deal with this eventuality. Although staff and patients had to be evacuated, some patients were able to move with their fellows left the health centres. However, later with rising tensions between patients and increasing damage to the structure of the hospital, it had been done slowly. After this earthquake, all affected patients were moved to Tabriz, and only one hospital was fully evacuated. In addition, the patients' documents also needed to be incorporated into the evacuation and eventual relocation. Therefore, it is not only the coordination of patients but also of their records that needs attention during and after a disaster, particularly when undertaking evacuations [36].

4.2 Lack of communications management

When a natural disaster occurs, some communication tools will be suspended, such as Wi-Fi signals, mobile and local telephone networks, and internet signals; for example, following the

Herris and Ahar earthquakes local telephones failed, and there was also an issue with hospital phones in Tabriz. Furthermore, although the mobile signal was not cut off, the signal was weak in most of areas due to a failure in the centre and a spike in call traffic. Nevertheless, the issue was resolved after 1 day, which demonstrates a rapid response to the issue of communication access. [23].

4.3 Lack of coordination

A further challenge following the earthquake was the lack of proper coordination of relief, which can be classified into two categories: intra coordination, and inter sectoral coordination. In fact, a lack of coordination affects all activities, and this and the shortage of equipment are significant barriers to an effective service delivery, which in the case of the Bam earthquake impacted as follows: coordination problems with volunteers who came to assist; the lack of coordination among health centres, and among officials of different hospitals; the lack of a hospital incident command system and a failure to perform sufficiently frequent examinations of some of the injured; confusion between personnel and officials; the lack of coordination between prehospital emergency officials and hospitals; staff disregard of officials' orders; the intractable performance of tasks by staff; the absence of a command unity and unit commander fragmentation; unwarranted interventions of irrelevant people; the absence of official accountability; and disaster management interventions by county headquarters [36].

4.4 Lack of preparedness

The lack of preparedness of disaster managers can have a significant impact in providing an effective service to hospital patients in a natural disaster event. With a lack of previous staff training, a lack of initial planning to address issues arising from disaster events, no attention to previous experience, no anticipation of demands, a failure to meet demands, and no attention to the preparedness plans, the impact of natural disasters can be increased [24].

4.5 Logistics deficiencies

Other challenges facing disaster managers in such events can include logistic deficiencies, which can involve; a lack of funds for emergencies, challenges to security management, issues with donations management, inappropriate places to provide medical services, and difficulties with human resources management. Both hospitals in the North-West of Iran were destroyed after earthquake happened; however, they had only prepared one location in their hospital campuses to assign their health services, which were not adequate in for the numbers of personnel and injured people. For instance, there were inadequate sanitation facilities, and an issue with the cooling and heating systems [30]. Nevertheless, although other health centres were not damaged, they experienced the same problems. Furthermore, there were not a lot of donations in the hospital and no funding was predicted for such urgent moments, which also caused issues for the disaster recovery of the hospital and its continued operations. Although there were no security issues, people were rushing to get first aid treatment and medicines. Staff members complained about the temporary settlements in the hospital campuses in that there was fear about incoming injured people, and the security of the buildings, due to a lack of sustainability in the temporary constructions [31].

5 DISCUSSION

In this study, the concept of disasters, the importance of hospitals in such events, and the impact of natural disasters on hospitals were explored. In addition, some major challenges facing hospital disaster managers were extracted. The earthquake within the North-West region of Iran provided an example that illustrated the impact of previous disaster management failures and highlighted key areas of concern. As health facilities, such as hospitals, are fundamentally important in mitigating the risk of worsening health impacts from natural disaster events, they should be a priority for disaster management planning. The response to injured people should be planned within an appropriate location and in the minimum time possible. However, the physical damage to the hospital buildings and equipment following the earthquake had a direct impact on its service and ability to meet the urgent demands at the time [31]. According to the Ahar earthquake lack of preparedness was one of the major challenges in face of disaster managers because of lack of previous training of personnel and training programs, lack of prior planning for disaster situations.

Also failure to anticipate and meet demands. In addition lack of coordination was the other challenge, due to coordination issues with volunteers who were referred to help, lack of

Table 2: Summary of challenges in face of Disaster Managers [36].

Lack of preparedness

- No previous training of personnel and lack of training programs
- Lack of prior planning for disaster situations
- Failure to anticipate and meet needs
- Lack of attention to the experiences and lessons of previous disasters

Communication and information management

- Contact with the media
- Communication within the hospital
- Out-of-hospital communications
- Management of very important people and visitors

Logistics challenges

- Inappropriate places for providing services to the injured
- Management of donations
- No emergency fund
- Security management
- Human resources management

Technical challenges

- Evacuation of hospitals
- Patient security
- Admission
- Entry and exit management and discharging of injured
- Triage and prioritization of patients

Lack of coordination

- Coordination problems with volunteers who were referred to help
- Lack of coordination among hospital officials
- Lack of coordination among the authorities in different hospitals
- Lack of coordination among the prehospital emergency and hospital authorities
- No Incident Command System and not running if there was any
- Disobeying the orders of officials by personnel
- Intractable performance of tasks by staff
- Absence of command unity and single commander
- Frequent examinations of some injured
- Bewilderment of personnel and officials
- Fragmentation and repetition
- Inappropriate interventions of unrelated individuals

coordination among hospital officials, lack of coordination among the authorities in different hospitals, lack of coordination among the prehospital emergency and hospital authorities, and failure incident command system made difficult situation for disaster managers at that event. In spite of lack of coordination and preparedness, there were other challenges such as logistic, and technical challenges because of failure management of donations, lack of emergency fund, issue with security management human resources management admission, and entry and exit management and discharging of injured people. Accordingly, accreditation organisations now require hospitals to have a preparation plan and demonstrate their preparedness for a disaster response. Thus, hospital preparation and identifying the need for disaster management is necessary in many countries that are susceptible to natural disasters. Nonetheless, the results of this research indicate that many health centres are often not ready for such eventualities [36].

Timely and effective responses to disasters require an organised disaster response system, which can provide the appropriate aid. Therefore, with all those challenges it is significantly important to prepare hospitals and get lessons with previous experience.

For instance providing disaster training for staff, equipping health care facilities in disaster events, and allocating building code can be productive solution to mitigate the risk of disasters in hospitals. In the following table, a summary of challenges is illustrated, taken from the existing cases that hospital disaster managers should be aware of.

6 CONCLUSION

In conclusion, preparing a hospital for a natural disaster event is important as it is likely to face high demand from injured people. Whether disasters are from flooding, hurricanes, severe heat waves, droughts, ice storms, tornadoes, typhoons, or fires are less significant than the need to respond quickly, appropriately, and with authority. The main issues are that firstly, there is no clear way to predict when a natural event will occur, and secondly that it will be difficult to determine about any potentially destructive trajectories. Nevertheless, it is important to prepare a hospital by maintaining its condition and learning from previous disaster events in order to mitigate the risk of hazards. In conclusion, it is crucial to have right disaster management plan in place to avoid and mitigate the risk of a disaster incapacitating a hospital. Hospital disaster managers have a critical role in keeping hospitals in a good condition as they provide the most strategic location in terms of relief in natural disaster events. Disaster managers may face a lack of coordination, and communication and logistics issues, which, as well as addressing at the time, should be considered as valuable learning for future disaster management plans. The next phase of the research will involve the collection and analysis of primary data from expert interviews with a view to capturing in depth knowledge on natural disasters and the challenges of hospital disaster managers.

REFERENCES

[1] UNISDR, *UNISDR terminology for disaster risk reduction*. United Nations International Strategy for Disaster Reduction (UNISDR) Geneva, Switzerland, 2009.

[2] Chen, J., Modern disaster theory: evaluating disaster law as a portfolio or legal rules. *Emory International Law Review*, **25**, p. 1121, 2011.

[3] Vale, L.J. & Campanella, T.J., *The resilient city: How modern cities recover from disaster*. Oxford University Press, 2005.

[4] Biswas, B.C. & Choudhuri, S.K., Digital information resources for disaster management of libraries and information centres. *Bangladesh Journal of Library and Information Science*, **2**(1), pp. 12–21, 2012. https://doi.org/10.3329/bjlis.v2i1.12915

[5] Shaluf, I.M., Disaster types. *Disaster Prevention and Management: An International Journal*, **16**(5), pp. 704–717, 2007. https://doi.org/10.1108/09653560710837019

[6] Lin Moe, T. & Pathranarakul, P., An integrated approach to natural disaster management: public project management and its critical success factors. *Disaster Prevention and Management: An International Journal*, **15**(3), pp. 396–413, 2006. https://doi.org/10.1108/09653560610669882

[7] Sabharwal, M. & Swarup, A., The implementation of disaster management by Indian banks. *International Journal on Arts, Management and Humanities*, **1**(1), pp. 73–80, 2012.

[8] Masellis, M., Thermal agent disaster and fire disaster: definition, damage, assessment and relief operations. In The management of mass burn casualties and fire disasters, M. Masellis & S.W.A. Gunn (eds.). Springer, pp. 7–12, 2014.

[9] Orlando, S., Danna, D., Giarratano, G., Prepas, R. & Johnson, C.B., Perinatal considerations in the hospital disaster management process. *Journal of Obstetric, Gynecologic, & Neonatal Nursing*, **39**(4), pp. 468–479, 2010. https://doi.org/10.1111/j.1552-6909.2010.01158.x

[10] WHO, Guidelines for drinking-water quality. *World Health Organization*, **216**, pp. 303–304, 2011.

[11] Eybpoosh, M., Dikmen, I. & Talat Birgonul, M., Identification of risk paths in international construction projects using structural equation modeling. *Journal of Construction Engineering and Management*, **137**(12), pp. 1164–1175, 2011. https://doi.org/10.1061/(asce)co.1943-7862.0000382

[12] Ardalan, A., Masoomi, GR., Goya, MM., Ghaffari, M., Miadfar, J., Sarvar, MR., Soroush, M., Maghsoodi, A., Holakouie Naieni, K., Kabir, MJ. & Khankeh, HR., Disaster health management: Iran's progress and challenges. *Iranian Journal of Public Health*, **38**(1), pp. 93–97, 2009.

[13] Djalali, A., Castren, M., Hosseinijenab, V., Khatib, M., Ohlen, G. & Kurland, L., Hospital incident command system (HICS) performance in Iran; decision making during disasters. *Scandinavian Journal of Trauma, Resuscitation and Emergency Medicine*, **20**(1), p. 14. https://doi.org/10.1186/1757-7241-20-14

[14] Salamati Nia, S.P., Kulatunga, U. & Thayaparan, M., *The importance of disaster management and impact of it into hospitals*. Paper presented at the SPARC 2016, Media city UK.

[15] Barbera, J.A., Yeatts, D.J. & Macintyre, A.G., Challenge of hospital emergency preparedness: analysis and recommendations. *Disaster Medicine and Public Health Preparedness,* **3**(S1), S74–S82, 2009. https://doi.org/10.1097/dmp.0b013e31819f754c

[16] Ardalan, A., Masoomi, G.R., Goya, M.M., Ghaffari, M., Miadfar, J., Sarvar, M.R., Soroush, M., Maghsoodi, A., Holakouie Naieni, K., Kabir, M.J. & Khankeh, H.R., Disaster health management: Iran's progress and challenges. *Iranian Journal of Public Health*, **38**(1), pp. 93–97, 2009.

[17] Eybpoosh, M., Dikmen, I. & Talat Birgonul, M., Identification of risk paths in international construction projects using structural equation modeling. *Journal of Construction Engineering and Management*, **137**(12), pp. 1164–1175, 2011. https://doi.org/10.1061/(asce)co.1943-7862.0000382
[18] Kenny, C., Disaster risk reduction in developing countries: costs, benefits and institutions. *Disasters*, **36**(4), pp. 559–588, 2012. https://doi.org/10.1111/j.1467-7717.2012.01275.x
[19] Jaiswal, P. & van Westen, C.J., Use of quantitative landslide hazard and risk information for local disaster risk reduction along a transportation corridor: a case study from Nilgiri district, India. *Natural Hazards*, **65**(1), pp. 887–913, 2013. https://doi.org/10.1007/s11069-012-0404-1
[20] Yarmohammadian, M.H., Atighechian, G., Haghshenas, A. & Shams, L., Establishment of Hospital Emergency Incident Command System (HEICS) in Iranian hospitals: a necessity for better response to disasters. *Iranian Red Crescent Medical Journal*, **15**(12), 2013. https://doi.org/10.5812/ircmj.3371
[21] World Health Orgainzation. Guidelines for drinking-water quality. *World Health Organization*, **216**, pp. 303–304, 2011.
[22] Noralfishah, S., Thayaparan, M., Haidaliza, M. & Kulatunga, U., Impact of flood disaster in Malaysia: a case study on public hospitals. *Special Themed Session: Cities, Infrastructure and Cascading Natural Disasters*, 28–33, 2015.
[23] Ardalan, A., Mowafi, H. & Khoshsabeghe, H.Y., Impacts of natural hazards on primary health care facilities of Iran: a 10-year retrospective survey. *PLOS Currents Disasters*, 2013.
https://doi.org/10.1371/currents.dis.ccdbd870f5d1697e4edee5eda12c5ae6
[24] Nakhaei, M., Khankeh, H., Masoumi, G., Hosseini, M. & ParsaYekta, Z., Health management in past disasters in Iran: a qualitative study. *Health in Emergencies and Disasters*, **1**(2), pp. 107–115, 2015.
[25] Shaluf, I.M., Disaster types. *Disaster Prevention and Management: An International Journal*, **16**(5), pp. 704–717. https://doi.org/10.1108/09653560710837019
[26] Guha-Sapir, D., Hargitt, D. & Hoyois, P., *Thirty years of natural disasters 1974–2003: the numbers*. Presses universityu de Louvain, 2004.
[27] McEntire, D. A. a., *Disaster response and recovery: strategies and tactics for resilience* (2nd ed.). John Wiley & Sons, Hoboken, NJ, 2015.
[28] Lin Moe, T. & Pathranarakul, P., An integrated approach to natural disaster management: public project management and its critical success factors. *Disaster Prevention and Management: An International Journal*, **15**(3), pp. 396–413, 2006. https://doi.org/10.1108/09653560610669882
[29] Rotimi, J.O., Wilkinson, S., Zuo, K. & Myburgh, D., Legislation for effective post-disaster reconstruction. *International Journal of Strategic Property Management*, **13**(2), pp. 143–152, 2009.
https://doi.org/10.3846/1648-715x.2009.13.143-152
[30] Boin, A. & McConnell, A., Preparing for critical infrastructure breakdowns: the limits of crisis management and the need for resilience. *Journal of Contingencies and Crisis Management*, **15**(1), pp. 50–59, 2007.
https://doi.org/10.1111/j.1468-5973.2007.00504.x

[31] Alexander, D.E., *Principles of emergency planning and management.* Oxford University Press on Demand, 2002.
[32] DeRoeck, D., Bawazir, SA., Carrasco, P., Kaddar, M., Brooks, A., Fitzsimmons, J. & Andrus, J., Regional group purchasing of vaccines: review of the Pan American Health Organization EPI revolving fund and the Gulf Cooperation Council group purchasing program. *The International Journal of Health Planning and Management*, **21**(1), pp. 23–43, 2006.
https://doi.org/10.1002/hpm.822
[33] Babaie, J., Ardalan, A., Vatandoost, H., Goya, M. M., & Akbarisari, A., Developing a performance assessment framework and indicators for communicable disease management in natural disasters. *Prehospital and Disaster Medicine*, **31**, pp. 27–35, 2016.
https://doi.org/10.1017/s1049023x15005452
[34] Francescutti, L.H., Sauve, M. & Prasad, A.S., Natural disasters and healthcare: Lessons to be learned. In *Healthcare Management Forum 2017* (Vol. 30, No. 1, pp. 53–55). Sage Publications, Los Angeles, CA, 2017.
[35] Razzaghi, M.S. & Ghafory-Ashtiany, M., *A preliminary reconnaissance report on August 11th, 2012, Varzaghan-Ahar twin earthquakes in NW of Iran.* Report of International Association of Seismology and Physics of the Earth's Interior, 2012.
[36] Pouraghaei, M., Jannati, A., Moharamzadeh, P., Ghaffarzad, A., Far, M.H. & Babaie, J., Challenges of hospital response to the twin earthquakes of August 21, 2012, in East Azerbaijan, Iran. *Disaster Medicine and Public Health Preparedness*, pp. 1–9, 2017.

DE-ICING OF AIRCRAFT: INCORPORATING BUSINESS RISKS AND OCCUPATIONAL HEALTH AND SAFETY

S. NADEAU & FRANÇOIS MORENCY
Mechanical engineering, École de technologie supérieure, Canada.

ABSTRACT
Airliner maintenance is a high-risk sector from an occupational health and safety perspective. In the current context of increasing air traffic worldwide and in view of the economic and safety issues associated with aircraft maintenance, it is imperative that companies providing this service receive support in their quest to reconcile operational performance with OHS. A survey of interdisciplinary research published from 2004 to 2014 allowed analysis of current thought in industrial engineering, human factors engineering and aeronautical de-icing, thus revealing a need for the design of aircraft de-icing activities that are sustainable, holistic and integrated. In response to market evolution, aeronautics will have to offer aircrafts that are greener, meaning safe for all users (including maintenance workers) and the environment throughout the product lifecycle.
Keywords: aeronautics, aircraft maintenance, aviation, de-icing, human factor engineering, integrated risk management, occupational health and safety, sustainable development

1 INTRODUCTION

The aim of sustainable development is to meet present needs of populations throughout the world without compromising the capacity of future generations to do so. In view of this, the International Labour Office (ILO) upholds the right of employees to work under conditions that do not threaten wellbeing, meaning that adequate measures are in place to prevent accidents and occupational illnesses. Whether in construction, manufacturing or other sectors, several researchers have concluded that achieving true sustainability will require an approach that is more holistic than that being taken in most cases (Hinze *et al.* [1], Haslam & Waterson [2], Demirel & Duffy [3], Lozano & Huisingh [4]). This means more precisely that pro-active preservation of all resources including human capital throughout the lifecycle of an economic activity, be it mining, air transport or other, is now a pre-requisite for future development (Dul *et al.* [5]).

The aviation sector foresees a major increase in future air traffic: 5–10% per year, depending on the continent (OACI [6]). Boeing estimates that 36,000 passenger aircraft will be in service in 2025 (Boeing [7]). One hour of flight requires 12 hours of aircraft maintenance (Rashid *et al.* [8]). A quadratic relationship exists between air traffic volume and accident rate, which explains the recommendations to increase safety levels by a factor of three by 2020 and by ten on the long term if air traffic should triple (Vogt *et al.* [9]). According to airliner incident and accident reports, the principal cause is improperly completed procedure (Drury and Johnson [10]), information corroborated by an Australian study (Hampson *et al.*, 2012 [88]). Maintenance activities represent 10–15% of airline budgets, two thirds of which are for labor costs in the case of heavy maintenance burdens (McFadden and Worrells 2012 [11]). In the case of commercial flight, the proportion of accidents related to airliner maintenance is 12–20% (Rashid *et al.* [8]). In Europe, 10–20% of workplace accidents and 10–15% of deaths in all industrial sectors are associated with maintenance activities (EU-OSHA [12]). In 2002, there were no fatalities in the Québec air transport maintenance sector, although numerous accidents and occupational illnesses were reported, leading the province's occupational health and safety commission to rate this sector as high risk.

DOI: 10.2495/SAFE-V7-N2-247-266

The literature is replete with studies describing the associations between deficient measures for accident or occupational illness prevention and drops in productivity, losses of business competitive advantage and even drops in national economic growth and competitiveness (HESAPRO [13]) and individual and corporate social wealth (Dul *et al.* [5]). Furthermore, organizational innovation through rapid and flexible adoption of innovative approaches to the design and management of production systems appears to favor sustainable business models (HESAPRO [13]). Market globalization, technological change and business regulation is leading to profound, rapid and continuous changes in corporations, notably in the workplace (Kristensen [14]). Production systems are becoming more integrated and complex, characterized by greater process interdependence, greater numbers of variables and arbitrations to be managed (Costella *et al.* [15]), focus on high-value-added products, and design based on cutting-edge technologies and knowledge. A sustainable business must place increased emphasis on balancing its performance imperatives with the wellbeing of its workers (Westgaard and Winkel [16]). We examine the question of how this might be achieved in the airliner maintenance sector, and more specifically in the context of de-icing operations, based on the available literature on integrated risk management, on maintenance operations and on ground de-icing. We analyze this knowledge from the perspective of justifying the need for modeling as means of developing sustainable and holistic de-icing activities that will ultimately balance worker performance with worker wellbeing. Further researches should be conducted to develop technology to reduce human-factor associated variability in de-icing activities. The models could help improve the integration of ergonomics in the de-icing procedure, through the development of task specific documentation and training. Tools to integrate variability into the risk evaluations of critical task are also needed and could eventually lead to design improvements at the aircraft level.

2 AERONAUTICAL DE-ICING

Icing of aircraft surfaces cannot be avoided completely and remains a significant cause of accidents. It can occur during flight or on the ground, whenever the aircraft encounters frosting conditions. Accumulation of ice on the wings of an aircraft increases the stalling speed and modifies the flight envelope, which may lead to accidents (Brown [17]). Certification authorities therefore define an envelope of the most probable icing conditions, and aircraft manufacturers must demonstrate that their planes fly safely under these conditions. For an aircraft to obtain authorization for flight through icing conditions, the manufacturer must install anti-icing systems, which are often based on heating of the wings (Hannat and Morency, [18]). When icing conditions arise on the ground before takeoff, the aircraft must be de-iced and protected against subsequent frost deposition using de-icing liquids.

Aircraft de-icing shares most of the common maintenance problems related to deregulation in the aircraft industry. In Canada and USA, the air carrier must have a ground de-icing program approved by the authorities, but outsourcing of aircraft ground de-icing services is becoming the norm (FAA, [92]). Although new policies have been under development over the last 10 years, no procedure yet exists for the service provider to have an approved de-icing program, as in some other aircraft maintenance activities. The service provider has to comply with every air carrier de-icing program, as many as 25 different air carriers (EASA, [94]). Another particularity is the fact that offshore de-icing activities are inevitable, as de-icing must be done at the aircraft departure airport if icing conditions occur. Contrary to most maintenance activities (Quinlan, 2013 [91]), human errors in de-icing have immediate effects, either by exposing workers to injury or by jeopardizing aircraft takeoff. Most airports nowadays have a single service provider dedicated to de-icing activities, potentially resulting in a growing level of experience and higher level of service (EASA, [94]). The typical centralized service provider is well organized with good training, experienced supervisors and management systems, avoiding some previously identified failures in other aircraft maintenance activities (Quinlan, 2014 [90]). The typical employee, however, has been with the company for less than 2 years, almost always does part time and seasonal work, and often it is his/her first exposure to aircraft maintenance work (EASA, [94]).

When conducting de-icing tasks on aircraft, maintainers are exposed to many risks as falls from heights of ten meters or more, splashes or run-offs of fluids from the aircrafts, the platform colliding with the aircraft, being sucked towards the intake of running motors during audit of small aircraft, slipping on the ground during audit of small aircraft, direct exposure to de-icing/anti-icing fluids, direct exposure to conditions on the airport tarmac (cold temperature, wind-chill factor, noise, glare (passing from darkness to brightness), musculoskeletal lesions associated with working postures and the weight (about 3 to 5 kg) of the de-icing nozzle being handled, fatigue (Torres *et al.*, [96]) and others. A de-icing ground crew sprays these liquids and inspects critical surfaces to ensure that no ice remains (Ayache *et al.*, [19]). De-icing operations on the ground must be conducted as quickly as possible not to cause undue delays in flight schedules.

The protection of aircraft against frost build-up is often divided into two aspects, namely ground and in-flight. However, in accidents or incidents due to frost build-up, very often both aspects are involved to varying degrees. For example, in the case of the crash of a Cessna 208B Caravan on Pelée Island in Ontario in January 2004, ice began to build up on the wings while the plane was on the ground and the process continued in flight (Transportation Safety Board of Canada [20]). From 2006 to 2010, the National Safety Board (NSTSB) has documented 228 accidents associated with ice formation in the USA (Appiah-Kubi *et al.* [21]). These accidents include all those caused by ice in the broad sense, be it frost build-up on aircraft surfaces or ice in engines under certain weather conditions, in flight or on the ground.

Frost formation occurs on aircraft on the ground when the temperature is near or below the freezing point of water. Frost may build up even when ground conditions are not 'frosty', since aircraft surfaces may become very cold during a long flight at high altitude and remain so for considerable time because of thermal inertia, thereby causing ground-level atmospheric moisture to condense (Transport Canada, 2004 [22]). Frost build-up may occur in flight during the summer as well, when the aircraft encounters a thunderstorm cell (Honomichl *et al.* [23]).

3 METHODOLOGY

This critical review of literature of interdisciplinary writings was conducted in two phases:

1. A search from an industrial engineering and human factors engineering perspective. The databases consulted were PubMed/Medline, Compendex/Engineering Village, and Web of Science. The keywords used were (in French and English) *occupational health and safety, ergonomics, human factors, integrated risk management, lean, Toyota, Honda, Volvo, Volkswagen, autonomous work, production planning, continuous improvement*;
2. A search from an on-ground de-icing and human factors engineering perspective. The databases consulted were Scopus, Compendex/Engineering Village, and Web of Science. The keywords used were (in French and English) *aircraft ground deicing, aircraft icing, human factors, human engineering, aircraft icing accident, Canada, aircraft icing statistic, aircraft maintenance, human error*;

For both searches, only documents available in the public domain, published in peer-reviewed journals or as peer-reviewed conference proceedings were retained. The timeframe was 2009 to 2014. Titles and abstracts of papers found in the databases were screened according to their relevance to the subject. Full papers were then read and quality assessed according to classical epistemology rules. Only relevant and high-quality papers were retained for critical analysis. By the snowball effect, flagship studies published prior to the targeted years were also identified.

4 RESULTS

4.1 Integrated risk management, a promising avenue

Many researchers, research organizations, practitioners and regulatory agencies consider integrating interventions in occupational health and safety with normal business processes to be a very promising focus for scientific advancement in the field of sustainable development. Contributions to development and exchange of knowledge in this sector have evolved from demonstrations of its relevance in the 1990s to more recent proposals and evaluations of models of integration. An integrated approach to occupational health and safety improves working conditions, the surrounding social climate and organizational processes notably by:

1. Improving the corporate brand image and hence labor recruitment, loyalty and satisfaction, improving system quality, reliability, availability and maintainability, increasing productivity and innovation, yielding better organizational decisions;
2. Decreasing direct costs, overhead, redesign, personnel turnover, absenteeism, slacking, inefficiencies, errors and system failures.

Integrated risk management focuses on complex sociotechnical systems involving numerous interdependent risks of diverse nature associated with a diversity of disciplines and in which uncertainty and the expectations of several parties must be taken into consideration. Its ultimate aim is to rank risks in terms of the various issues and constraints in play, and to guide choices among the various means available for managing these risks (avoidance, mitigation, deflection, contingence, retention) (Aubert & Bernard, 2004).

To achieve this, we need: (1) to understand and formalize the functioning and dysfunction of the system under study; (2) to model the system; (3) to quantify/qualify the model; (4) to apply the model in the field and validate its results. The principal scientific challenge consists of integrating risks characterized at different levels of sophistication (Aubert and Bernard [24]), for example risks for which it is not possible to identify the events that could result (level 0), risks for which there are definable event scenarios (level 4) and finally risks for which the uncertainty can be estimated in quantitative terms (level 6). It is also necessary to materialize this integration in interdisciplinary or even inter-sectorial contexts, which poses the challenge of determining the applicability of definitions and evaluation methods to the uncertainty concept (Péry *et al.* [25]). The degree of integration of risks will also depend on the size, structure and competitive positioning of the business concerned (Gangolells *et al.* [26]). Characteristics such as these dictate which management tool to adopt (Rocha [27]).

We note at this point that in conventional management systems, integration of risk is achieved through the application of standardization frameworks on a voluntary basis (Fernandez-Muniz *et al.* [28]). American Society for Testing and Materials International, (ASTM E2350-07 [29]), Occupational Health and Safety Assessment Series, (OHSAS

18001–2007), American National Standards Institute/American Industrial Hygiene Association, (ANSI/AIHA Z10-2005), Canadian Standards Association (CSA Z1004-12), British Standards (BS 8800:1996) or British Standards European Norm International Organization for Standardization (BS EN ISO 14001) are the standards most widely applied to occupational health and safety. Adoption of occupational health and safety certification supports the process of getting workers involved in prevention, can lead to improvements or innovations going beyond legal or regulatory requirements and facilitates transparency of business audits and contractual decisions/relations (Kristensen [14]) but does not support the advanced practices of organizational learning (Granerud and Rocha [30]). Standards dealing with management of quality, the environment and occupational health and safety are all based on the Deming wheel. Their integration can lead to overlapping of tasks involving different human resources, and hence challenges of coordination and alignment with business strategy and the potential for financial and social gains (Santos *et al.* [31]; Zeng *et al.* [32]). Initial set-up costs are high and the continuous updating of all the documentation can be tedious (Santos *et al.* [31], Zeng *et al.* [32]). However, if integration is successful, internal audits become unified, external audits are simplified, employees are better trained, responsibilities are better defined, documentation is clearer, communications are improved, and the entire management system is more effective and efficient, including supply management (Rebelo *et al.* [33], Oliveira [34], Santos *et al.* [31]). For project-type processes involving much technical variability, subsystems, temporary locations (e.g. construction, mining, maintenance activities) integration is difficult (Gangolells *et al.* [26]). The challenges stem from deficiencies in understanding how to integrate these standards at the planning stage and in understanding of risk identification, evaluation and control, and from gaps in support from certification agencies and poor suitability of the technical guides they offer (Gangollels *et al.* [26]). Some are attempting to meet these challenges for conventional processes in the manufacturing sector (Rebelo *et al.* [33], Oliveira [34]). Solutions have been explored in the construction and mining sectors (Badri *et al.* [35], Badri *et al.* [36, 37], Badri *et al.* [38]).

Integrated risk management may also be achieved through non-certified management systems, in particular those based on a philosophy of continuous improvement of quality and productivity, such as the so-called Total Quality Management (TQM), lean, six sigma, lean sigma or fit sigma production systems (Yang & Yang, [97]). Studies examining the occupational health and safety impact of systems called 'lean' indicate positive effects such as increased involvement of workers in control over tasks, greater focus on quality and improved training, task diversification, more teamwork with greater autonomy, a system that treats workers as equals, but also negative effects such as more tasks and stress, less autonomy, more intense and complex work, more intrusive supervision and frequent overtime on short notice (Westgaard and Winkel [16], Dul and Neumann [39], Saurin and Ferreira [40]). These contrasts stem from organizational culture issues, from the maturity of the production philosophy being introduced, from the socioeconomic context of the business and from the level of worker participation (Saurin and Ferreira [40]). We note that the literature provides little support for the tools of such philosophies and their level of maturity, making the scientific implications of the studies difficult to describe and practically impossible to extrapolate to other contexts.

Techniques such as functional analysis, failure mode effects analysis (FMEA), hazard and operability studies (HAZOP), cause/failure trees, the bow-tie model, human error analysis, or other risk analysis techniques (Zahra, J.T. *et al.* [41], Tixier *et al.* [42]) in laboratory experiments (Dempsey *et al.* [43]) or using digital human modeling (Illmann *et al.* [44], Demirel and Duffy [3], Fritzsche *et al.* [45], Neumann and Medbo [46], Cimino *et al.* [47]) have been used to complete the comprehension, formalizing and modeling of the decisional subsystem of the system under study. Quantitative or qualitative characterization of such a model can be achieved using Bayesian methods (Knegtering and Pasman [48], Hu *et al.* [49]) or multi-criteria methods such as AHP (analytical hierarchy process) (Saaty, [98]) or ELECTRE, when specific events can be associated with risks and probabilities of their occurrence can be estimated. When probabilities of occurrence cannot be estimated, consideration must be given to using combinatory or continuous optimization algorithms, depending on the nature of the system under study. Combinatory optimization can be achieved using exact methods (e.g. Markov chains) or approximate methods such as meta-heuristics or techniques of artificial intelligence (Hu *et al.* [49]). In all cases, the qualitative characterization of a model will necessarily involve elicitation of experts (Knegtering and Pasman [48], Zahra *et al.* [41]). The advantage of using a decision aid is to be able to use a model repeatedly to make decisions, choose options or set diagnostics where consequences are major and possibly spread out over time and past experience is of little help, all while minimizing the introduction of bias. These are simplified representations of reality, allowing decision-making of which the quality (level of performance of the resulting actions) depends on the model chosen. A decisional process must be in place to guide any decision aid that might be used. The literature identifies explicitly the category of personnel for whom the decisional processes are intended. The question of the resources at their disposition (time, information, finances) and the constraints under which the decisions must be made nevertheless remains crucial.

At the operational level, integrating occupational health and safety into the design and management of a production system means, among other things, integration of these risks into the techniques and tools used to determine production system capacity, to study and measure work, factory layout, stock management, production planning, raw material needs, scheduling, quality management, equipment and facility management and supply management. A few of the avenues that have been explored in published studies are listed in Table 1. Research in this field nevertheless remains emergent and faces the challenges of quantifying empirical relationships, integrating them into mathematical models currently used in operations/industrial engineering management and demonstrating to practitioners the usefulness of these new integrated models (Neumann and Dul [50]). This involves demonstrating their robustness in the face of uncertainty and under the changing conditions of business. The research must also provide assurance that these tools all can be fit harmoniously into a system of strategic management (Asif *et al.* [51]).

4.2 Human factors engineering of airliner maintenance and de-icing

The scientific literature on ergonomics and human factors engineering in aviation has been growing significantly since the Second World War. Questions regarding sustainable human factors engineering in this sector nevertheless remain to be examined (Haslam and Waterson [2]). Among these questions, some are directly related to maintenance workers: Is the green and safe label valid for aircraft in view of the accident and illness statistics indicating that aircraft maintenance is a high-risk occupation? What occupational health and safety risks associated with airliner maintenance work have the greatest influence on the operational performance of airlines? How can these considerations and risks be integrated into the lifecycle of an aircraft? Is the 'Design for Maintainability' approach sufficient? Or is an entirely new systemic engineering model needed? These questions are of crucial importance in the aviation sector, in view of the considerable impact of maintenance worker health and safety on the safety of passengers and the public (Vogt *et al.* [9]).

Table 1: Avenues of integration of occupational health and safety at the operational level

Research topic	Author	Goal	Methodology	Contribution
Integration of ergonomics and measurement of work	Dempsey, P. *et al.* [43]	To determine the influence of work autonomy on productivity and the associated physiological response.	23 subjects Assembly task in a conventional linear system, 36 cycles. Imposed pace (MOST) versus moderate fixed pace and totally free psychophysically determined pace.	Negative effects of free pace: reduced idle off time, chosen pace similar to both preceding assembly conditions, pre-programing of movement or psychophysical instructions?
	Caragnano and Lavatelli [52]	Integration of increases in biomechanical fatigue into a predetermined motion time system (MTM).	European Assembly Worksheet (EAWS).	Predetermined times modified as a function of the German estimator EAWS of physical load.
Integration of ergonomics and process balancing	Di Benedetto and Fanti [53]	Integration of musculoskeletal lesion risk analysis process balancing.	OCRA Index.	Software tool for balancing production lines involving repetitive tasks soliciting the upper limbs, based on the Italian estimator of risks OCRA.
	Otto and Scholl [54]	Integration of ergonomic risk estimation methods and process balancing models.	Pose the balancing optimization problem while considering ergonomic risks as measurements in the objective function; Solve the NP-hard optimization problem using simulated annealing technique.	Proposal of a model for optimizing the capacity of linear production systems incorporating ergonomic risk.
	Choi [55]	Integration of physical workload into process balancing.	Modeling by the goal programming technique.	Proposal of a model for optimizing process balancing, incorporating risk associated with physical workload.
Integration of lockout/ tag-out into production management	Emami-Mehrgani *et al.* [56]	Integration of lockout/ tag-out into production management in a flexible manufacturing system.	Two machines working in passive redundancy and producing one type of part. The system is subject to random failures and repairs. It deteriorates with machine	Passive redundancy optimizes production and maintenance costs while enhancing OHS.

(*Continued*)

Table 1: (*Continued*)

Research topic	Author	Goal	Methodology	Contribution
			age, as a function of its total output since the previous repair or servicing. Markov chain and homogeneous Markov chain processes.	
	Emami-Mehrgani *et al*. [57]	Integration of lockout/tag-out into management of production capacity in a flexible manufacturing system.	Three machines working (two in passive redundancy and one in series) and producing one type of part. The system is subject to random failures and repairs. It deteriorates with machine age, as a function of its total output since the previous repair or servicing. Analytical control approach (Markov chain), experimental design method and simulation using ARENA.	Passive redundancy improves production costs and worker safety. Integration of mean time to lockout/tag-out into management of production capacity.
Integration of ergonomics with scheduling	Lodree *et al*. [58]	Integration of physical and cognitive ergonomic considerations into production scheduling.	Review of the literature and construction of an integration framework.	Proposal of a scheduling model incorporation ergonomic risks and questions of productivity.
Integration of ergonomics and quality	Erdinç and Yeow [59]	Impact of the integration of ergonomics on quality.	Five field studies drawn from the manufacturing sector Critical analysis: internal/external validity.	Ergonomics has positive effects on quality (leading to financial gain for the business) in the manufacturing sector, more specifically through reduction of human error associated with man-machine interfaces (working posture, vision). Employee participation is crucial. Ergonomics methods should be incorporated into quality improvement tools.

Maintenance activities in the conventional sense are carried out in a wide variety of industrial sectors and workplaces, which implies a wide variety of risks: physical, chemical, biological, ergonomic, machine safety and psychosocial. The activities are not routine in nature and take place in changing work environments (EU-OSHA [60]). Maintenance workers operate under highly restrictive time constraints and communication between the actors involved or affected by their work is often deficient (EU-OSHA [61]). This combination of risks and time constraints may lead to circumvention or deactivation of machine safety devices and hence increased accident rates among maintenance workers (Papadopoulos *et al.* [62], Reason 2016). These time constraints and communication deficiencies are respectively the second and third causes of incidents in airliner maintenance (Warren *et al.* [63], Suzuki *et al.* [64]). Distribution of maintenance tasks is increasing rapidly in many industrial sectors (EU-OSHA [12]) and aviation is certainly no exception. According to the FAA, 64% of aircraft maintenance activities were distributed in 2007 (McFadden and Worrells [11]). Subcontracting (and sometimes offshoring) these activities may lead to challenges associated particularly with lack of knowledge about the environment and the risks present, increased complexity of work organization (coordination and supervision of fragmented tasks, among others), definition of roles and responsibilities of the various actors involved, and with implementation of occupational health and safety measures, to name only a few (EU-OSHA [60, 61], Papadopoulos *et al.* [62], Hampson *et al.*, 2012 [88], Quinlan *et al.*, 2013 [91], Gregson *et al.*, 2015 [89]).

Aicraft de-icing activities exist potentially at all the airports of the world where below freezing temperature are likely to occur. According to a study of the United States Environmental Protection Agency (EPA, 2000), around 200 airports in United States performed de-icing operation typically from October through May. In the case of Montréal, Canada, the season extends mainly from November through March. Since aircraft de-icing is a seasonal maintenance activity, the workers are often temporary employees on call. There is much occupational health and safety literature showing an association between job insecurity and increases in work-related accidents (for example Benavides *et al.* [65], Papadopoulos *et al.* [62]). The mechanisms explaining this are still being studied. Two factors, namely poor work conditions and lack of experience in and knowledge about the work environment (Benavides *et al.* [65]), could be significant for de-icing personnel. Fatigue also plays a frequent role in incidents associated with aircraft maintenance. Adequacy of rest, work proficiency, communication and coordination, and work shift are the major factors determining the fatigue of individuals (Wang & Chuang [66]). Communication in particular appears as important as time constraints in de-icing operations on the ground (Von Thaden [67]). Although now more than 15 years old, the Von Thaden study sums up very well the various human factors that make de-icing on the ground difficult. It emphasizes the impossibility for the pilot without help from inspectors outside to be certain that no ice has accumulated on the aircraft prior to takeoff and still respect the time constraints imposed by the airport.

In the world of aviation, the problem of human-factor-associated variability is often solved by introduction of technology (Vogt *et al.* [9]), imposing written procedures and documented work codes or training (Cromie *et al.* [68]). Several regulatory organizations including the ICAO and IHST are nevertheless promoting proactive safe design (Rashid and Braithwaite [8]). We therefore propose that to be sustainable and balance performance criteria with worker wellbeing, the airliner maintenance sector and in particular de-icing operations should aim for systemic design throughout their activity lifecycles.

5 DISCUSSION

5.1 Introduction of technology to reduce human-factor-associated variability in aircraft de-icing

The use of decision-aid tools is one approach studied to help humans make decisions under time pressure. The majority of modern aircraft are equipped with detectors that warn the flight crew that external conditions allow ice formation and in some aircraft activate anti-frost protective systems automatically (Federal Aviation Administration, [69]). These detectors only signal the presence of the conditions and do not propose action. Decision support systems (DSS) have been developed and tested by pilots in flight simulators to determine whether or not they bring any significant benefit. One so-called 'smart' system detects frost build-up in flight by monitoring changes in aircraft aerodynamic performance (Si-Bin *et al.* [70]). It has been shown that when the information obtained from a DSS is accurate, pilots' management of ice formation is improved, but when the information is inaccurate, their management is poorer than it is without the DSS (Sarter and Schroeder, [71]). As an alternative to this reactive type of decision aid, proactive decision-aid systems could be developed. Instead of waiting for ice to accumulate on the aircraft, the danger of this occurrence could be predicted in advance, based on aircraft flight and atmospheric conditions (Zeppetelli and Habashi, [72]). Most of the research effort has been devoted to aiding pilots, with little attention to decision aids for de-icing personnel on the ground, who in fact can only advise pilots. The most promising systems for ground-level detection of ice formation on wings use cameras and image processing. The current versions detect ice reliably but too slowly for use in airports (Terrace *et al.* [73]).

5.2 Implementation and documentation of procedures designed to reduce human-factor-associated variability in aircraft de-icing

Studies have focused on delays caused by de-icing operations. Various strategies that take into consideration interactions between airports and airlines have been proposed, including at the de-icing station design stage, to shorten the line-up of aircraft at airports that have multiple de-icing bays (Wu *et al.* [74, 75]). Reduced line-ups would be beneficial if they allowed reductions in the time constraints imposed on de-icing ground crews.

5.3 Development of training to reduce human-factor-associated variability in aircraft de-icing

Various types of training have been devised to reduce the risk associated with frost both in flight and on the ground. For example, pilots practice in flight simulators the actions to undertake to escape from stalling caused by ice. The principal difficulty associated with this approach has been the design of a calculation algorithm that allows the simulator to reproduce realistic behavior of the aircraft (Crider [76]). This is a daunting challenge, since the behavior of a stalling aircraft is difficult to predict and training on a simulator that behaves unrealistically could worsen decision-making in real situations. Meanwhile, de-icing ground workers receive training prior to each de-icing season. The service provider is responsible for this training, but it remains the duty of the air carriers, the certificate holder, to ensure that the service provider knows their approved de-icing procedure. The certificate holder must provide a manual with the procedures for each type of aircraft operated (FAA, [93]). Observations on the field are presently in progress to look at the tasks of de-icing ground workers and identify the gap in the training. These observations will lead to training improvements by enabling additional task-oriented formation content.

5.4 Sustainable, holistic and integrated risk management in aircraft de-icing

Although human error plays a major role in accidents and incidents associated with aircraft maintenance, various situations are the root cause of such errors: workers inadequately qualified, poor inspections, incorrect replacement of components, erroneous entries in the logbook, and over-extended task distribution (Geibel *et al.* [77]). Application of the Human Factors Analysis and Classification System (HFACS) to the study of accidents allows consideration of specificities of flight crews, the environment, supervision and organizational factors (Shappell *et al.* [78]). The AHP can be used to create a weighted model of human factor involvement in aeronautical accidents associated with frost build-up. The AHP requires the creation of a judgment matrix and allows consideration of the interactions between human factors, equipment, management and the environment, which has revealed that human factors and the environment are the dominant factors in accidents associated with frost build-up (Lijuan & Shinan, [79]).

From human factors research, the trend in aviation has evolved towards the development of tools for anticipating the risks stemming from human error during the performance of critical tasks (Kunlun *et al.* [80]). One of the difficulties here is that the system variables, in particular the environment and the conditions of use, evolve over time, introducing variability into the probability of accidents. One method proposed for taking this variability into consideration allows identification of critical events that mark the boundary between safety and risk of accident (Luo and Hu [80]). However, this does not address the essential step of identifying the root cause of human error (Rashid and Braithwaite [81]), which in the case of maintenance, repair and overhaul organizations, may occur in design practices, during manufacturing processes, or at later stages due to organizational, individual or workplace conditions (Rashid *et al.* [82]).

However, ‘In the tradition of product ergonomics, user involvement is considered essential for the development of user-friendly products and services, and the participatory design methods and tools that have been developed in ergonomics could be useful for linking ergonomics to product innovation’ (Dul and Neumann [39]). There is now general agreement that products incorporating occupational health and safety concerns will have a positive impact on cost reduction (Rose *et al.* [83]).

Integration of occupational health and safety concerns could be achieved through the development of wing profiles that are less sensitive to frost build-up, thus reducing the deterioration of aircraft performance (Ghisu *et al.* [84]) and making de-icing less critical. However, integration of these more robust wing profiles into initial aircraft design is a slow process, if it is occurring at all. For the existing fleet however, only specific training or technological decision aid for de-icing personnel could enable integration of occupational health and safety concerns. For example, the wing of the Cessna 208B Caravan is notoriously sensitive to frost build-up and this aircraft has been involved in numerous accidents due to frost and ice (FSF Editorial Staff [85]). The solution proposed to allow this aircraft to continue to receive clearance to fly under frosting conditions has been primarily to improve pilot training

(Turner [86]). This measure has proven to be only somewhat effective, since accidents continue to occur (Transportation Safety Board of Canada [87]).

A systemic and integrated design of de-icing operations (Asif *et al.* [51]) throughout the lifecycle starting at aircraft and anti-icing system design and continuing all the way to elimination of frost, ice or snow on aircraft surfaces is necessary. Such an integrated design does not exist yet and further researches are needed to find appropriate solutions for the aircraft manufacturers, the aircraft carriers and service provider. Decision tools that enable integration of both quantitative and qualitative measurements must be developed. These decision tools should take into account variability for anticipating the risk of accidents. For example, it could be interesting to know how a wing design could impact aircraft ground de-icing activities.

6 CONCLUSION

From a sustainable development perspective, proactive support for occupational health and safety throughout human activity lifecycles is essential. Airliner maintenance is a high-risk sector. In the current context of worldwide increases in air traffic, in view of the economic and safety issues stemming from maintenance, balancing of the imperatives of business performance and occupational health and safety can no longer be put off.

Integrated management of business risks and occupational health and safety is a promising avenue. The scientific literature proposes developments as much in the macro realm of business (conventional or non-certified systems management, decisional subsystems) as in the micro realm (operational systems). Airliner maintenance activities and de-icing in particular are fraught with risk, variability and uncertainties and are subject to exacting time constraints. They involve numerous actors and are subject to distribution of tasks.

The aviation industry has attempted to solve the problem of human-factor-associated variability by introducing technology (e.g. decision aids. frost-detecting cameras), imposition of procedures (e.g. to manage line-ups at de-icing stations) and documented work codes, and by training. Modeling and resolution of accidents/incidents via the control of human error run into numerous limitations recognized in the literature, particularly in the case of complex systems such as aircraft de-icing. Integration of occupational health and safety, as an important strategic tool in the business processes of airliner maintenance companies and their quest for better control over business risks, can be achieved only through sustainable, holistic and integrated design of airliner maintenance systems (de-icing activities in the present specific case).

ACKNOWLEDGMENTS

The authors thank Natural Sciences and Engineering Research Council of Canada (NSERC), the Ministère de l'Économie, de la Science et de l'innovation – Québec (MDEIE), École de technologie supérieure (ÉTS), Fonds de recherche Santé – Québec (FRQS), Fonds de recherche Nature et technologies – Québec (FRQNT) and Fonds de recherche Société et culture – Québec (FRQSC) for financial contributions and support for the development of our interdisciplinary and inter-sector research programs through their research network program.

REFERENCES

[1] Hinze, J., Godfrey, R. & Sullivan, J., Integration of construction worker safety and health in assessment of sustainable construction. *Journal of Construction and Engineering and Management*, **6(139)**, pp. 594–600, 2013. https://doi.org/10.1061/(asce)co.1943-7862.0000651

[2] Haslam, R. & Waterson, P., Ergonomics and sustainability. *Ergonomics*, **3(56)**, pp. 343–347, 2013.
https://doi.org/10.1080/00140139.2013.786555
[3] Demirel, H.O. & Duffy, V.G., A sustainable human centered design framework based on human factors. In Duffy, V.G. DHM/HCII Part 1. *Springer-Verlag*, pp. 307–315, 2013.
[4] Lozano, R. & Huisingh, D., Inter-linking issues and dimensions in sustainability reporting. *Journal of Cleaner Production*, **2–3(19),** pp. 99–107, 2011.
https://doi.org/10.1016/j.jclepro.2010.01.004
[5] Dul, J., Bruder, R., Buckle, P., Carayon, P., Falzon, P., Marras, W.S., Wilson, J.R. & van der Doelen, B., A strategy for human factors/ergonomics: developing the discipline and profession. *Ergonomics*, **55(4)**, pp. 377–395, 2012.
https://doi.org/10.1080/00140139.2012.661087
[6] OACI (Organisation de l'Aviation Civile Internationale. L'OACI prévoit que le trafic continuera d'augmenter jusqu'à, 2015. available at: http://www.icao.int/Newsroom/Pages/FR/ICAO-predicts-continued-traffic-growth-through-2015_FR.aspx. (Accessed on: 18 April, 2014).
[7] Boeing. Long-term market. Current market outlook 2013–2032. available at: http://www.boeing.com/boeing/commercial/cmo/. (Accessed on: 2014.)
[8] Vogt, J., Leonhardt, J., Köper, B. & Penning, S., Human factors in safety and business management. *Ergonomics*, **2(53)**, pp. 149–163, 2010.
https://doi.org/10.1080/00140130903248801
[9] Drury, C.G. & Johnson, W.B., Writing aviation maintenance procedures that people can/will follow. *Proceedings Human Factors and Ergonomics Society 57th Annual Meeting*, **57**, pp. 997–1001, 2013.
https://doi.org/10.1177/1541931213571223
[10] McFadden, M. & Worrells, D.S., Global outsourcing of aircraft maintenance. *Journal of Aviation Technology and Engineering*, **2(1)**, pp. 63–73, 2012.
https://doi.org/10.5703/1288284314659
[11] EU-OSHA, 2010. Factsheet 90 – *Maintenance and OSH – A Statistical Picture. Office for Official Publications of the European Communities*, Luxembourg, available at: http://osha.europa.eu/en/publications/factsheets/en.90.pdf/view
[12] HESAPRO, The link between productivity and health and safety at work. *Background research paper*, April, 32 pages, 2013.
[13] Kristensen, P.S., Managing OHS: A route to a new negotiating order in high-performance work organisations? *Safety Science*, **7(49)**, pp. 964–973, 2011.
https://doi.org/10.1016/j.ssci.2011.02.001
[14] Costella, M.F., Saurin, T.A. & De Macedo Guimaraes, L.B., A method for assessing health and safety management systems from resilience engineering perspective. *Safety Science*, **8(47)**, pp. 1056–1067, 2009.
https://doi.org/10.1016/j.ssci.2008.11.006
[15] Westgaard, R.H. & Winkel, J., Occupational musculoskeletal and mental health: significance of rationalization and opportunities to create sustainable production systems a systematic review. *Applied Ergonomics*, **2(42)**, pp. 261–296, 2011.
https://doi.org/10.1016/j.apergo.2010.07.002
[16] Brown, A.P., Aircraft performance degradation - the effects of inflight icing upon lift, drag and propulsive efficiency. *SAE Technical Paper* 2011-38-0073, 2011.

[17] Hannat, R. & Morency, F., Numerical validation of conjugate heat transfer method for anti-/de-icing piccolo system. *Journal of Aircraft,* **1(51)**, pp. 104–116, 2014. https://doi.org/10.2514/1.c032078
[18] Ayache, A., Morency, F. & Nadeau, S., Évaluation de la fatigue dans l'activité de dégivrage d'avions au sol et ses effets sur la performance. *Canadian Aeronautics and Space Institute 60th Aeronautics Conference AERO 2013*, Toronto, Ontario, Canada, April 30, 2013.
[19] Transportation Safety Board of Canada 2006. Aviation investigation report. *Loss of Control,* Georgian Express Ltd. Cessna 208B Caravan C-FAGA Pelee Island, Ontario January 17, 2004.
[20] Appiah-Kubi, P., Martos, B., Atuahene, I. & William, S., U.S. inflight icing accidents and incidents. *IIE Annual Conference and Expo 2013*. 2006 to 2010, 2013.
[21] Transport Canada. When in Doubt... Small and Large Aircraft. *Aircraft Critical Surface Contamination Training For Aircrew and Groundcrew*, report number TP 10643E, Ottawa, Chapter 2, 2014.
[22] Honomichl, S.B., Detwiler, A.G. & Smith, P.L., Observed hazards to aircraft in deep summertime convective clouds from 4–7 km. *Journal of Aircraft,* **50(3)**, pp. 926–935, 2013. https://doi.org/10.2514/1.c032057
[23] Aubert, B. & Bernard, J.G., *Mesure intégrée du risque dans les organisations*, Presses de l'Université de Montréal, 524 pages, 2004.
[24] Péry, A.R.R., Schüürmann, G., Ciffroy, P., Faust, M., Backhaus, T., Aicher L., Mombelli, E., Teddy, C., Cronin, M.T.D., Tissot, S., Andres, S., Bringnon, J.M., Frewer, L., Georgiou, S., Mattas, K., Vergnaud, J.C., Peijnenburg, W., Capri, E., Marchis, A. & Wilks, M.F., Perspectives for integrating human and environmental risk assessment and synergies with socio-economic analysis. *Science of the Total Environment*, **456–457**, pp. 307–316, 2013.
https://doi.org/10.1016/j.scitotenv.2013.03.099
[25] Gangollels, M., Casals, M., Forcada, N., Fuertes, A. & Roca, X., Model for enhancing integrated identification, assessment, and operational control of on-site environmental impacts and health and safety risks in construction firms. *Journal of Construction Engineering and Management*, **2(139)**, pp. 138–147, 2013.
https://doi.org/10.1061/(asce)co.1943-7862.0000579
[26] Rocha, R.S., Institutional effects on occupational health and safety management systems. *Human Factors and* Ergonomics *in Manufacturing & Services Industries*, **3(20)**, pp. 211–225, 2010.
https://doi.org/10.1002/hfm.20176
[27] Fernandez-Muniz, B., Montes-Péon, J.M. & Vasquez-Ordas, C.J., Occupation risk management under the OHSAS 18001 standard: analysis of perceptions and attitudes of certified firms. *Journal of Cleaner Production*, **24**, pp. 36–47, 2012.
https://doi.org/10.1016/j.jclepro.2011.11.008
[28] ASTM, Standard Guide for Integration of Ergonomics/Human Factors into New Occupational Systems. E235007, 10 pages, 2013.
[29] Granerud, L. & Rocha, R., Organisational learning and continuous improvement of health and safety in certified manufacturers. *Safety Science*, **7(49)**, pp. 1030–1039, 2011.
https://doi.org/10.1016/j.ssci.2011.01.009

[30] Santos, G., Mendes, F. & Barbosa, J., Certification and integration of management systems: the experience of Portuguese small and medium enterprises. *Journal of Cleaner Production*, **17–18(19),** pp. 1965–1974, 2011.
https://doi.org/10.1016/j.jclepro.2011.06.017
[31] Zeng, S.X., Tam, C.M. & Tam, V.W.Y., Integrating safety, environmental and quality risks for project management using a FMEA method. *Economics of Engineering Decisions*, **21(1)**, pp. 44–52, 2010.
[32] Rebelo, M.F., Santos, G. & Silva, R., A generic model for integration of quality, environment and safety management systems. *The TQM Journal*, **26(2)**, pp. 143–159, 2014.
https://doi.org/10.1108/tqm-08-2012-0055
[33] De Oliveira, O.J., Guidelines for the integration of certifiable management systems in industrial companies. *Journal of Cleaner Production*, **57**, pp. 124–133, 2013.
https://doi.org/10.1016/j.jclepro.2013.06.037
[34] Badri, A., Nadeau, S. & Gbodossou, A., A new practical approach to risk management for underground mining project in Quebec. *Journal of Loss Prevention in the Process Industries*, **26(6)**, pp. 1145–1158, 2013.
https://doi.org/10.1016/j.jlp.2013.04.014
[35] Badri, A., Nadeau, S. & Gbodossou, A., Occupational health and safety risks: towards the integration into project management. *Safety Science*, **50(2)**, pp. 190–198, 2012.
https://doi.org/10.1016/j.ssci.2011.08.008
[36] Badri, A., Nadeau, S. & Gbodossou, A., Proposal of a risk-based analytical approach for integrating occupational health and safety into project risk evaluation. *Accident Analysis and Prevention*, **48**, pp. 223–234, 2012.
https://doi.org/10.1016/j.aap.2011.05.009
[37] Badri, A., Nadeau, S. & Gbodossou, A., Integration of OHS into risk management in an open-pit mining project in Quebec (Canada). *Minerals*, **1(1)**, pp. 3–29, 2011.
https://doi.org/10.3390/min1010003
[38] Dul, J. & Neumann, P., Ergonomics contributions to company strategies. *Applied Ergonomics*, **40(4)**, pp. 745–752, 2009.
https://doi.org/10.1016/j.apergo.2008.07.001
[39] Saurin, T.A. & Ferreira, C.F., The impacts of lean production on working conditions: a case study of a harvester assembly line in Brazil. *International Journal of Industrial Ergonomics*, **39(2)**, pp. 403–412, 2009.
https://doi.org/10.1016/j.ergon.2008.08.003
[40] Zahra, J.T., Seyyed, S.H. & Aziruddin, R., Risk assessment of ergonomic risk factors at construction sites. *Applied Mechanics and Materials*, **330**, pp. 857–861, 2013.
https://doi.org/10.4028/www.scientific.net/amm.330.857
[41] Tixier, J., Dusserre, G., Salvi, O. & Gaston, D., Review of 62 risk analysis methodologies of industrial plants. *Journal of Loss Prevention in the Process Industries*, **15(4)**, pp. 291–303, 2002.
https://doi.org/10.1016/s0950-4230(02)00008-6
[42] Dempsey, P., Mathiassen, S.E., Jackson, J. & O'Brien, N., Influence of three principles of pacing on the temporal organisation of work during cyclic assembly and disassembly tasks. *Ergonomics*, **53(11)**, pp. 1347–1358, 2010.
https://doi.org/10.1080/00140139.2010.520745

[43] Illmann, B., Fritzsche, L., Leidholdt, W., Bauer, S. & Dietrich, M., Application and future developments of EMA in digital production planning and ergonomics. In Duffy, V.G., DHM/HCII, Part II, LNCS 8026, pp. 66–75, 2013.
[44] Fritzsche, L., Leidholdt, W., Bauer, S., Jäckel, T. & Moreno, A., Interactive production planning and ergonomic assessment with digital human models – introducing the editor for manual work activities (ema). *Work*, **41**, pp. 4428–4432, 2012.
[45] Neumann, W.P. & Medbo, P., Integrating human factors into discrete event simulations of parallel flow strategies. *Production Planning & Control*, **20(1)**, pp. 3–16, 2009. https://doi.org/10.1080/09537280802601444
[46] Cimino, A., Longo, F. & Mirabelli, G., A Multi-measure-based methodology for the ergonomic effective design of manufacturing system workstations. *International Journal of Industrial Ergonomics*, **39(2)**, pp. 447–455, 2009. https://doi.org/10.1016/j.ergon.2008.12.004
[47] Knegtering, B. & Pasman, H., The safety barometer. How safe is my plant today? Is instantaneously measuring safety level Utopia or realizable*? Journal of Loss Prevention in the Process Industries*, **26(4)**, pp. 821–829, 2013. https://doi.org/10.1016/j.jlp.2013.02.012
[48] Hu, J., Zhang, L., Ma, L. & Liang, W., An integrated safety prognosis model for complex system based on dynamic bayesian network and ant colony algorithm. *Expert Systems with Applications*, **38(3)**, pp. 1431–1446, 2011. https://doi.org/10.1016/j.eswa.2010.07.050
[49] Neumann, P. & Dul, J., Human factors: spanning the gap between OM and HRM. *International Journal of Operations & Production* Management, **30(9)**, pp. 923–950, 2010. https://doi.org/10.1108/01443571011075056
[50] Asif, M., de Bruijn, E.J. & Fisscher, O.A.M., Meta-management of integration of management systems. *TQM Journal*, **22(6)**, pp. 570–582, 2009. https://doi.org/10.1108/17542731011085285
[51] Caragnano, G. & Lavatelli, I., ERGO-MTM model: an integrated approach to set working times based upon standardized working performance and controlled biomechanical load. *Work*, **41**, pp. 4422–4427, 2012.
[52] Di Benedetto, R. & Fanti, M., An integrated tool to support engineers for WMSDs risk assessment during the assembly line balancing. *Work*, **41**, pp. 2329–2333, 2012.
[53] Otto, A. & Scholl, A., Incorporating ergonomic risks into assembly line balancing. *European Journal of Operational Research*, **212(2)**, pp. 277–286, 2011. https://doi.org/10.1016/j.ejor.2011.01.056
[54] Choi, G., A goal programming mixed-model line balancing for processing time and physical workload. *Computers & Industrial Engineering*, **57(1)**, pp. 395–400, 2009. https://doi.org/10.1016/j.cie.2009.01.001
[55] Emami-Mehrgani, B., Nadeau, S. & Kenné, J.P., Lockout/tagout and operational risks in the production control of manufacturing systems with passive redundancy. *International Journal of Production Economics*, **132(2)**, pp. 165–173, 2011. https://doi.org/10.1016/j.ijpe.2011.03.028
[56] Emami-Mehrgani, B., Kenné, J.P. & Nadeau, S., Lockout/tagout and optimal production control policies in failure-prone non-homogeneous transfer lines with passive redundancy. *International Journal of Production Research*, **51(4)**, pp. 1006–1023, 2013. https://doi.org/10.1080/00207543.2012.662305

[57] Lodree E.J. Jr., Geiger, C.D. & Jiang, X., Taxonomy for integrating scheduling theory and human factors: review and research opportunities. *International Journal of Industrial Ergonomics*, **39(1)**, pp. 39–51, 2009. https://doi.org/10.1016/j.ergon.2008.05.001

[58] Erdinç, O. & Yeow, P.H.P., Proving external validity of ergonomics and quality relationship through review of real-world case studies. *International Journal of Production Research*, **49(4)**, pp. 949–962, 2011. https://doi.org/10.1080/00207540903555502

[59] EU-OSHA. Factsheet 63 – Procurement of Maintenance Services and Health and Safety at Work: Office for Official Publications of the European Communities, Luxembourg, available at: http://osha.europa.eu/en/publications/factsheets/en.63.pdf/view

[60] EU-OSHA, Factsheet 62 – Safe Maintenance: Working with Contractors and Subcontractors: Office for Official Publications of the European Communities, Luxembourg, 2012. Available at: http://osha.europa.eu/en/publications/factsheets/en.62.pdf/view

[61] Papadopoulos, G., Georgiadou, P., Papazoglou, C. & Michaliou, K., Occupational and public health and safety in a changing work environment: an integrated approach for risk assessment and prevention. *Safety Science*, **48(8)**, pp. 943–949, 2010. https://doi.org/10.1016/j.ssci.2009.11.002

[62] Warren, W.R., Blickensderfer, B., Cruit, J. & Boquet, A., Shift turnover strategy and time in aviation maintenance. *Proceedings Human Factors and Ergonomics Society 57th Annual Meeting*, **57(1)**, pp. 46–50, 2013. https://doi.org/10.1177/1541931213571013

[63] Suzuki, T., von Thaden, T.L. & Geibel, W.D., Coordination and safety behaviors in commercial aircraft maintenance. *Proceedings of the Human Factors and Ergonomics Society 52nd Annual Meeting*, **52(1)**, pp. 89–93, 2008. https://doi.org/10.1177/154193120805200120

[64] Benavides, F.G., Associations between temporary employment and occupational injury: what are the mechanisms? *Occupational Environmental Medicine*, **63**, pp. 416–421, 2006. https://doi.org/10.1136/oem.2005.022301

[65] Wang, T.C. & Chuang, L.H., Psychological and physiological fatigue variation and fatigue factors in aircraft line maintenance crews. *International Journal of Industrial Ergonomics,* **44(1)**, pp. 107–113, 2014. https://doi.org/10.1016/j.ergon.2013.11.003

[66] Von Thaden, T.L., Air carrier deicing: human factors approach for building redundancy and commonality in standard operating procedures. *Transportation Research Record*, **1567**, pp. 41–48, 1997. https://doi.org/10.3141/1567-06

[67] Cromie, S., Liston, P., Ross, D., Corrigan, S., Vani, L., Lynch, D., Demosthenous, S., Leva, C., Kay, A. & Demosthenous, V., Human and organisational factors training as a risk management strategy in an aviation maintenance company. *Chemical Engineering Transactions*, **33**, pp. 445–450, 2013.

[68] Federal Aviation Administration. *Aviation maintenance technician handbook—airframe,* **(1)**, FAA-H-8083-31, USA, chapter 15, 2012.

[69] Si-Bin, Y., Tong, G. & Jian-Liang, A., H parameter identification and H2 feedback control synthesizing for inflight aircraft icing. *Journal of Shanghai Jiaotong University (Sci.),* **3(18)**, pp. 317–325, 2013. https://doi.org/10.1007/s12204-013-1401-6

[70] Sarter, N.B. & Schroeder, B., Supporting decision-making and action selection under time pressure and uncertainty: The case of in-flight icing. *Human Factors,* **43(4)**, pp. 573–583, 2001.
https://doi.org/10.1518/001872001775870403
[71] Zeppetelli, D. & Habashi, W.G., In-flight icing risk management through computational fluid dynamics-icing analysis. *Journal of Aircraft,* **49(2)**, pp. 611–621, 2012.
https://doi.org/10.2514/1.c031609
[72] Terrace, S.M., Bender, K.D., Sierra, E.A., Marcil, I., D'Avirro, J., Pugacz, E. & Eyre, F., Comparison of human ice detection capabilities and ground ice detection system performance under post deicing conditions. *Proceedings of the Human Factors and Ergonomics Society,* **50(7)**, pp. 2051–2055, 2006.
https://doi.org/10.1177/154193120605001772
[73] Wu, M., Li, L. & Chen, B., Optimization research of the aircraft deicing problem. *Advanced Materials Research,* **616–618**, pp. 1926–1929, 2013.
https://doi.org/10.4028/www.scientific.net/amr.616-618.1926
[74] Wu, M., Chen, B. & Xing, Z., Analysis of centralized aircraft deicing process. *Applied Mechanics and Materials,* **101–102**, pp. 779–782, 2011.
https://doi.org/10.4028/www.scientific.net/amm.101-102.779
[75] Crider, D.A., Upset recovery training: lessons from accidents and incidents. *Aeronautical Journal*, **1160(114)**, pp. 629–636, 2010.
https://doi.org/10.1017/s0001924000004103
[76] Geibel, W.D., Von Thaden, T.L. & Suzuki, T., Issues that precipitate errors in airline maintenance. *Proceedings of the Human Factors and Ergonomics Society*, **52(1)**, pp. 94–98, 2008.
https://doi.org/10.1177/154193120805200121
[77] Shappell, S., Detwiler, C., Holcomb, K., Hackworth, C., Boquet, A. & Wiegmann, D.A., Human error and commercial aviation accidents: An analysis using the human factors analysis and classification system. *Human Factors,* **49(2)**, pp. 227–242, 2007.
https://doi.org/10.1518/001872007x312469
[78] Lijuan, C. & Shinan, C., An approach of AHP for human factors analysis in the aircraft icing accident. *Procedia Engineering*, **17**, pp. 63–69, 2011.
https://doi.org/10.1016/j.proeng.2011.10.008
[79] Kunlun, S., Yan, L. & Ming, X., A safety approach to predict human error in critical flight tasks. *Procedia Engineering*, **17**, pp. 52–62, 2011.
https://doi.org/10.1016/j.proeng.2011.10.007
[80] Luo, P. & Hu, Y., System risk evolution analysis and risk critical event identification based on event sequence diagram. *Reliability Engineering and System Safety,* **114**, pp. 36–44, 2013.
https://doi.org/10.1016/j.ress.2013.01.002
[81] Rashid, H.S.J., Place, C.S. & Braithwaite, G.R., Eradicating root causes of aviation maintenance errors: Introducing the AMMP. *Cognition, Technology and Work,* **16(1)**, pp. 71–90, 2014.
https://doi.org/10.1007/s10111-012-0245-4
[82] Rose, L.M., Orrenius, U.E. & Neumann, W.P., Work environment and the bottom line: survey of tools relating work environment to business results. *Human Factors and Ergonomics in Manufacturing & Service Industries*, **23(5)**, pp. 368–381, 2011.
https://doi.org/10.1002/hfm.20324

[83] Ghisu, T., Jarrett, J.P. & Parks, G.T., Robust design optimization of airfoils with respect to ice accretion. *Journal of Aircraft,* **48(1)**, pp. 287–304, 2011. https://doi.org/10.2514/1.c031100

[84] FSF Editorial Staff, Investigators urge restrictions on cessna 208 icing operations. *Flight Safety Foundation Accident Prevention*, **2(63)**, 2006.

[85] Turner, T.P., Saving The Cessna Caravan. *Aviation Safety Magazine.* Belvoir Media Group, 2008.

[86] Transportation Safety Board of Canada 2014. *Loss of control and collision with terrain gogal air services limited Cessna 208B*, C-GAGP Snow Lake, Manitoba, November 18, 2012.

[87] Reason, J., *Managing the risks of organizational accidents*. Taylor and Francis, UK, 2016.

[88] Hampson, I., Junor, A., & Gregson, S., Missing in action: aircraft maintenance and the recent 'HRM in the airlines' literature. *International Journal of Human Resource Management*, **23**(12), pp. 2561–2575, 2012. https://doi.org/10.1080/09585192.2011.633278

[89] Gregson, S., Hampson, I., Junor, A., Fraser, D., Quinlan, M. & Williamson, A., Supply chains, maintenance and safety in the Australian airline industry, *Journal of Industrial Relations*, **57**(4), pp. 604–623, 2015. https://10.1177/0022185615582234

[90] Quinlan, M., Hampson, I. & Gregson, S., Slow to learn: Regulatory oversight of the safety of outsourced aircraft maintenance in the USA. *Policy and Practice in Health and Safety*, **12**(1), pp. 71–90, 2014. https://doi.org/10.1080/14774003.2014.11667798

[91] Quinlan, M., Hampson, I. & Gregson, S., Outsourcing and offshoring aircraft maintenance in the US: Implications for safety. *Safety Science*, **57**, pp. 283–292. https://doi.org/10.1016/j.ssci.2013.02.011

[92] Federal Aviation Administration, Adjustment of Compliance Date for Review of Outsourcing of (Third Party Provider) Aircraft Ground Deicing at Foreign Locations, FAA NOTICE – N 8900.26, available at: https://www.faa.gov/documentLibrary/media/Notice/n8900_33.pdf. (Accessed on: 2007.)

[93] Federal Aviation Administration, Ground deicing and anti-icing program, *Document Information*, AC 120-60B, available at: https://www.faa.gov/regulations_policies/advisory_circulars/index.cfm/go/document.information/documentID/23199. (Accessed 20 Děcember 2004).

[94] European Aviation Safety Agency. Regulation of ground de-icing and anti-icing services in the EASA Member States, *Interim Report*. EASA.2009.OP 21, available at: https://www.easa.europa.eu/system/files/dfu/EASA.2009_4%20Interim_Report.pdf (Accessed 2009).

[95] United States Environmental Protection Agency, Preliminary Data Summary. *Airport Deicing Operations (Revised)*, EPA-821-R-00-016, available at: https://www.epa.gov/sites/production/files/2015-06/documents/airport-deicing-pds-2000.pdf. (accessed August 2000).

[96] Torres, Y., Nadeau, S. & Morency, F., Study of fatigue and workload among aircraft deicing technicians. *Occupational Ergonomics*, **13**(2), pp. 79–90, 2016. https://doi.org/10.3233/oer-160240

[97] Yang, C.-C & Yang, K.-J, An integrated model of the toyota production system with total quality management and people factors. *Human Factors and Ergonomics in Manufacturing & Service Industries*, **23**(5), pp. 450–461, 2013. https://doi.org/10.1002/hfm.20335

[98] Saaty, T.L., Relative measurement and its generalization in decision making; why pair-wise comparisons are central in mathematics for the measurement of intangible factors. *The Analytic Hierarchy/Network Process, Royal Academy of Sciences, Spain, Series A. Mathematics,* **102(2)**, pp. 251–318, 2008.
https://doi.org/10.1007/bf03191825

A PATHWAY TO INITIATE BOTTOM-UP COMMUNITY-BASED DISASTER RISK REDUCTION WITHIN A TOP-DOWN SYSTEM: THE CASE OF CHINA

TIMOTHY SIM[1], LENA DOMINELLI[2] & JOCELYN LAU[1]
[1]Department of Applied Social Sciences, The Hong Kong Polytechnic University, Hung Hom, Kowloon, Hong Kong.
[2]School of Applied Social Sciences, Durham University, Durham, UK.

ABSTRACT

In China, as in other countries, inadequate knowledge of local vulnerability and hazard characteristics, and a rapidly industrialising society render enhancing resilience to natural disasters particularly challenging. This is particularly evident in rural areas with limited human and financial resources available for disaster risk reduction initiatives. The Chinese government institutionalized a top-down community-based disaster risk reduction (CBDRR) system to ensure that the capacity of communities would be enhanced effectively. In the long run, a top-down management style often undermines local capacities and vernacular DRR (disaster risk reduction) knowledge. There is a need to recognize the importance of communities as complex and dynamic entities in reducing disaster risks. Adopting participatory action research (PAR), this in-progress exploratory study examines a pathway to initiate bottom-up CBDRR within China's top-down institutional setting. Through PAR, the study of a rural village in Shaanxi Province shows that bottom-up initiatives can complement the existing system. Its current progress demonstrates the potential for using a transdisciplinary perspective to initiate CBDRR in China, where both top-down and bottom-up approaches, come together alongside different disciplines to increase a rural community's disaster resilience.

Keywords: community, community-based disaster risk reduction (CBDRR), disaster risk reduction (DRR), Sendai Framework, rural China, top-down and bottom-up approaches to CBDRRR, participatory action research, natural disaster

1 INTRODUCTION

The *Sendai Framework for Disaster Risk Reduction 2015–2030* acknowledges that 'disaster risk reduction' requires engagement and partnership with all-of-society because everyone is situated in a 'socio-cultural system' where rich people and poor, governments and societies alike, are affected when a disaster strikes. Responding to these events requires empowerment, inclusivity, accessibility, and non-discriminatory participation that pays special attention to the people disproportionately affected by disasters, especially the poorest ones [1]. The literature that explores the 'differentiated experiences of disaster' reveals that the basis for addressing the specific needs of groups whose vulnerabilities and capacities for resilience varies [2]. Moreover, disasters arise when a socio-cultural system fails to protect its population from external and internal vulnerability [3–6].

China is a signatory to the Sendai Framework. As the world's most populated and also a geographically stretched country, China is among the most vulnerable countries prone to natural disasters. Natural disasters in China are diverse, have high frequency, wide geographic distribution, significant damage and high disaster risks [7]. In parallel with advocacy on DRR by the United Nations Office for Disaster Risk Reduction (UNISDR), China has endorsed community-based disaster risk reduction (CBDRR) as a key component in its national strategies, e.g., the 11th, 12th and 13th Five Year Plans, for comprehensive disaster prevention and reduction.

DOI: 10.2495/SAFE-V7-N3-283-293

CBDRR represents a suite of methods embedded within communities to provide an operational process for supporting community resilience [8]. In the Western paradigm, DRR initiatives and knowledge are generated and 'owned' by communities [9] wherein local actors are at the heart of decision-making and implementation [10]. In contrast, CBDRR in China follows a top-down, government-led approach whereby local governments and communities are not involved in decision-making processes.

The Chinese government-led CBDRR strategy remains significant, and has produced some remarkable results. For instance, the low death toll during the magnitude 6.1 Ludian earthquake in Yunnan Province in 2014 has been attributed to this strategy. Nevertheless, this top-down disaster management is not without problems. It undermines local capacities and vernacular DRR knowledge; and underplays the heterogeneity of communities. In the long run, an alternative form of CBDRR initiatives is required in China to bridge the gap between the 'top' and the 'bottom' layers of society. In the face of marginal and limited discussions on bottom-up initiatives in China, this article considers what bottom-up CBDRR approaches are feasible within the predominantly top-down 'context' of China. What steps could be taken to mobilize communities and local actors to take a pro-active role in initiating CBDRR activities considering this national context?

This paper discusses an exploratory project in response to the above questions. Our research project has been running since 2012, but this article focuses on a pilot site located in a rural village in Shaanxi Province. We reflect upon iterative processes of discovery in the search for a complementary model that will effectively enhance the resilience of populations exposed to natural disasters by the end of the project in July 2017. We begin with a discussion of the concept of CBDRR, paying particular attention to its application in China. We then describe our journey of identifying a pathway to initiate CBDRR in this pilot site using participatory action research (PAR) methods.

2 COMMUNITY-BASED DISASTER RISK REDUCTION (CBDRR)

Community-based disaster risk reduction (CBDRR) is a process aimed at transforming passivity and powerlessness into action and resilience [11]. Despite a myriad of intrinsic elements and features, CBDRR consists of self-developed, culturally and socially acceptable, economically and politically feasible ways of coping with and avoiding crises related to natural hazards [11,12]. The Asian Disaster Preparedness Centre (ADPC), one of the earliest institutes to promote the practice of CBDRR through its training programmes, defines it as:

> 'a process in which at-risk communities are actively engaged in the identification, analysis, treatment, monitoring and evaluation of disaster risks in order to reduce their vulnerabilities and enhance their capacities. This means that people are at the heart of decision-making and implementation of disaster risk management activities. The involvement of [the] most vulnerable social groups is considered as paramount in this process, while the support of the least vulnerable groups is necessary for [its] successful implementation' [10].

At the core of this conceptualization, is the assumption that local communities are capable of responding to natural hazards on their own, as long as they are empowered and have adequate resources [8, 12, 13].

To date, vulnerability is increasingly understood as socially-constructed [4, 6, 8] and disasters are recognized as local events that first and foremost affect local communities [13]. It is logical to suggest that nobody is more interested in reducing risks of disasters than local people whose lives and livelihoods are threatened. Having more incentive to act, and the potential to mobilize and unlock the application of resources and know-how at the local level [12], local people can and should do more than simply react and respond to disasters. They should also be involved in preparing, planning and acting so that disaster risk is reduced [13].

Driven by international advocacy for a more people-centered approach to CBDRR, the Chinese government has drawn on the experience of other countries, and officially introduced the idea of community-based disaster risk reduction (*shequ jianzai*) in the National Plan, initially in the Comprehensive Disaster Reduction element of the Eleventh Five-Year Period in 2007. This emphasis on strengthening community capacity received institutional support through Chinese laws and regulations [14–18]. Specifically, the Chinese government perceived that the main methods for improving community capacity are by accurately assessing the levels of disaster risk at the community-level, improving the capacity of community engagement, and enhancing the capabilities of a community in disaster preparedness, reduction, response, and recovery [7].

The Chinese authorities introduced a series of national campaigns to implement CBDRR in China. For example, the Comprehensive Disaster Reduction Demonstration Community and Safe Community programmes were to be implemented by local governments to mainstream in their current practices the tasks of perfecting contingency plans, organizing emergency drills, and improving monitoring and early warning capacities [18]. These tasks entailed an explicit focus for enhancing community capacity on disaster prevention and reduction, and improving disaster management systems at community level [16]. All in all, the key objective for launching community-based disaster risk reduction in China by the authorities was to address effectively disaster risks and strengthen local capacities. These contemporary disaster risk reduction strategies also coincide with the goals of the Chinese developmental agenda.

However, unlike the trajectory of CBDRR implementation in other parts of the world, China applies it in a top-down fashion. The label of a top-down CBDRR system has thus emerged. Through it, the national government assumes key responsibilities in institutionalizing the concept of CBDRR and designing and applying the methodologies of community-based disaster risk assessment. The tasks of developing CBDRR by means of national campaigns such as the Demonstration Communities or Safe Communities are composed and assigned by central government to a community, level by level. Whereas the national and provincial governments introduce specific policies on CBDRR and assign tasks to their lower levels, local governments are responsible for undertaking the key tasks for developing CBDRR initiatives handed down by central government. Local governments are also responsible for all the expenditures incurred during the process of their implementation [18].

Others may argue that the outcome of such 'top-down' CBDRR remains uncertain as the effort of implementation is heavily dependent upon commitments and the financial capacities of each level of government within the overall administrative management system. Optimistically, initiating disaster risk activities in a top-down manner enables the concept to be put into action and sustained across the country quickly and efficiently. Tasks outlined through CBDRR campaigns could also materialize through standardization and become nationally applicable. In addition, by means of this management system, both local governments and communities would be given access to the necessary information and resources to actualize CBDRR. This, in turn, would empower not only communities, but also local governments involved in managing and reducing disaster risks in a given area [18].

In the long-run, the underlying weaknesses of a 'top-down' CBDRR strategy become apparent. Its tendency to assume that 'one size fits all' may fail to take into account local capacities and vernacular knowledge. For the same reason, this strategy is unable to recognize communities as complex and dynamic entities. Instead of being homogenous, communities are made up by diverse groups with different vulnerabilities, capacities and needs [13, 19]. Vulnerability to disasters is differentiated according to age, disability, gender, ethnicity and other social divisions, including class expressed through poverty [20, 21]. People's exposure to vulnerabilities and experiences to disasters in China are particularly varied, as a result of the rapid pace of urbanization, and the expansion of the income gap across regions. These socio-economic conundrums indicated that the 'one-size-fits-all strategy' is not a sustainable solution for addressing disaster risks in China.

Besides, it is increasingly recognized that simple one-way transfers of 'expert' information are inadequate. In this globalized world, effective communication is at the core of all activities in which experts are be involved, including in reducing disaster risks [22]. To cope better with the increasingly uncertain, complex and ambiguous risks and hazards, exchanges of knowledge, experiences and concerns among actors of various backgrounds and at different levels become important and constitutive of meaningful and effective communication [22–24]. Being more attentive to local capacities and vernacular knowledge regarding CBDRR is in line with the contemporary need for a substantial reduction of disaster risk and that requires 'perseverance and persistence' [1].

While the Chinese top-down CBDRR system is effective in putting the concept of CBDRR into practice, in the long-run, it is necessary to spell out more explicitly the role of multiple stakeholders in framing and implementing disaster risk reduction and resilience building endeavours to plug the gaps left by a 'top-down' approach to CBDRR. Given its strong 'top-down' capacity to coordinate and distribute resources quickly within disaster-stricken areas, China has a great potential in facilitating the successful integration of the 'top-down' and 'bottom-up' approaches. This could involve a sharing of 'responsibilities among central governments and relevant authorities, sectors and stakeholders, as appropriate to the national circumstances and systems of governance' [1].

In light of the limited discussion on the process and effect of initiating bottom-up CBDRR endeavours within a top-down context, this exploratory study aims to shed light on the feasibility of implementing bottom-up CBDRR efforts in China, where the government is actively involved in advancing the DRR agenda. Specifically, we employ participatory action research (PAR) strategies, with the aim of bridging the gaps between scientific knowledge and vernacular knowledge; between theoretical and practical CBDRR; and between top-down interventions and bottom-up initiatives [25].

3 PARTICIPATORY ACTION RESEARCH (PAR)

As we explore a bottom-up approach to CBDRR in China, we shall delineate our aims for 'bottom-up' CBDRR in our project. In the short-run, a bottom-up approach to CBDRR would actively engage local people in the identification, analysis, treatment, monitoring and evaluation of disaster risks to reduce their vulnerabilities and enhance their resilience capacities. This means that people, particularly vulnerable social groups, are at the heart of decision-making and implementation of disaster risk management activities [23]. In the long-run, the DRR knowledge and capacities of local communities, households, authorities and other local stakeholders would engage in DRR incorporated by and legitimized within the Chinese disaster management framework. That is, we anticipate a CBDRR model with Chinese characteristic that has both the community and government integrated within one holistic framework.

Participatory action research (PAR) was adopted as the guiding framework for this exploratory study to examine the pathway of getting the local actors, namely, the local village head, secretary, and residents to take a pro-active role in initiating CBDRR. PAR methods have a long history in the social sciences and have been used extensively for DRR planning and implementation around the world. It significantly enhances the disaster resilience of local stakeholders via the co-production of 'hybrid' knowledge that combines local and outside specialist knowledge [25]. On the other hand, instead of theoretical discussions, PAR encourages communities, researchers and local authorities to work together in all stages of disaster risk reduction [25, 26] to find solutions to their problems and develop action plans owned by the community [21].

With PAR, research is no longer seen as solely a means of creating and disseminating knowledge. Instead, it is also a process of education, a development of consciousness and a call to action for DRR [26–31]. Although the participants in this study are residents from a rural village with relatively low literacy rates, we regard them as expert community-based 'knowers' and valorize their knowledges and experiences. The researchers and village participants enter into a reciprocal relationship in a research process wherein external researchers do not presume to have a superior perspective. This creates a safe environment that fosters bottom-up initiatives. Therefore, in the pilot site, the rural villagers are full partners in the research processes and they, like us, are referred as co-researchers [26, 29, 31].

4 PATHWAY TO INITIATING BOTTOM-UP CBDRR

4.1 Working together: Integrate DRR and local community development programme

The interdisciplinary research team, consisted of four overseas researchers (two social workers, one seismologist, one social policy scholar and one research assistant), who worked closely with an established local non-governmental organization (NGO) in China. This NGO has focused on community development and promoted earthquake disaster resilience in Shaanxi province since 2013. The province experienced the Huaxian Earthquake in 1556 that killed more than 830,000 people. We identified C County with the help of the local seismological bureau to consider known geophysical hazards in Shaanxi. C County has been constantly affected by hailstorms and drought, but has not faced earthquakes in recent decades. The research team aimed to increase the earthquake resilience of C County by integrating DRR with a community livelihood development project run in C Village in Shaanxi Province under the auspices of the local NGO. The livelihood project focused on promoting tourism in the village given its strategic location near several famous historical sites and plantations of apples which would enable villagers to attract tourists from towns and cities. The NGO has been working closely with C Village since 2014 and has successfully engaged the local village committee, local leaders and residents in these activities.

4.2 Baseline study: Gaining an understanding of the local context

DRR planning and implementation require salient community-based information. It seeks to identify potential coping capacities and resources, the availability of support networks, and local knowledge of their vulnerabilities and the hazards to which they are exposed [25]. The next step we took was to conduct a baseline study of C Village.

Prior to the baseline study, desk research of local and international tools was conducted to identify a CBDRR baseline tool which was appropriate for the rural Chinese context. This enabled us to identify a gap, namely that there are few accessible CBDRR baseline tools in the Chinese literature. We then embarked on developing a toolkit relevant for investigating local disaster risk reduction capacities and villagers' perceptions of different types of natural disasters. Once the toolkit was designed, feedback from social scientists, natural scientists, local NGOs and our co-researchers both in China and overseas were sought and the tool was revised accordingly. Ethical approval was provided by the Hong Kong Polytechnic University via the Human Subjects Ethics Application Review System. The date of approval was 3 May 2015 and the reference number is HSEARS20150503001.

The baseline study was conducted between 4 and 14 March 2016. Five trained field staff went door to door to visit every household in C Village to interview all available villagers aged 13 or above. In the end, 230 villagers took part in the baseline study where 219 questionnaires were valid for analysis. In brief, we found that:

- Participants were most concerned about hailstorms, followed by drought, then earthquake hazards;
- Participants with primary school education or below have a significantly weaker preparedness for natural disasters than those with education level of junior high and above; and
- Children have the strongest preparedness for disasters, followed by heads of household (usually men), the spouses of heads of households and parents of the heads of households.

These findings above suggested that older people and women were the groups more vulnerable to disasters in C Village. Overall, the baseline study enabled us to: (a) engage with our co-researchers; (b) understand the ways in which different elements within the community have adapted to the risk contexts; and (c) the way CBDRR knowledge is understood by different co-research groups. This step provided a preliminary understanding of the complex interplay of resilience and vulnerability to disasters in the village, and how these are embedded in other aspects of everyday living. These form important knowledge for co-researchers to educate each other, and inform the next steps in the co-production processes that lead to action.

4.3 In-depth interview: Interacting with 'representatives' of co-researchers

The analysis of the baseline study produced a surprise for the authors. We learned that although our pilot site is not far from the epicenter of the 1556 Shaanxi earthquake, the study revealed that our co-researchers were not only unfamiliar with that earthquake, but they also had limited experiences with earthquakes in general, despite the recent 2008 Wenchuan earthquake which did have a severe impact on some counties nearby [32].

In view of the knowledge gap identified by the baseline study, we sought to understand further what life narratives, histories, cultural contexts, socio-economic conditions and governance structures existed in the village, and what impact they may have had on the issues of resilience, vulnerability, and risk.

Therefore, an indicative in-depth interview was conducted with a specific focus on two identified vulnerable groups: middle-aged women (aged between 40 and 60) and older people (aged 60 or above). In a nutshell, these interviews highlighted that:

- Hailstorms were the natural disaster most often experienced in C Village;
- Recovery was conducted independently on a household basis after previous hailstorms (versus a neighbourhood/community approach whereby members mutually helped one another);

- Village norms and insufficient resources were the main reasons for the lack of support between neighbours;
- Availability of funding was instrumental in determining a family's preparedness for hailstorms;
- Villagers were not ignorant about the impact of earthquake-related hazards;
- Villagers have an exclusive focus on engineering, especially housing, for earthquake preparedness; and
- Villagers are keen to learn about CBDRR.

In addition, three specific types of needs for enabling our co-researchers to carry out CBDRR in the village were identified: financial support from the government, CBDRR education and the availability of hail cannons (to fire into the sky and prevent hailstones from dropping on their crops).

4.4 An interactive field study: knowledge acquisition through action learning

Based on the findings above, the authors were delighted to find out that the villagers were keen to learn about CBDRR. To our knowledge, although bottom-up CBDRR initiatives are rare in China, there are NGOs that have been working closely with local governments on CBDRR. After consulting with the villager committee members and some of the villagers, a field study was organized for them to learn about the CBDRR experience of Y County in Shaanxi province. This field study served as an action learning process to expose co-researchers and other key local stakeholders to CBDRR practices; and for them to make direct observations and learn lessons about the process and possible products of CBDRR in a rural setting within their own province.

In this field study, 12 co-researchers were exposed to DRR drill demonstrations that were targeted at various hazard types, ranging from household fires to landslides, in two rural villages in Y County. Both formal and informal interactions between the co-researchers and NGOs representatives, government officials, were facilitated during the visit. Informal interactions occurred over meals and breaks. During the formal discussion session, the co-researchers of Y County, including local government officials were invited to share their views and experiences of developing CBDRR initiatives. Co-researchers from C Village were also engaged in the debriefing session later when they reflected on the CBDRR knowledge that they had gained through this visit, and were enabled to explore potential facilitators and obstacles in implementing CBDRR activities appropriate to their own village. The impact of this interactive field study was apparent immediately, especially considering their insight and motivation. The co-researchers pointed out:

- the importance and necessity of governmental support and involvement in their CBDRR activities;
- the need to build a comprehensive local CBDRR system, which involves the building of a CBDRR committee, and a coordination and communication mechanism.

4.5 Consultation: Action prioritization

After the field trip, the research team collaboratively ranked the key CBDRR activities appropriate for C Village with the co-researchers and NGO staff who have been working in C

Village. The objective was to facilitate and guide local stakeholders in producing a detailed plan outlining the content and processes for CBDRR implementation.

Owing to the low literacy rate of some of the co-researchers, pictures were utilized to help explain the types of activities that were identified based on the findings and discussions from the previous steps in engaging them. Specifically, co-researchers were asked to rank independently the CBDRR activities to be carried out according to their needs and those of their family members.

For the next step of this ongoing project, the authors and local NGO staff will be helping the co-researchers to set up their own CBDRR system with reference to the national frameworks, followed up by designing with them the CBDRR activities that they deem relevant. CBDRR experts from local universities and relevant government departments within China's DRR system will also be invited to join the consultations around designing and carrying out these activities where relevant.

5 DISCUSSION

Maskrey [12] highlighted the essence of CBDRR is 'neither specifically, nor particularly about implementing mitigation measures exclusively at the community level of action, but the progressive activation of all the levels, starting with the community itself'. We believe there is value in going beyond 'persuading the community to participate in government-based mitigation programs, and to persuade government at all levels to participate in communities' own programmes'[12]. This will form the basis of the integration of a top-down with a bottom-up approach.

On the other hand, although the origin of the CBDRR concept emerged in the 1970s as an alternative paradigm to challenge the predominantly top-down strategy of disaster management then practiced worldwide [11, 13], our exploration for a bottom-up CBDRR initiative in China does not advocate such a polarized vision. On the contrary, we seek to explore a complementary approach to plug the gaps left by top-down approaches that ignore community needs and capacities. Bottom-up CBDRR initiatives in China can utilise the heterogeneity of a community and integrate the disparate elements of local CBDRR knowledge into the existing top-down CBDRR system.

Although this is only an exploratory study, we are able to layout a pathway that introduces a bottom-up CBDRR approach in the Chinese context through the use of a PAR framework. Whether the approach is top-down or bottom-up, the ultimate goal should be to enhance effectively the resilience of communities to disasters. Future studies on CBDRR can, therefore, learn from this iterative process whereby villagers are invited as co-researchers to work with a range of stakeholders from different disciplines and sectors (governmental and non-governmental) and locations (other counties and overseas) to co-design the research agenda throughout the process of increasing the disaster resilience of village co-researchers in a community development project.

6 CONCLUSION

It is recognized that CBDRR initiatives often consist of self-developed, culturally and socially acceptable, economically and politically feasible ways of coping with and avoiding crises related to natural hazards [11, 12]. By adopting the participatory action research framework, this study provided multiple opportunities to empower practitioners and co-researchers to construct vernacular knowledge and integrate CBDRR practices in ways that are unique and practical to the context of China.

This ongoing study attempts to identify a pathway to initiate bottom-up CBDRR endeavours in a rural village setting in China. The authors began by pulling the available resources to work together to integrate CBDRR into the local community development programme. Next, we sought to understand the local context and local knowledge about hazard characteristics, vulnerability of people and their ability to prepare and cope with disasters. We then acquired more in-depth information about the vulnerable groups' coping mechanisms and adaptation, post-disaster mitigation and preparedness, lived experiences of disasters and specific CBDRR needs. Nine months after the baseline study, the co-researchers and local stakeholders were involved in an interactive field visit to other rural villages in Shaanxi, where there were some established CBDRR products and activities initiated by non-governmental bodies. Based on their observation and lessons learned, the co-researchers became more confident and capable of mapping out the CBDRR activities that are appropriate for their own village.

Overall, different skills and mixed methods were used for different steps involved throughout the exploratory activities. To learn about existing CBDRR capacity and the needs of the community, we employed PAR methods such as surveys, participant observation, and in-depth interviews. One of the characteristics of PAR is that data collection and analysis cannot be separated. Hence, we carefully documented each step we took. Moreover, we analyzed the data collected in an iterative fashion and constantly reflected upon the previous results. More importantly, information sharing and discussions with co-researchers on a constant basis in advance of planning out details and actions formed an integral part of our methodology. Our overall conclusion at this stage is that this exploratory study reflects that participatory action research is a feasible approach for identifying a pathway for initiating bottom-up CBDRR initiatives in the context of China because its processes serve as a foundation whereby to complement the current top-down disaster management system.

ACKNOWLEDGEMENTS

This study is an integral part and funded by the Earthquakes without Frontiers project (Grant number: NE/J01995X/1). We would like to extend our thanks to the local collaborators: Shaanxi Seismological Bureau, Shaanxi Gender Development Solution, World Vision International – China Office, and local government officials for their support. We would also like to thank the co-researchers, i.e., the village committee members and residents of C Village who kindly agreed to engage fully in this study.

REFERENCES

[1] United Nations Office for Disaster Risk Reduction (UNISDR). *The Sendai Framework for Disaster Risk Reduction 2015–2030*, 2015.

[2] Dominelli, L., Environmental Justice at the heart of social work practice: greening the profession. *International Journal of Social Welfare*, **22**(4), pp. 431–439, 2013. https://doi.org/10.1111/ijsw.12024

[3] Barton, A.H., *Social Organization Under Stress,* National Research Council, National Academy of Sciences: Washington, DC, 1963.

[4] Dynes, R.R., Coming to terms with community disaster. In *What is a Disaster: Perspectives on the Question*, ed. E.L. Quarantelli, Routledge: London, pp. 109–126, 1998.

[5] Bates, F.L. & Peacockm, W.G., *Living Conditions, Disasters and Development*, University of Georgia Press: Athens, 1993.

[6] Oliver-Smith, A., Global challenges and the definitions of disaster. In *What is a Disaster: Perspectives on the Question,* ed. E.L. Quarantelli, Routledge: New York, pp. 177–194, 1998.
[7] Shi, P.J., Wang, M. & Ye, Q., Achievements, experiences and lessons, challenges and opportunities for china's 25-year comprehensive disaster reduction. In: *Planet @ Risk,* 2(5), Special Issue for the Post-2015 Framework for DRR: pp. 353–358, Global Risk Forum GRF Davos, Davos, 2014.
[8] Wisner, B. & Kelman, I., *Community Resilience to Disasters.* In *International Encyclopedia of the Social and Behavioral Sciences,* 2nd edn., ed. E.L. Quarantelli, Elsevier: Oxford, vol 4, pp. 354–360. 2015.
https://doi.org/10.1016/B978-0-08-097086-8.28019-7
[9] Amaratunga, C.A., Building community disaster resilience through a virtual community of practice (VCOP). *International Journal of Disaster Resilience in the Built Environment,* **5**(1), pp. 66–78. 2014.
https://doi.org/10.1108/IJDRBE-05-2012-0012
[10] Abarquez, I. & Zubair, M., *Community-Based Disaster Risk Management: Field Practitioners' Handbook,* ADPC (Adapted from CBDRM-II course materials), 2004.
[11] Luna, Emmanuel, M., Community-based disaster risk reduction and disaster management. In *Disaster Management: International Lessons in Risk Reduction, Response and Recovery,* eds A. López-Carresi, B. Wisner, I. Kelman & J.C. Gaillard, Routledge: New York, pp. 43–63, 2013.
[12] Maskrey, A., Revisiting community-based disaster risk management. *Environmental Hazards,* **10**(1), pp. 42–52, 2011.
https://doi.org/10.3763/ehaz.2011.0005
[13] Delica-Willison, Z. & Gaillard, J.C., Community action and disaster. In B. Wisner, I. Kelman & J.C. Gaillard, *Handbook of Hazards and Disaster Risk Reduction,* Routledge: New York. pp. 711–722. 2012.
[14] Jiang, L.I., Implementation of disaster reduction measures and enhancement of integrated risk governance in China. *International Journal of Disaster Risk Science,* **4**(2), pp. 101–104, 2013.
https://doi.org/10.1007/s13753-013-0011-0
[15] Zhai, G.F., Li, S.S. & Chen, J., Reducing urban disaster risk by improving resilience in China from a planning perspective. *Human and Ecological Risk Assessment: An International Journal,* **21**(5), pp. 1206–1217, 2015.
https://doi.org/10.1080/10807039.2014.955385
[16] Zhang, X., Yi, L.X. & Zhao, D., Community-based disaster management: a review of progress in China. *Journal of the International Society for the Prevention and Mitigation of Natural Hazards,* **65**(3), pp. 2215–2239, 2013.
https://doi.org/10.1007/s11069-012-0471-3
[17] Zou, M. & Yi, Y., China's comprehensive disaster reduction. *International Journal of Disaster Risk Science,* **1**(1), pp. 24–32, 2010.
[18] Feng, X.J. & Wang, D.M., *Policy Analysis of Community-Based Disaster Risk Reduction in China,* Peking University Press: Beijing, 2014.
[19] Anderson, M.B. & Woodrow, P.J., *Rising From the Ashes: Development Strategies in Times of Disaster,* Lynne Rienner: London, 1989.
[20] Dominelli, L., Sim, T. & Ke, C., Community-based approaches to disaster risk reduction in China. In Report: *Pathways to Earthquake Resilience in China,* Overseas Development Institute, Oct: 27, 2015.

[21] Dominelli, L., *Green Social Work*, Cambridge: Polity Press, 2012.
[22] Renn, O., Klinke, A. & van Asselt, M., Coping with complexities, uncertainty and ambiguity in risk governance: a synthesis. *Ambio,* **40**, pp. 231–246, 2011. https://doi.org/10.1007/s13280-010-0134-0
[23] Rodriguez, H., Diatz, W., Santos, J.M. & Aguirre, B.E., Communicating risk and uncertainty: science, technology, and disasters at the crossroads. In *Handbook of Disaster Research,* eds H. Rodriguez, E.L. Quarantelli, R.R. Dynes, Springer: New York, pp. 476–488, 2007. https://doi.org/10.1007/978-0-387-32353-4_29
[24] Benn, S., Dunphy, D. & Martin, A., Governance of environmental risk: new approaches to managing stakeholder involvement. *Journal of Environmental Management,* **90**, pp. 1567–1575, 2009. https://doi.org/10.1016/j.jenvman.2008.05.01
[25] McCall, M.K. & Peters-Guarin, G., Participatory action research and disaster risk. In *Handbook of Hazards and Disaster Risk Reduction,* eds B. Wisner, I. Kelman, J.C. Gaillard, Routledge: New York, pp. 772–786, 2012.
[26] McIntyre, A., *Participatory Action Research,* Sage Publications, 2007.
[27] Park, P., People, knowledge, and change in participatory research. *Management Learning,* **30**(2), pp. 141–157, 1999. https://doi.org/10.1177/1350507699302003
[28] Reason, P. & Hilary, B., *The SAGE Handbook for Action Research: Participative Inquiry and Practice,* SAGE: London, 2008. https://doi.org/10.4135/9781848607934
[29] Small, S.A., Action-oriented research: models and methods. *Journal of Marriage and Family*, **57**(4), pp. 941–955, 1995. https://doi.org/10.2307/353414
[30] Kindon, S., Pain, R. & Kesby, M., (eds). *Participatory Action Research Approaches and Methods: Connecting People, Participation and Place*, Routledge: London, 2007.
[31] Schruijer, S.G.L., Research on collaboration in action. *International Journal of Action Research*, **2**(2), pp. 222–224, 2006.
[32] Wang, F., Cheng, Q.G. & Highland, L., Preliminary investigation of some of the large landslides triggered by the 2008 Wenchuan earthquake, Sichuan Province, China. *Landslides*, **6**, pp. 47–54, 2009. https://doi.org/10.1007/s10346-009-0141-z

A STUDY OF MOTIVATIONAL ASPECTS INITIATING VOLUNTEERISM IN DISASTER MANAGEMENT IN GERMANY

P. HOLWITT, S. STROHSCHNEIDER, R. ZINKE, S. KAISER, I. KRANERT, A. LINKE & M. MÄHLER
Department of Intercultural Studies and Business Communications, Friedrich-Schiller University, Jena, Germany.

ABSTRACT
Volunteer work is an important aspect that influences a society´s ability to cope with different kinds of security threats and disasters. However, the motivational and social sources that drive civil engagement in the field of disaster management are not understood very well. If societies want to preserve or increase their resilience and thus reduce their vulnerability to disasters, volunteerism is an important topic to look into. In order to foster voluntary engagement in disaster management, it is essential to both understand the motivational sources that drive volunteers and establish appropriate conditions for future voluntary engagement. In this article, motivations of volunteers in three regions of Germany are analysed using a theoretical model that builds on the works by psychologists Dörner and Bischof. The model considers volunteerism as a way of catering to three basic needs of human beings: the need for affiliation, the need for certainty and the need for control. This model is applied to data gathered from unstructured and semi-structured interviews with volunteers and professionals working in the field of disaster management in Germany.
Keywords: disaster management, Germany, motives, training, volunteerism.

1 INTRODUCTION

In recent years, the concept of volunteer work is undergoing fundamental changes: long-term commitment to a particular organization or team is decreasing and, at the same time, more flexible forms of voluntary engagement are emerging. These new forms of volunteerism often respond to single urgent crises and quickly disperse again. New modes of communication and coordination foster this incident-related volunteerism. Furthermore, volunteer work attracts only particular segments of society, while others are under-represented, particularly elderly people and migrants. These transformations of volunteerism and shortcomings of traditional modes of managing volunteers are challenges to organizations that operate in the field of disaster management and are hence asking for a rethinking of strategies.

In this article, we present a theoretical model that builds on psychological theory and can be used to understand the motivational sources of volunteers who work in the field of disaster management. We apply this model to study the motivational sources of volunteers who work in the field of disaster management in three regions of Germany. The model is intended to help understand the different motivations of volunteers and to account for their various reasons to take up voluntary work in the field of disaster management.

The study presented here was carried out within the collaborative Indo-German research project 'INitiate VOLunteerism to counter VulnErability' [1]. The focus of the German study was on sources and motivations of voluntary engagement in disaster management in selected German regions contributing to an increase of societal resilience. The project aims at generating knowledge about why and in which ways people engage in voluntary work in the field of disaster management and to improve ways of integrating volunteers into established structures of professional disaster management organ,izations based on these findings. It is yet an ongoing research project and the results presented in this article are still at a preliminary stage.

DOI: 10.2495/SAFE-V7-N3-294-302

The article is structured in the following way: it starts with a short review of the state of the art regarding works from various disciplinary backgrounds that engage with motivational sources of volunteerism. In accordance with the approach presented in this article, a special focus lies on psychological theories about volunteerism. After this brief overview, the theoretical model applied in the INVOLVE-project is presented. The next section explains the wider context of our data collection and the methods used. In the final section, the theoretical model is applied to data gathered in the context of the research project. The conclusion summarizes the preliminary results of the project and presents the implication of interventions in existing trainings and work routines of organizations working in the field of disaster management.

2 VOLUNTEERISM AND MOTIVATION

What can be accounted for as volunteerism is subject to controversies in the literature. Many authors even question the possibility and necessity of formulating a watertight definition of volunteerism [2, 3]. Consequently, the following working definition of volunteerism does not strive for universal validity, but represents the result of an attempt to circumscribe the phenomenon in a very open and inclusive manner while still allowing to distinguish it from closely related types of activities like hobbies or paid work. For these purposes, volunteerism is understood as *actions taken up by people based on a voluntary decision, which do not primarily aim at generating economic profit and which are supposed to influence particular as well as vaguely defined social circumstances.*

Scholars from different disciplinary backgrounds have studied volunteerism in quite distinct ways and with different research interests. The topic has stirred interest among scholars of such diverse backgrounds as psychology, sociology and political sciences. Accordingly, research questions, methods, theories and concepts applied to the topic of volunteerism vary significantly. For example, while psychologists have mostly studied volunteerism as an individual decision based on particular personality traits, sociologists – while also interested in the question who volunteers and why – have considered volunteerism first and foremost as a social phenomenon and asked what benefits volunteers receive from volunteerism [3]. Political scientists consider volunteerism as a cornerstone for civic society and democratic participation and study the ways in which it establishes particular societal values among citizens.

Although in the framework of the INVOLE-project, the overall aim is the combination of psychological and sociological perspectives in a multi-level theoretical model, this article focuses on the psychological aspect of this work. This perspective strives to explain volunteerism on the level of the individual actor and his or her motivations to volunteer. In the following section, important existing work in this field is reviewed in order to explain how it is built upon and further developed in the INVOLVE-project.

2.1 Psychological perspectives on motivation to volunteer

Various works in the field of psychology have attempted to conceptualize volunteerism. Particularly, the works of Clary *et al.* [4, 5] and their model of the Volunteer Functions Inventory (VFI) have received attention for providing a useful theoretical framework to conceptualize motivations of volunteers. In the VFI, volunteerism claims to serve six main functions or motivations, which are values, understanding, enhancement, career, social and protective. Of these, values, understanding and enhancement are claimed to be of greater importance than career, social and protective [5]. Value-driven volunteerism is defined as a way of expressing and acting on important values like humanitarianism, understanding-driven volunteerism is defined as a way to learn more about the world or to exercise skills that are often unused and enhancement-driven volunteerism is defined as a way to grow and develop psychologically. Among the lesser important motivations in the VFI, career-driven volunteerism is defined as a way to gain career-related experience, social-driven volunteerism is defined as a way to strengthen the volunteer´s social relationships and protective-driven volunteerism is defined as a way to reduce negative feelings, such as guilt, or to address personal problems.

Although the VFI conceptualizes the different motivations as separate entities, the authors stress the multi-motivational nature of volunteering. This means that a person´s decision to take on volunteer work is driven by an amalgamation of motivations rather than by a single one. Put differently, the VFI proclaims that volunteers pursue different goals when engaged with one and the same volunteer work.

The VFI provides one way of conceptualizing volunteers´ motivations and is a useful starting point to think about why people volunteer in the first place. However, the model can be criticized on various accounts: It is rather eclectic and its theoretical basis is not clearly defined. Most importantly, it ignores existing, more elaborate models about motivation in psychological research. Taking these shortcomings of the VFI into account, the following section introduces a slightly different model that builds upon the main insights of the VFI but integrates them into more general and encompassing psychological theories of motivation.

2.2 Integrative model for motivation to volunteer

The work of Dörner [6–8], Bischof [9, 10] and Strohschneider [11, 12] helps to integrate a concept of motivation to volunteer into a general psychological theoretical framework for motivation. Based on these works, the proposed model identifies three main needs that are addressed by voluntary engagement in disaster management. These are the need for affiliation, the need for certainty and the need for control.

The *need for affiliation* connotes a need of belonging to a form of community. It is satisfied by signals that acknowledge an individual´s membership to a group. These signals can involve particularly meaningful gestures and symbols that communicate membership to outsiders as well as insiders, including types of clothing typical for the group or participation in seemingly banal rituals like communal gatherings and group meetings. In the context of disaster management, such signals can be found in uniforms that are typical for a certain group and widely known. Other nonverbal signals of affiliation involve touch and close spatial proximity to each other, which is why for example firefighters have been described as prototypical 'touching cultures' [13]. Coordinated group activities also work as important means to foster a strong bond among members and provide effective affiliation signals to members. Last but not least, affiliation is of course also expressed verbally in everyday encounters between members. Taken together, these different ways of expressing a notion of belonging form an organization´s language of affiliation that is important for the smooth collaboration within the group and that works to integrate members in a subtle way.

The second need that drives people to volunteerism is the *need for certainty*. It is associated with an interest in understanding the world and a desire for predictability and order, fairness and reciprocity. This need is particularly triggered by unexpected, potentially dangerous and disturbing events that are typical for the context of disaster management. It is satisfied by actions (and information resulting from those actions) that re-establish a sense of order in afflicted individuals and render the world calculable again. People who are driven by a strong need for certainty can be expected to prefer individual, often short-term, event-related forms of voluntary engagement that do not necessarily involve long-term commitment to a group or organization, in contrast to volunteers who are predominantly driven by a need for affiliation.

Lastly, the *need for control* denotes a need to actively shape the world, to influence one´s surroundings and immediately recognize the effects of one´s actions. People with a prominently developed need for control can be expected to feel frustrated by circumstances in which they are not able to recognize the immediate results of their work or in which they have to deal with uncertainty and ambiguity. They strive to ´do things´ in a rather straightforward manner and prefer clear task designations and a short delay between action and outcomes. These outcomes are not restricted to the primary agenda of an organization but can also be competences that the volunteer acquires during the course of his or her work. Taken together, if people who are driven by a need for control are able to recognize a particular effect of their engagement, their motivation is sustained.

These basic needs are more or less prominent in every human being, although they are developed to different degrees. They never occur in a singular manner, but people exhibit a mixture of needs that are hierarchized. Similar to the VFI, this model rests upon the claim that it is possible to identify different personal motivations without assuming that these exclude each other.

Furthermore, these needs have to be understood in a bipolar manner in the sense that there can be an abundance of a certain emotional stimulation, which people experience as too much. For example, if a person feels bored by his or her daily work routine, this could be understood as an abundance of certainty that results in a need for uncertainty. Taking up volunteer work can provide that person with a means to achieve just that and satisfy a lust for adventure and excitement. Too much affiliation can trigger a need for de-affiliation and as a result, people might feel an urge to engage in voluntary work on their own or in a smaller group. Likewise, certain work can be experienced as too controlled, which could result in a person losing motivation due to feeling unchallenged in his/her work. In this way, an overstimulation of particular needs can lead to new needs. This makes a careful balancing of volunteers´ needs important.

3 RESEARCH CONTEXT AND METHODS

Disaster management in Germany lies within the realm of responsibility of the federal and local administrative bodies. Only in the case of war and aggression is the national level responsible, here for the purpose of civil protection and defence [14]. A more detailed comparison of national civil security systems of European states was done in the European research project ANVIL [15].

With respect to German disaster management it is necessary to emphasize that any action carried out in disaster response highly depends upon the engagement of (mostly) unpaid volunteer workers. The greater portion of these volunteers is engaged in disaster management organizations. Overall numbers vary but estimates suggest that approximately 1.5 to 2 million people engage in volunteer work in organizations that are integrated with disaster response. Out of these volunteers, roughly 1.3 million are part of voluntary fire brigades. Approximately 50.000 are part of the only partially nationally funded organization THW (Technisches Hilfswerk), providing technical support in disasters, while the remaining volunteers are organized in various non-profit relief organizations [16]. Throughout recent disaster incidents, e.g. floods, a noteworthy number of volunteers engaged free of organizational affiliation. Since these volunteers are more difficult to coordinate outside an organization´s structure, relief organizations of disaster response are trying to become more attractive to these volunteers too.

The authors conducted *semi-structured and unstructured interviews* [17] with volunteers as well as employees of the above-mentioned disaster management organizations and also municipal institutions who work with volunteers. The interviews with volunteers were deliberately designed in an unstructured, open-ended way to allow interviewees to lead the course of the conversation and reflect on their biographical experiences with voluntary engagement. By contrast, the interviews with people who worked with volunteers were semi-structured and followed a stricter protocol with guiding questions.

In total, a number of 40 interviews were conducted with volunteers from different regions of Germany. All of them were active in local-level institutions of various disaster relief organizations, including the Red Cross Society (Deutsches Rotes Kreuz), the Workers´ Samaritan Federation (Arbeiter Samariter Bund) and the German Lifeguard Association (Deutsche Lebens-Rettungs-Gesellschaft). Although these institutions work in several regions across Germany, the interview partners were chosen only from urban areas. One reason for this choice of sampling was that disaster sites in densely populated areas potentially affect a high number of people at the same time, thereby creating a need for broad support in disaster management. The interview partners came from the comparably large cities of Berlin and Leipzig, and the smaller town of Jena.

The main aim of these interviews was to collect data on why volunteers engage in the field of disaster management in Germany. Moreover, the study was supposed to generate insights into reasons for volunteers to engage in particular kinds of voluntary work and specific relief organizations. Another interest was in the knowledge of the people working with volunteers with respect to perceived reasons of voluntary engagement and the devised strategies based upon that to attract volunteers.

Interviews were conducted from June 2015 to July 2016, with each interview lasting approximately 90 to 120 minutes. All interviews were digitally recorded, transcribed word by word and analysed using the software tool MAXQDA. Due to the relatively small number of interviews, the analysis dispenses with calculating statistical significances; conclusions were derived qualitatively by in-depth analysis. Since statements of the interviewees are highly valued to infer the motivational basis for their involvement the results presented in this article include a number of verbal quotes. For reasons of confidentiality, no names of interviewees are stated.

4 FINDINGS

4.1 Need for affiliation

The *need for affiliation* was at the core of a number of interviewees´ statements. A woman in her mid-twenties told that she had started her voluntary work in school when she participated in a project group with a group of four fellow students. She stated that it was the camaraderie within this group that kept her motivated and led to her engagement in the Youth Red Cross (Jugendrotkreuz) after she had completed school. Summarizing her motivation to volunteer, she stated: 'What really keeps me there are the people. I don't really care what it is that I am doing with them. The main thing is that I am in the company of those people'. For her, volunteerism first and foremost was a means to develop a strong bond with other people and maintain it through regular meetings with a group of other volunteers and colleagues within the organization. Although she also stated that she was interested in learning about administering first aid, she did not consider this as what kept her motivated to continue her engagement. Instead, she used volunteer work primarily to satisfy her need for affiliation.

This need also resounded in statements that highlighted the importance of regular group meetings. Another woman who worked in the same youth organization of the Red Cross Society in Jena stated: 'Well, you meet every Monday. Really, every Monday we have a team meeting. That creates a very unique feeling of community. Everything is being done in common'. Such regular group meetings are a very common form of satisfying a need for affiliation in volunteers and many other interviewees also mentioned this way of interacting with fellow people being an important factor that kept them motivated to continue their volunteer work over the years.

Other statements also stressed the importance of common meetings and allowed to differentiate the interlocutor´s affiliation-centred motivation from other motivations even more clearly. For example, a man in his mid-forties who worked in the DLRG in Berlin stated that what was important for his decision to engage in this kind of voluntary work was the fact that his current engagement was not as strict as other forms of voluntary work in the field of disaster management. He said: 'Well, we don´t have such a strict training and because the atmosphere is a bit different from other organizations. After our monthly meeting, when we clean up material and vehicles and prepare things for the next meeting, well, then we also share a beer or two'. In this statement, the interlocutor quite blatantly put the aspect of common leisure upfront and did not refer to more ideological or value-centred motivations as the main reason for him to engage in volunteer work. For him, voluntary work primarily provided a means to meet people he was familiar with and enjoy their company in a context that blurred the boundaries between work and leisure. He also admitted that he was not that interested in trainings or other ways to further expand his practical knowledge. In this way, his statements were in line with the basic assumptions about volunteers who are primarily motivated by a need for affiliation and less so by a need for certainty or control.

4.2 Need for certainty

The *need for certainty* was also prominent in a number of statements. It featured as a main motivation in the statements of a middle-aged woman with a doctoral degree who worked as a voluntary dog trainer in a dog rescue group [German: Rettungshundestaffel] of the local Red Cross Society in Jena. She stated: 'The love for dogs is a part of my motivation and being a part of the team, but the most decisive factor is my enthusiasm for work that contributes to the common good'. When asked if she could imagine other outlets for this motivation, she responded that she also considered the food bank and a recently established refugee camp in Jena as possible places where she could pursue voluntary work. Her statements did not put group cohesion and personal sympathies towards colleagues at the centre, but gravitated towards particular values and ideals to explain her engagement. To her, these values could just as well be realized in other groups and organizations than the ones where she was active already. In other words, she did not engage in voluntary work to satisfy a need for affiliation, but a need for certainty that puts particular emphasis on the creation of a just and fair world. Consequently, she did not link her engagement to a particular group of people – as do people with a more pronounced need for affiliation – but considered the possibility of shifting to other forms of volunteerism in the future.

Another interviewee who displayed a strong need for certainty was a man who worked for the medical service (Sanitätsdienstliche Einsatzgruppe) of the Red Cross Society in Jena. He had started volunteerism in disaster management while he was in primary school when he had enrolled in the Youth Red Cross and continued his engagement since then in various departments of the Red Cross Society. When asked for the reason why he devoted that much time to voluntary work, he replied: 'I don´t want to get any money for my work here. I do it out of some kind of altruism. Ever since I was a child I have some sort of protective instinct. It is simply in my nature to help others'. He identified very passionately with the Red Cross Society´s guidelines and ideology and saw it as a way to change the world. He said: 'I know that I alone cannot change the world but I would like to contribute to that with my work'. To achieve this goal, it was important to him to take on responsibility in the organization. As the following quote shows, he would not have been satisfied with what he regarded as menial tasks. He said: 'When you say that you want to help people then it is not done by distributing and buttering bread at a blood donation event'.

This statement shows a clear demarcation between the need for certainty and the need for control. It was important to our interlocutor that he perceived the work he did as relevant and as `making a difference´ in the larger scheme of things. It had to be consistent with his world-view and did not need to foster immediate results.

4.3 Need for control

The *need for control* also provides a strong incentive for many people to engage in volunteer work. One interviewee – the training supervisor of the local department of the Workers´ Samaritan Federation – very concisely captured this type of motivation in the expression 'Immediately after you help, you see that you did help'. He described the situation of the floods in Jena in 2013 and claimed that the immediate visibility of results contributed immensely to a heightened availability of volunteers during that catastrophic event. In contrast, he stated that the recent refugee situation with its much less `controllable´ timeframe and unforeseeable development was a less attractive field for possible volunteers, because it did not promise immediate, tangible results. Therefore, one could argue, it was much less attractive for volunteers whose engagement is motivated by a strong need for control and more attractive for those who strive to satisfy a need for certainty.

A 19-year old Law student from Leipzig who had volunteered in various disaster management organizations since his teens and had started volunteering as a paramedic for the Workers´ Samaritan Federation half a year ago spoke about the differences between volunteerism and work as a student as follows: 'Well, it is a different kind of work. You have something practical, you see results. With every patient whom I treated, I get a practical result in a rather short amount of time. And that is a different way of interacting with people'. In this statement, the interviewee drew a clear distinction between the rather `theoretical´ work as a student and the `practical´ work as a voluntary paramedic. Translated into the conceptual language of the model for motivations presented here, voluntary work as a paramedic fulfiled the student´s need for control much better than did his work at the university. It provided him with tangible results of his work that he felt lacked from his studies. Furthermore, he could experience these results without any delay and in the form of immediate human interaction, another crucial difference to his work as a student. His statement thus resonated a need for control that he could fulfil with his voluntary work.

4.4 Further implications

Certain responses suggest a processual character of motivations for voluntary engagement, as also stated in the VFI [5]. One interviewee from Berlin explained his changing reasons for volunteerism with the words 'You come because of the issue and stay because of the people'. This statement suggests a shift from a need for certainty to a need for affiliation as the dominant motivation to engage in volunteer work. Therefore, it is important to keep in mind that the needs identified in the model outlined in this article can shift over the course of time. Some needs can become more important or even replace others that were dominant at an earlier stage of voluntary engagement. In this sense, voluntary engagement has to be understood not as stable but as a dynamic process.

Furthermore, as stated above, needs do not exclude each other, but are often complementary. For example, the woman quoted above who exhibited a strong need for affiliation also expressed a need for certainty in the same interview. She stated: 'I also come here because I believe in a certain world view and that is simply manifested in the guidelines of the Red Cross Society'. Based on her quotes, it is impossible to decide if her decision to take up volunteerism was primarily rooted in a need for affiliation or a need for certainty or in both in equal measure. Therefore, one has to keep in mind that in most cases of volunteerism a mixture of motivations occurs.

5 CONCLUSION

People engage in voluntary work for very different individual reasons. The psychological model presented in this article is intended to further a better understanding of the multitude of volunteers´ motivational sources. This model looks at volunteerism in the field of disaster management at the level of the individual actor and reads this phenomenon as a means to address basic psychological needs of human beings. It offers a conceptual tool to categorize these needs and to integrate them into a sophisticated theoretical framework of motivation.

The model proposes three basic human needs that drive volunteers. These are the need for affiliation, the need for certainty and the need for control. These needs do not exist in a singular manner, but are developed to different degrees in every human individual. They also vary according to particular situations and circumstances and can change over time. Traumatic episodes that are typical for contexts of disasters can trigger certain needs – for example a need for certainty – and needs that cause an initial decision to take up voluntary work can become replaced by other needs at later stages in the volunteers´ life. Taking this dynamic nature of needs into account, the model allows to conceptualize volunteerism and to understand individual motivations to volunteer from a psychological perspective.

These findings have ramifications on how organizations that work in the field of disaster management can address volunteers´ needs. In order to improve the integration of volunteers into their structures, organizations should first identify what kind of volunteers they want to attract and design their inter- as well as extra-organizational communication, their trainings and their everyday work routines according to volunteers´ needs. This can help both organizations and volunteers working in the field of disaster management to improve their cooperation and set their expectations towards each other in the most appropriate way.

ACKNOWLEDGEMENTS

This research is funded by the German Ministry for Education and Research (BMBF) contract No. 13N13028 (project: INVOLVE).

REFERENCES

[1] Involve, available at www.involve-project.com.

[2] Wilson, J., Volunteering. *Annual Review of Sociology,* **26**, pp. 215–240, 2000. https://doi.org/10.1146/annurev.soc.26.1.215

[3] Hustinx, L., Cnaan, R.A. & Handy, F., Navigating theories of volunteering: A hybrid map for a complex phenomenon. *Journal for the Theory of Social Behavior,* **40**(4), pp. 410–434, 2010. https://doi.org/10.1111/j.1468-5914.2010.00439.x

[4] Clary, E.G., Snyder, M., Ridge, R.D., Stukas, A.A., Copeland, J. & Miene, P., Understanding and assessing the motivations of volunteers: a functional approach. *Journal of Personality and Social Psychology,* **74**(6), pp. 1516–1530, 1998. https://doi.org/10.1037//0022-3514.74.6.1516

[5] Clary, E.G. & Snyder, M., The motivations to volunteer: theoretical and practical considerations. *Current Directions in Psychological Science,* **8**(5), pp. 156–159, 1999. https://doi.org/10.1111/1467-8721.00037

[6] Schaub, H. & Dörner, D., Eine synthetische Theorie der Motivation. In*Bericht über den 40. Kongreß der deutschen Gesellschaft für Psychologie in München 1996*, ed. H. Mandl, Hogrefe: Göttingen, pp. 855–858, 1997.

[7] Dörner, D., *Bauplan für eine Seele*, Rowohlt Verlag: Reinbek, 1998.

[8] Dörner, D., Bartl, Ch., Detje, F., Gerdes, G., Halcour, D., Schaub, H. & Starker, U., *Die Mechanik des Seelenwagens. Eine neuronale Theorie der Handlungsregulation*, Huber: Bern, 2002.

[9] Bischof, N., Das Rätsel Ödipus. Die biologischen Wurzeln des Urkonfliktes von Intimität und Autonomie, Piper: München, 1985.

[10] Bischof, N., Untersuchungen zur Systemanalyse der sozialen Motivation I: Die Regulation der sozialen Distanz – Von der Feldtheorie zur Systemtheorie. *Zeitschrift für Psychologie,* **201**, pp. 5–43, 1993.

[11] Strohschneider, S., Kompetenzdynamik und Kompetenzregulation beim Planen. In *Ja, mach nur einen Plan: Pannen und Fehlschläge – Ursachen, Beispiele, Lösungen*, eds S. Strohschneider & R. von der Weth, 2nd Edn., Huber: Bern, pp. 35–51, 2002.

[12] Strohschneider, S., Ja, mach´ nur einen Plan. In *Panne – Irrtum – Missgeschick*, ed. B. Boothe, Huber: Bern, pp. 127–144, 2003.

[13] Dibiase, R. & Gunnoe, J., Gender and culture differences in touching behaviour. *The Journal of Social Psychology,* **144**, pp. 49–62, 2004. https://doi.org/10.3200/socp.144.1.49-62

[14] The Crisis Management System in Germany; Federal Ministry of the Interior, Division KM 1 – Crisis Management Coordination Centre, Alt Moabit 140, 10557 Berlin, available at https://www.bmi.bund.de/SharedDocs/Downloads/EN/Broschueren/2012/system_krisenmanagement_en.pdf?__blob=publicationFile

[15] ANVIL, available at http://anvil-project.net/

[16] Hegemann, H. & Bossong, R., Country study: Germany, project report. *Analysis of Civil Security Systems in Europe [ANVIL].* University of Hamburg: Germany, available at http://anvil-project.net/wp-content/uploads/2013/12/Germany_v1.0.pdf

[17] Lamnek, S., *Qualitative Sozialforschung*, 5th Edn., Beltz: Weinheim, 2010.

PERVASIVE TECHNOLOGIES FOR THE REDUCTION OF DISASTER CONSEQUENCES: OPPORTUNITIES AND QUESTIONS

VINCENZO FERRARA
Department of Information Engineering, Electronics and Telecommunications (DIET)
Sapienza University of Rome, Italy.

ABSTRACT

The paper aims at put in evidence as spreading of a better and efficient technology can decrease the extreme consequences of a disaster in scenarios such as avalanches or collapsed buildings. Therefore, it is an opportunity that increases the probability of successful rescue. At the same time involves questions that may have consequences on the same mode of operation of the measuring instruments, practically by hiding or preventing their capability of detection. A review of detection methods belonging to emerging technology shows these aspects along with the questions that arise. On the other hand, rescuers need specialized instrumentations, which are available to all times, easily accessible and manageable, and that meet stringent requirements in terms of detection accuracy, quick location and reduction of false alarms. An improved technology allows reaching a satisfactory quality level of the detection method, but it must be obtained limiting spending, because a more expensive price generally decreases the availability of that instrumentation for rescuers. This aspect of availability is to be emphasized because it could create delays in the performance of the rescue practices. Recent disasters that occurred in Italian region show this as a critical point. Other important aspects are the manageability and easy operation. In order to solve efficiently the detection, actual systems integrate different techniques, often based on hybrid technologies: electronics, electromagnetics, and so on. These detection systems need an accurate management of single parts and of their mutual interaction. Complexity of technical solution must not get worse the practical use. Moreover, size and weight reduction of electronic components, with which the device is built, increase the transportability of detection system, e.g. on board of an unmanned aerial vehicle. This has consequence on an augmentation of their employment, with the opportunity to use them in scenarios characterized by high-risk rescue operations.
Keywords: disaster monitoring and mitigation, emergency preparedness, preparedness and training.

1 INTRODUCTION

Technological progress makes available the tools that offer more valid and effective solutions in cases of post-disaster interventions. Above all, may increase the rate of survivors allowing faster and more precise interventions, an opportunity that increases the probability of rescue. There are now available tools that allow various environmental analysis, by checking the status of real-time areas, as well as other sensors allow you to monitor the vital signs of trapped individuals. The quantity and quality of information take a specific weight, according to the shortness of the time available to carry out the rescue of people in danger of life. In fact, while the detection of vital signs is evaluated positively when it has available a relatively long time in order to intervene, the same one is a given that many would not be used when the survival time is very short, and there may be uncertainty of the data detected. In any case, it is true the rescuers need specialized instrumentations, available to all times, easily accessible and manageable, and that meet stringent requirements in terms of detection accuracy, quick localization, and reduction of false alarms.

In the second chapter, we introduce different technologies and methods for detecting buried victims during the emergencies. The chapter examines two major categories of rescue scenarios and then of detection methods that must be distinguished in the case of presence of:

DOI: 10.2495/SAFE-V7-N3-303-312

active or passive targets. That is, the detection by using active devices and detection of passive targets.

The third chapter examines some aspects of the dissemination of technology.

2 TECHNOLOGY FOR DETECTING BURIED VICTIMS IN EMERGENCY

The trapped victims and/or buried by rubble because of a natural or manmade disaster are among the most disturbing events and strong emotional impact. The organisation of the rescue has as priority the detection and saving of victims. First between all targets. The relief efforts must still be made in safety, to prevent accidents to rescuers, and this may cause a slowdown of operations. Unsafe building structures as a result of an earthquake, jeopardizing the lives of rescuers, prevent de facto checking of relevant areas to search for victims. At the same mode, the motion along a mountain slope run over by avalanche is particularly arduous and dangerous, due to the presence of snow mixed with bulky slabs of ice. Often, fastest results could be obtained if the rescuers have available more efficient instrumentations, suitable for the particular scenario. Although a specific instrumentation can be useful for most of emergencies, to be an effective detection in all of particular scenarios, should at least change the detection procedure.

However, there are many examples of different detection methods that use the same basic technology. Usually the scenario is the one that imposes a particular detection technique. Two major categories of rescue scenarios and then of detection methods must be distinguished: the presence of active or passive targets. They are considered active targets all those objects that produce a steady flow of information, detectable in a consistent manner from suitable sensors. The nature of the 'message' can be different: electromagnetic, sound, and so on. Belong to this category: transceivers, geophones, etc. Also fall into this category all those electronic appliances not directly designed for the detection, but that can be localized thanks to their electromagnetic emissions.

In any case, situations that are more frequent show passive targets that not transmit any useful data for the detection. Or rather, we need to organize the detection without the aid of radiation sources, worn by humans or in their proximity. Therefore, after a list of principal issues and relative solutions generally adopted for detecting in the two situations, we will discuss principally the case of passive state of the targets.

2.1 Detection by using active devices

The technologies based on the use of active devices are more frequent when persons are conscious of to what the environment where they live is dangerous. This is typical of many outdoor sports activities, such as off piste skiing or mountain hike in winter. But a similar methodology is adaptable for localizing objects used daily by persons, like smartphones, electronic watches, and all electronic or electromagnetic devices that should modify in a controlled mode the electromagnetic radiation that comes from a transmitter managed by a rescuer. Examining at first the case corresponding to the use of devices dedicated to perform the localization of avalanche victims, every skier or hiker brings with it a transceiver.

2.1.1 Avalanche beacon

Normally the device, named Avalanche Beacon or ARTVA, works as transmitter which operates at 457 kHz ± 80 Hz international standard frequency, ETSI standard EN 300 718. The transmitting frequency is suitable for penetrating through some meters of snow layers, sufficient to localizing persons in the most of these disasters.

Digital beacons that show higher performance are also equipped with another transmitter at a frequency referred to as W-Link. This second transmitter is for resolving of multiple, complex burial situations by better differentiating individual transceivers. It helps the operational management of rescue by avoiding that more operators focus simultaneously on the identification of a single subject and they forget about the remaining buried victims. Moreover, the W-link allows to transmit and to receive additional data including wearer's vital signs or identification. The transmitting frequency is different for region. European countries adopt a frequency fixed to 869.8 MHz, United States and Canada establish frequencies in the range 916 ÷ 926 MHz.

The wave at the frequency of 475 kHz is about but not perfectly omnidirectional and consists of a signal arranged as a train of low-power pulses. The pulse width is in the interval 70–900 ms, with a duty-cycle in the range 10%–69%. The ARTVAs worn by hikers buried under snow, after they are run over by avalanche, continue to transmit. The remaining other hikers, which were not affected by the same avalanche, switch the ARTVA device in the receiver mode, and then they start to locate each buried person. Instrument determines the direction of a transmitter evaluating the maximum value of power receiving signal. That is, identifying the exact path by means of the antenna radiation pattern, designed to be selective in angle, i.e. suitably directional. The detectable signal is a function of the used ARTVA, analogue or digital, in that the two models have a different limit on maximum distance allowable between the two transmitting and receiving antennas, for a proper detection. Generally, for a digital device the maximum distance is about 10 m. Moreover, digital devices use three antennas in order to manage the detection of multi targets, evaluating for each both direction and distance of the buried transmitters. Wireless data of the W-link channel, which come from other ARTVAs switched in search mode, contribute to determine location of buried person faster and with greater precision.

When all skiers bring Avalanche Beacon, each of them could assume the function of rescuer, with the advantage to start immediately the research. In effect, the survival time of the victims under avalanche decreases quickly from 90% to 30% if rescuers remove a person after one hour. In order to complete the search on entire area, where the avalanche has invested people, it suggests a specific procedure, which takes into account the useful link budget NLOS (no line of sight) of the sensors. That is, the distance within which we have certainty of the detection, even in the worst orientation of buried transmitter. Such a search procedure is differentiated by number of researchers at work and for advancement phases of the localization, coarse and precise. Figure 1 shows the suggestion for the optimum path during the coarse phase in the case of only one researcher (a) and of many (b). The path must trace lines characterized by mutual distance equal to double of the useful link budget NLOS (d). A minor distance increases total time of the monitoring. Using a distance greater, the risk of not detecting some people is large.

An alternative to ARTVA transceivers usage is to wear passive reflectors, named Recco® reflectors, which operate as tuned resonant circuits, when an electromagnetic radiation affects them. Normally a skier wears two reflectors, by including them into each jacket's sleeve or inside the two boots. The main advantage is the absence of battery. The more evident issue is the uncertainty that the few skiers fully equipped of necessary transceiver can be between the survivors. Otherwise, there will be a big delay for the start of the rescue. Although this alternative uses passive reflectors, the basis functioning principle is the same of previous one, which adopts ARTVA, but with lower efficiency. The frequency of transmitted signal is equal to 917 MHz. Passive reflector retransmits the wave doubling its frequency, and consequently the receiver of the transceiver operates at a frequency of 1834 MHz.

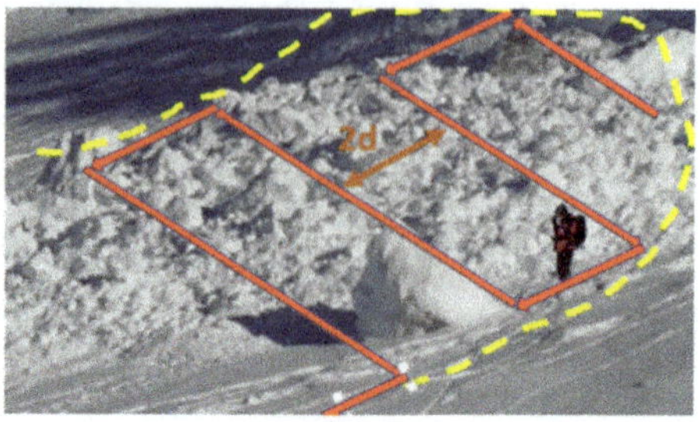

(a)

(b)

Figure 1: Suggestion for the optimum path, during the coarse phase, in the case of only one researcher (a) and of many (b).

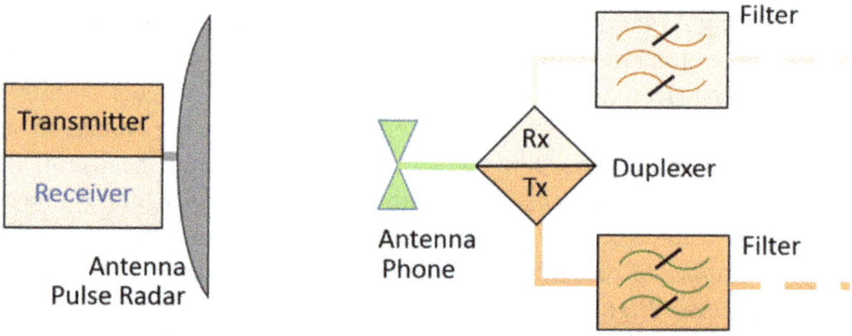

Figure 2: General setup for the detection of RF-devices even if defective or broken off.

2.1.2 Radio frequency device detection

In all different situations from the activities who plan to face high-risk environments, people do not wear special devices to ensure an efficient localization, in anticipation of a possible disaster. The search for persons that bring personal electronic devices, like smartphone, radio controlled watches, and so on, i.e. all electronic device that operate at radio frequency (RF), can be assisted with pulse radar systems [1]. As the Recco® reflector, such devices can be exposed to a radar pulse electromagnetic wave. If this pulse is near to resonance frequency of filters included in the RF-devices, the generated echo could be used for locating the owner person. In the particular case of defective smartphone, and with reference to the Fig. 2, a radar pulse is sent to buried and broken device, the antenna of phone receives impulse, and signal is forwarded through the duplexer, a high-Q filter used to separate the transmit and receive frequencies, to the filter that selects among the frequencies contained in the signal.

If the spectrum of pulse is in the range of RF filter bandwidth, the energy is stored in the filter. When it switches off the transmitting signal, i.e. the radar pulse is no longer transmitted, the stored energy generates a RF echo that propagates over duplexer and to transmitting antenna of smartphone. Finally, the radar can detect this RF echo in the same tuned band.

Unfortunately, it is not certain that the victims bring these electronic devices or objects are close to the body. However, it is a valid alternative when other methods fail for adverse environmental conditions. Better, this technology should be integrated and to cooperate together with different detecting systems in order to resolve many hard situations.

2.2 Detection of passive targets

As emphasized previously, when as a result of a disaster you must locate buried victims, the most common situation is that of individuals without electronic devices worn, neither those designed for the purpose of their detection, nor the generic RF devices, usable for the occasion like pseudo-coherent re-transmitters. Their position will only happen if one or some physical, biological or biomechanical characteristics of buried people are clearly identifiable with respect to similar physical and mechanical characteristics of the environment that surrounds them. Technology must accentuate the differences and/or have a high sensitivity to be able to identify individuals and avoid false alarms.

2.2.1 Geophones

A possibility is to detect the sounds emitted for example by the persons calling help. There are examples in the literature of sound detection systems manufactured using a network of geophones [2] or a single tri-axial geophone [3], in order to detect vital signs as heartbeat, breathing, and footstep. Actually, most of these systems are dedicated to discover persons hidden in closed environments, e.g. clandestine immigrants concealed in truck trailers. In effect, the main use is in the field of security, in the surveillance tasks. The main issue of this detection technique is the requirement of a very quiet environment, hard to find during rescue and emergency activities.

2.2.2 Thermography

The thermal inspections could be another alternative. It uses infrared (IR) radiations, in the range of wavelength around from one to ten micron. At the basis of this method, the well-known law of Stefan-Boltzmann, concerning black body emittance. Equation 1 is its extension to real body, where q, ε, σ, and T are respectively: the emittance of a black body, emissivity of real body, Stefan Boltzmann constant, equal to $5.67 \cdot 10^{-8}$ W m^{-2} K^{-4}, and the absolute temperature.

$$q = \varepsilon \cdot \sigma \cdot T^4. \qquad (1)$$

This method, also used in civil engineering, for the state of building structures surveys, provides results that are strongly influenced by environmental conditions, such as sunlight, rain and wind. For an acceptable quality of the detection images it is also necessary that there is a minimum of thermal difference between target and objects of the surrounding environment that are at direct contact, a minimum temperature difference of the order of 10°C. These conditions are best found in the cases of people buried under rubble. Unfortunately, the depth of penetration through the debris layers is very limited and this constitutes a major impediment to the detection. Variants that make use of the stress of the material layers by means of microwaves to raise the contrast of the image photogrammetric [4], through analysis of the evolution of temperature profiles during the heating and cooling phases, are normally used to locate cold objects, and not the human beings.

Moreover, in the mountain environment should be observed that you add to the adverse environmental conditions, the physiological mechanism of the rapid reduction of the temperature of the human body in the process of frostbite under the snow and ice. Note, since the radiation manifests a poor depth of penetration, the result of detection normally has origin from heat of the subcutaneous layers, and minimally by the deeper organs.

2.2.3 Ground penetrating radar

We can summarize the operation of a GPR system as follows:

- GPR system transmits an electromagnetic wave into the depths of the soil or other structure that is obstacle to the detection of objects buried or hidden;
- the different value of permittivity (ε_r) between adjacent layers changes the wavelength of the radiation and its propagation velocity in the medium;
- also the relative magnetic permeability (μ_r) and the electrical conductivity (σ) have influence on the characteristics of the electromagnetic wave parameters;
- the change in the electromagnetic properties between two contiguous layers generates a new scattered wave that can reach the receiving antenna of the GPR system;
- the received wave allows to calculate the distance of the surface that produced the scattered wave;
- other assessments, such as the relative movements between GPR and target, can be obtained by using specific types of GPR systems.

However, we should emphasize that the structures mainly made of metallic materials or reinforced concrete slabs are electromagnetic transmission media with losses, which reduce the depth of penetration achievable and create deleterious multiple reflections.

GPR systems operate with frequencies in the range from a few tens of MHz to some GHz, i.e. they show a larger wavelength than the IR radiation, and penetrate more deeply into the materials. Therefore, they are more suitable for the purpose of the research of buried individuals. The depth reached depends on the type of GPR system and by the material that constitutes the transmission medium. In general, a frequency increase improves resolution but reduces the penetration depth. The GPR systems are primarily distinguished by the domain design choice: in time or frequency, see Fig. 3. In time domain, the operating

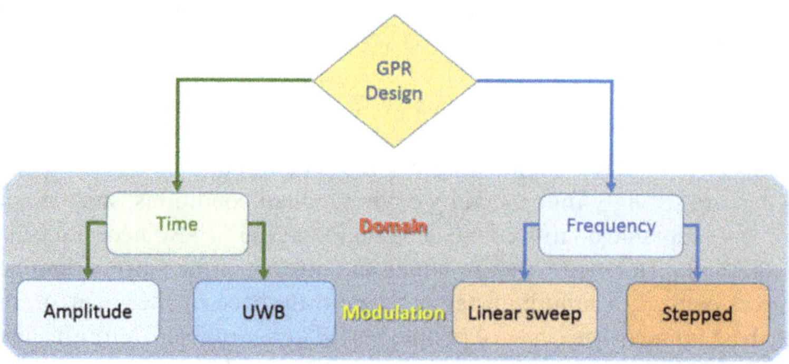

Figure 3: Different GPR designs.

principle is to send a pulse into the ground and then listen to the backscattered echoes. Main categories of time domain systems are two: the amplitude modulated and the carrier free or UWB GPR. Actually, the term UWB (Ultra-Wide Band) slightly differs from that of the conventional air radar systems, defining UWB GPR those systems with signals that have a fractional bandwidth ($Fract_{BW}$) of eqn. 2, where BW is the bandwidth and f_{centr} the central frequency, larger than 100%.

$$Fract_{BW} = \frac{BW}{f_{centr}} \cdot 100\% \geq 100\%. \tag{2}$$

The first type, i.e. the amplitude modulation, sends a pulse with the carrier frequency, which in turn is modulated by a square wave. The pulse width determines the vertical resolution: more the pulse is narrow, the better the resolution. Note the UWB GPR has been developed to get systems with widths of much larger bands.

In the frequency domain, you modulate a continuous wave (CW) by main two modes: linear sweep and stepped. A linear sweep generates frequency-modulated continuous-wave (FMCW), and you can carry out different shapes of the modulation such as at saw tooth and triangular. Obviously, if you change the frequency of the CW in fixed steps, the modulation is named stepped frequency continuous wave (SFCW).

The FMCW setup of Fig. 4 demonstrates how the distance between target and GPR is evaluated. A control voltage with ramp shape modifies the frequency of the oscillator (VCO). A splitter divides amplified waveform by sending two identical signals, one toward transmitting antenna and other to mixer. The scattered wave from the target, with a fractional amount of energy, arrives to the receiving GPR antenna. A low noise amplifier (LNA) increases the signal magnitude at proper value for the mixing operation. The mixer generates output signals of frequency equal to the sum and to the difference of the frequencies of input. The filters select the signals with frequency difference, which are then amplified. These signals are significant of the distance that exists between target and GPR system, like eqn. 3 expresses. In the equation, D is the searched distance, v is velocity of electromagnetic wave in the transmission medium, Δf is frequency difference between transmitted and received signals, and *df/dt* is frequency shift per unit time.

$$D = \frac{v \cdot \Delta f}{2 \cdot \left(\frac{df}{dt}\right)}. \tag{3}$$

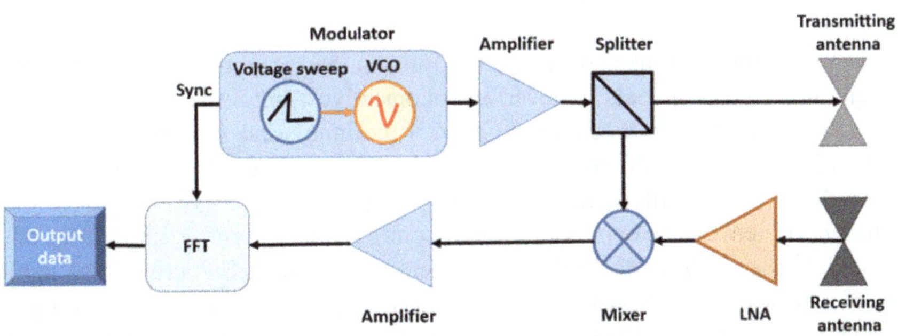

Figure 4: Block diagram of FMCW GPR.

In the case of triangular shape, the measure of Doppler frequency f_D can be implemented by evaluating two frequencies difference Δf_1 and Δf_2 at the rising and falling edges respectively, as you can see from eqn. 4. Doppler frequency is useful for detecting movement of a target, e.g. vital signs of a buried person.

$$f_D = \frac{|\Delta f_1 - \Delta f_2|}{2}. \tag{4}$$

Examples of cheap design of FMCW GPR are in literature [5].

From the point of view of the false alarms, the main source is the non-stationary clutter. However, there are other different noise sources, such as stationary clutter, internal noise, random jitter, and narrowband jamming. Some researchers proposed interesting solutions that solve problems arising from non-stationary clutter. In particular, Zaikov and Sachs designed a prototype UWB radar that allows the removal of that clutter together with signal to clutter separation, by principal component analysis [6]. The problem of false alarms is an important issue, which need to be resolved especially for the case emergency scenario. This is because, in such a scenario are more frequent the noises as motion of objects located near the GPR, and rescue engines, together with radio interferences added, like smartphones [7].

3 A WIDESPREAD TECHNOLOGY SUCH AS EFFECTIVE SOLUTION

All the technologies examined in the previous chapter have been selected for their non-invasive feature. An essential quality for not exposing the rescue workers to the dangers, which generally follow a disaster. However, among the listed technologies, those that can be used for detecting and locating people buried without personal devices, dedicated to the detection and/or exchange of information with the search system, are of major interest for personnel rescue. In fact, rescuers work directly in the field, in various situations and environmental conditions and prefer to use stand-alone instruments, showing a direct result, not conditioned by the presence/absence of devices owned of the victims. Among these technologies, those using the GPR systems result from what has been said the most appropriate in the case of people buried or trapped. Both for the nature of the electromagnetic radiation, which allows a greater depth of penetration compared to that based for example on thermography, and for the possibility to operate in noisy environments, commons in emergency scenarios, choosing the most appropriate type of GPR, and realizing a suitable cancellation technique or at least a reduction of noisy effects. Therefore, in the following of the paper we will refer on GPR systems, to distinguish detection techniques and the methodologies you can use.

3.1 Dissemination of the technology

At first, it is appropriate to make some observations on the relationship between technological advancement, in terms of goals achieved and perspectives, and actual use of technologically advanced instruments. It often occurs that, even if the innovative instrument has passed the post prototypal phase, and could be available for use, it is rarely used. The origin of this problem is manifold, but essentially: cultural and economic.

The cultural problem can be expressed by a dissemination of scientific knowledge and innovation limited to the circle of the only researchers and experts. Sometimes the application versatility of new devices or algorithmic methods is not appreciated until after a long process. The awareness of this problem has reached the top institutions, particularly European ones, which have launched in recent years many international programs of collaboration, e.g. the

COST Actions. The main purpose of them is the exchange of technological information at various levels, university, industrial and state agencies. At corollary of these collaborations have been undertaken a series of initiatives aimed at the dissemination of core technologies and expertise to young people.

The economic aspect often represents a ballast to the spread of complete systems, or simple devices, technologically advanced in the institutions responsible for security and civil protection. Fortunately, in recent times it has been observed that the technological progress feeds himself. That is, the progress made in nanotechnology has led to a considerable reduction in costs. Thus the realization of a FMCW radar system at the ISM band performed using integrated RF components has a cost of implementation which is also less than a tenth of an equivalent realized employing components housed in rugged unibody chassis.

The considerable reduction in the size of the devices and consequently of the entire system has other important consequences: the reduction of the energy required for its operation, which is accompanied by the decrease of the voltage values involved, and the considerable weight reduction. The implications are different. The characteristic low voltage-low power increases the energy autonomy of the system that can best be designed for use in mobility. To reach even to the complete autonomy, using harvesting techniques in order to recover all the energy needed for its operation from the surrounding environment. The reduction of the weight means greater manageability of the system, but it is also a prerequisite to the design of higher performance systems that can solve difficult operational situations.

It has already been said that the characteristic of the non-invasive research systems is important for safety purposes. However, if the monitoring of an area run over by an avalanche is made by dragging of the detection system over the snow and the slabs of ice, the danger to the rescuers persists. Similarly, in the case of collapsed buildings, where to move among the rubble poised it is extremely dangerous. In such cases, the adoption of self-propelled robot carrying the sensor or the mounting on board UAVs of the same monitoring systems is possible and opportune. In the case of on board of UAV GPR, the monitoring technique is more complex [8] and requires the integration of the aircraft's trim measures. These ones first of all serve to correctly interpret the pointing direction of the GPR antenna and to calculate the position of the target detected by applying the necessary fixes to the aircraft coordinates. This arrangement provides greater monitoring speeds with all the benefits that come with it.

A policy of greater diffusion of technology, and technological culture, could allow the achievement of regional centres of more efficient monitoring, with powers to intervene faster and more targeted. The search for a solution, or a technological tool for control of the area at risk, it would not be the result of initiatives forced by the necessities of the last hour, close to the event disastrous. In fact, it should be taken into account even the appearance of bureaucratic slowness, which slow down exceedingly purchases and the installation of important systems and especially needed. In recent events in Italy, for example, only the good will and abnegation of researchers volunteers allowed the rescuers to operate in relative safety, by implanting a radar control to check on the looming danger of avalanches.

REFERENCES

[1] Loschonsky, M., Feige, Ch., Rogall, O., Fisun, S. & Reindl, L.M., Detection technology for trapped and buried people, *IEEE MTT-S International Microwave Workshop on Wireless Sensing, Local Positioning, and RFID*, pp. 1–6, 2009. https://doi.org/10.1109/imws2.2009.5307879

[2] Ascione, M., Buonanno, A., D'Urso, M., Vinetti, P., Angrisani, L. & Schiano Lo Moriello, R., A method based on passive acoustic sensors for detection of vital signs in closed structures. *Instrumentation and Measurement Technology Conference (I2MTC), IEEE International*, pp. 1764–1769, 2012.
https://doi.org/10.1109/i2mtc.2012.6229551

[3] Vinod Reddy, V., Divya, V., Khong, Andy W.H. & Ng, B.P., Footstep detection and denoising using a single triaxial geophone. *Circuits and Systems (APCCAS), 2010 IEEE Asia Pacific Conference on Circuits and Systems,* pp. 1171–1174, 2010.
https://doi.org/10.1109/apccas.2010.5775004

[4] Di Marzio, C.A., Rappaport, C.M., Wen, L. & Sauermann, G.O., Microwave-enhanced infrared thermography. *Proceedings of SPIE*, **3710**, pp. 173–179, 1999.

[5] Ferrara, V., Troiani, F., Frezza, F., Mangini, F., Pajewski, L., Simeoni, P. & Tedeschi, N., Design and realization of a cheap ground penetrating radar prototype @ 2.45 GHz. *IEEE Browse Conference Publications - 2016 10th European Conference on Antennas and Propagation (EuCAP)*, pp. 1–4, 2016.
https://doi.org/10.1109/eucap.2016.7482008

[6] Zaikov, E. & Sachs, J., *UWB radar for detection and localization of trapped people*, eds Boris Lembrikov, Ultra Wideband, InTech, 2010.
https://doi.org/10.5772/10004

[7] Sachs, J., Helbig, M., Herrmann, R., Kmec, M., Schilling, K., Zaikov, E. & Rauschenbach, P., Trapped victim detection by pseudo-noise radar. *Proceedings of ACWR '11, 1st International Conference on Wireless Technologies for Humanitarian Relief*, pp. 265–272, 2011.
https://doi.org/10.1145/2185216.2185289

[8] Fruehauf, F., Heilig, A., Schneebeli, M., Fellin, W. & Scherzer, O., Experiments and algorithms to detect snow avalanche victims using airborne ground-penetrating radar. *IEEE Transactions On Geoscience and Remote Sensing*, **47**(7), pp. 2240–2251, 2009.
https://doi.org/10.1109/tgrs.2009.2012717

SOCIAL MEDIA DURING MULTI-HAZARD DISASTERS: LESSONS FROM THE KAIKOURA EARTHQUAKE 2016

BRIONY GRAY, MARK J. WEAL & DAVID MARTIN
Department of Electronics and Computer Science, University of Southampton, United Kingdom.

ABSTRACT
Social media provides channels of communication during emergency events such as earthquakes. Such sites may be utilised for a range of emergency response strategies providing that data is processed rapidly and management strategies employed effectively. The processing of social media data presents many challenges for emergency responders: information overload, organisational communication and information reliability remain prevalent issues. Furthermore, there is a growing need to improve the management of multi-hazard disasters (sometimes referred to as 'cascading disasters') due to an increase in their frequency and severity, exacerbated by underlying global problems such as climate change. This is especially important to geographical regions that are prone to particular hazards – New Zealand for instance recorded nearly 33,000 earthquakes in 2016 alone. Similarly, there is an increasing need to evaluate developments in technology and social media sites themselves as they are progressively being relied upon during emergency events. In this study, we examine the crisis communications of the Kaikoura earthquake (New Zealand, 2016) using mainstream media content such as new stories, and online content such as Twitter data. A mixed method approach was employed, which combined content analysis with the application of a conceptual framework. The paper then presents (i) an analysis of crisis communications during the event, focusing on changes in media content and theme, (ii) the structure of online emergency response in the country and its affect on management and (iii) the barriers effecting emergency response in this case study.
Keywords: conceptual framework, content analysis, disaster management, earthquake, emergency response, multi-hazard disaster, social media.

1 INTRODUCTION

Social media are increasingly utilised for the dissemination of information during emergency situations [1–3]. Individuals may use social media as a source of information to make personal and complex safety critical decisions during such events [4]. Social media offer a range of supporting features which may reduce risk during crises: real-time monitoring and evaluation can be used for targeted action purposes [1], generalized monitoring and evaluation may support policy-making [5] and as a means to establish situational awareness [6]. Specific to disaster management, social media may provide warning systems throughout the disaster lifecycle phases [7], identify or track potential hazards or problems [8] and strengthen human interaction, coordination and crisis communications [9].

Using social media to disseminate time-critical information does however face obstacles. Information overload transpires when there are high volumes of accessible information which may not be relevant or useful [10]. Often this occurs when information is not entirely related to a situation, is not targeted at individuals or networks, or is simply outdated [11, 12]. Poor communication and uncoordinated dissemination from emergency responders may exacerbate these problems [13]. Despite a rise in global connectivity through developments in technology, such as the World Wide Web ('the Web'), accessibility to social media remains a significant obstacle [14]. Factors such as social class, gender, ethnicity, income and geographical location may all impact access to social media and other online resources [2, 15]. Reliability of information created or shared via social media during disasters can also be

DOI: 10.2495/SAFE-V7-N3-313-323

questionable as disseminating information requires little to no authentication, and often individuals do not site original sources of data [12, 16].

The challenges of using social media in emergency response evolve alongside the development of new applications, services and uses either provided by or taken up by social media themselves [10, 17]. Consequently, a number of emergency responders have sought to address such limitations. The American Red Cross for example launched a 'Digital Operations Center' dedicated to global humanitarian relief by utilising social media [18]. The uptake of crowdsourcing public information has in recent years offered emergency responders with valuable insight into disaster situations, as well as empowering the citizens who experience a disaster [8, 19]. In such situations the challenges of information reliability and credibility are prominent [20]. The inability to verify information during crisis situations can be highly dangerous: especially as this may form the basis for emergency decision making processes, both for individuals and organisations, where poorly planned consequences may potentially be dire to citizens [2].

Utilising social media for emergency response additionally faces challenges presented by complex, multi-hazard events. During such crises, it is even more difficult to provide reliable, tailored, and time-critical information due to more unpredictable nature of the disaster and subsequently triggered hazards. As a result, there is a growing need to analyse the causes of high magnitude multi-hazard disasters especially as they are becoming more common and devastating than previously expected [21–23]. New functionalities, services, software and tools are continually evolving to meet this demand, and often use social media as a lens for analysis. Twitter in particular has been a popular focus for research in crisis communications, and remains extensively used during emergency events by individuals and responders alike. Micro-blogging platforms such as this offer rapid situational awareness – an aspect that has become vital to many studies seeking to utilise online networks for effective information dissemination.

2 THE STUDY

2.1 The Kaikoura earthquake

According to the USGS Earthquake Hazards Program (2016) the Kaikoura earthquake occurred on the 14th of November, 2016, on the South Island of New Zealand. The geographical setting of the earthquake is illustrated in Figure 1 below. It recorded a magnitude of 7.8 and a duration of roughly two minutes, making it the second largest earthquake to be recorded in the country since its colonisation. The epicentre was recorded at 60 kilometres south-west of the town of Kaikoura, thus providing its name, and at a depth of around 15 kilometres. The intensity of the earthquake has partly been attributed to a series of ruptures occurring on multiple fault lines in a complex sequence. The initial earthquake hazard then triggered nearly 12,000 aftershocks along multiple fault lines, as well as causing a 5 m tall tsunami wave which affected thousands of kilometres of coastline.

3 METHOD

3.1 Twitter data

Twitter data was collected for the Kaikoura earthquake case study using the Twitter streaming API. Tweets were collected for featuring two or more of the terms 'Kaikoura', 'Earthquake', 'New Zealand' and 'Tsunami'. The data collection ran from the 13th of November 2016 (the

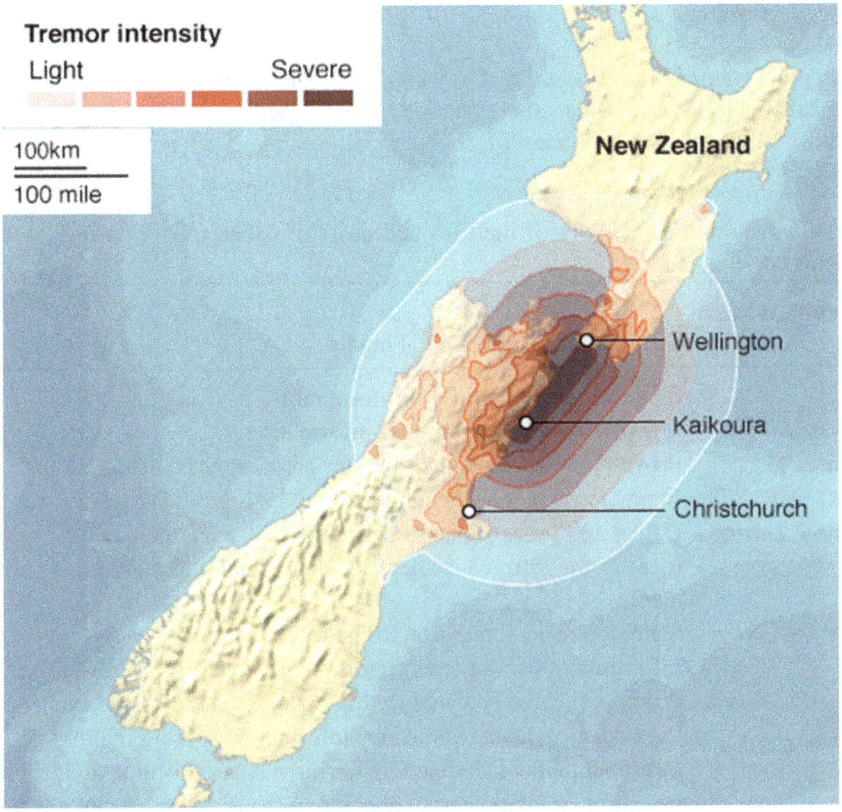

Figure 1: A map showing the epicentre and distribution of the Kaikoura earthquake, 2016. (*Source: USGS, 2016*).

day the earthquake occurred, GMT), and ended on the 13th of December 2016. This produced a data set of roughly 30,000 tweets. The dataset is hosted on the Southampton Web Observatory at a 100% sample size.

3.2 Content analysis

Manual content analysis was conducted on the dataset to indicate the most used hashtags, phrases, and popular re-tweets. This demonstrated how online media content changed over the duration of one month, and illustrated the larger picture of crisis communications on Twitter. Following this, tweets for a 24-hour period were then systematically selected at 7-day intervals. This formed four smaller datasets throughout the course of the event. These smaller datasets highlighted how specific hazards and micro-variations in hashtag and term trends manifested on the social media site, providing a more detailed insight to the event. Additionally, analysis of smaller datasets avoids common problems with the large size and velocity at which content is created and re-tweeted on Twitter during crises [24]. Similarly, manual content analysis was applied to indicate the most used hashtags, phrases and popular re-tweets, where each smaller dataset could be compared with one another to determine changes in content over time.

3.3 Conceptual framework

The conceptual framework proposed by Gray *et al.* [25] shown in Table 1 was applied to the systematically sampled datasets to show change in social media content over time. Application results indicated that the uses of social media, categorised from existing disaster literature, were present in the case study.

Table 1: A conceptual framework of the uses and users of social media during the disaster lifecycle phases.

Disaster lifecycle phase	The uses of disaster social media
All stages	Evaluate the reliability of information
	Identify and/or contain false information
Pre-event	Provide and seek general disaster preparedness information
	Provide and receive general national and regional disaster warnings
Pre-event → During	Detect and warn of disasters and specific hazards locally
	Identify the differences between actual and potential uses of social media
During event	Send and receive requests for help or assistance
	Inform others about ones condition and location
During → Post-event	Provide, receive and analyze big data generated by the event
	Provide, receive and encourage information sharing in multiple formats
	Document what is happening during a disaster online and offline
	Consume or create news coverage of the disaster
	Provide and receive location based real-time warnings
	Express public and/or individual emotion or empowerment; reassure others
	Raise and develop awareness; donate and receive donations; list ways to help or volunteer
	Seek to inform and support existing disaster management strategies
	Provide and receive specific disaster response, rescue and evacuation information
	Seek and assess mental, behavioral and emotional health support
	Filter, categorize critically analyze information
	Understanding how online and offline situations differ
	Provide and receive information regarding disaster response, recovery and rebuilding; tell and hear stories from the disaster
	Understand how ones access to the Web has had an effect on their experiences
Post-event	Discuss socio-political causes, implications and responsibility
	Re-connect community members
	Discuss the accessibility of the Web as an intermediary to social media
	Discuss the accessibility and reliability of specific social media; discuss perceptions
Post-event → Pre-event	Consolidate lessons learnt to develop new/improved social media applications

4 RESULTS

4.1 Crisis communications during the Kaikoura earthquake

Communications on Twitter during the Kaikoura earthquake, although generally following an expected pattern for communications previously recorded by other studies (see for example Houston *et al.* [26] official, and scientific literature was carried out in 2012–13 to develop a framework of disaster social media. This framework can be used to facilitate the creation of disaster social media tools, the formulation of disaster social media implementation processes, and the scientific study of disaster social media effects. Disaster social media users in the framework include communities, government, individuals, organisations, and media outlets. Fifteen distinct disaster social media uses were identified, ranging from preparing and receiving disaster preparedness information and warnings and signalling and detecting disasters prior to an event to (re) were more focused on the recording and spread of news relating to individual hazards within the overall, larger event. The hazard-specific news primarily referred to a tsunami which followed the initial earthquake some two hours later, a large aftershock recording a magnitude of 6.8, another aftershock at 6.2, and finally an aftershock at 5.8: all of which occurring within one month of the first quake. The changes in online content was noted in variations of popular hashtags, demonstrated in Table 2 below. The table compares 24 hour systematic samples beginning on the day of the first quake. Although the first three most popular terms in each dataset refer to generalised terms, the subsequent terms illustrate the changes in events that are both represented online as well as offline.

The term '#Tsunami' is more extensively used in the first dataset of Table 2 which corresponds with the occurrence of the hazard in the real world based on media content. Similarly, the term '#Argentina' is popular within the same sample. This refers to the fault triggering a similar earthquake with roughly the same magnitude within quick succession of the Kaikoura quake. In the second sample, '#HMCSVancouver' recorded a high volume, which referred to a series of offline events where HMCS Vancouver aided with unfolding of disaster management strategies in the local area. Similarly to Argentina, in the second sample '#Japan' become popular, representing a plethora of warnings through news and media channels warning of potential risk to the country based on the recent seismic movements. In the third sample, '#BREAKING' and '#waiau' were prevalent, referring to particular large aftershocks in the area. Finally, in the

Table 2: The most used terms during a 24 hour period during the Kaikoura earthquake.

14/11/2016	21/11/2016	28/11/2016	05/12/206
#NewZealand	#NewZealand	#NewZealand	#NewZealand
#earthquake	#Kaikoura	#Kaikoura	#earthquake
#nzearthquake	#earthquake	#Earthquake	#Kaikoura
#Argentina	#nznews	#NZ news	#geochat
#sup	#eqnz	#BREAKING	#nznews
#eqnz	#HMCSVancouver	#prophecy	#weather
#Tsunami	#imwithher	#Geology	#wx
#NuevaZelanda	#Japan	#Waiau	#cli
#NZQuake	#Erdquake2011	#Wellington	#Auckland
#supermoon	#Tsunami	#Kaikoura earthquake	#humanitarian

last sample '#geochat' illustrated the transition of online activities from being heavily centred on news stories, to shifting towards a lively online debate that considered the causes of earthquake and multiple hazards, as well as the socio-political and economic considerations.

4.2 The structure of emergency response in New Zealand

The Kaikoura earthquake experience relatively low levels of information overload. This may in part be attributed to the set-up of online media channels, as there are a number of dedicated news and media outlets specifically set up to distribute real-time information regarding earthquakes (for instance, @WeatherWatchNZ, @USGS, @LastQuake etc.). As this is most likely due to the fact that the country, well-practised in coping with seismic hazards as a result of their geographical location, has much higher preparedness and resilience strategies and infrastructure in place. This, combined with a comparatively strong global economic standing, means that a large proportion of the public are able to access the Web during times of crisis, access online resources, which include that of social media and are generally able to maintain cohesion between offline and online sites.

The structure of New Zealand's emergency response is arguably therefore well managed: online media, as well as the dedicated Twitter accounts for specific geographic information more often than not re-tweet and disseminate emergency warnings and responses from local and Governmental authorities, as opposed to creating their own. This means that information is often clear, structured to a particular platform, and automatically disseminated through automatic systems set-up by certain twitter accounts (i.e. @LastQuake make the most recent earthquake information publicly available as and when something occurs, adding other time-critical information which is dictated by Governmental bodies and emergency responders). The application of the conceptual framework has demonstrated how online content changed over the following four weeks of the disaster.

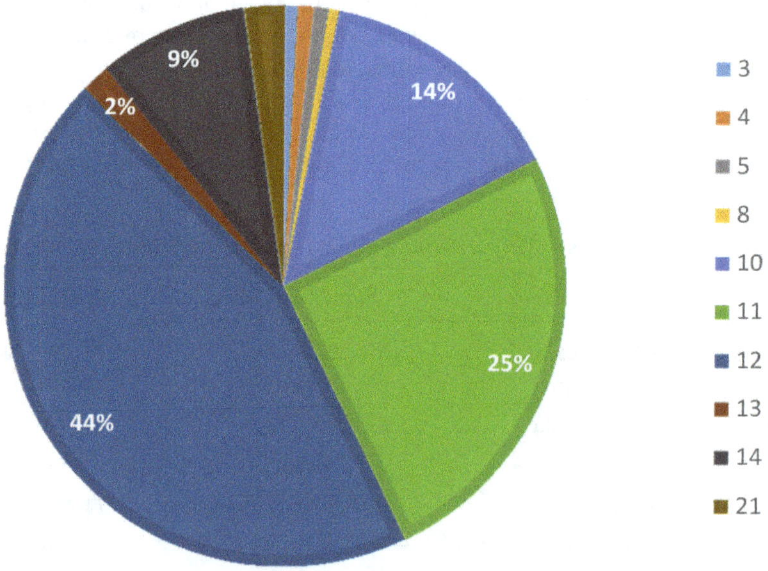

Figure 2: The count of framework categories during the first 24 hour sample of the Kaikoura earthquake (14/11/2016). Framework category numbers are listed on the right hand column.

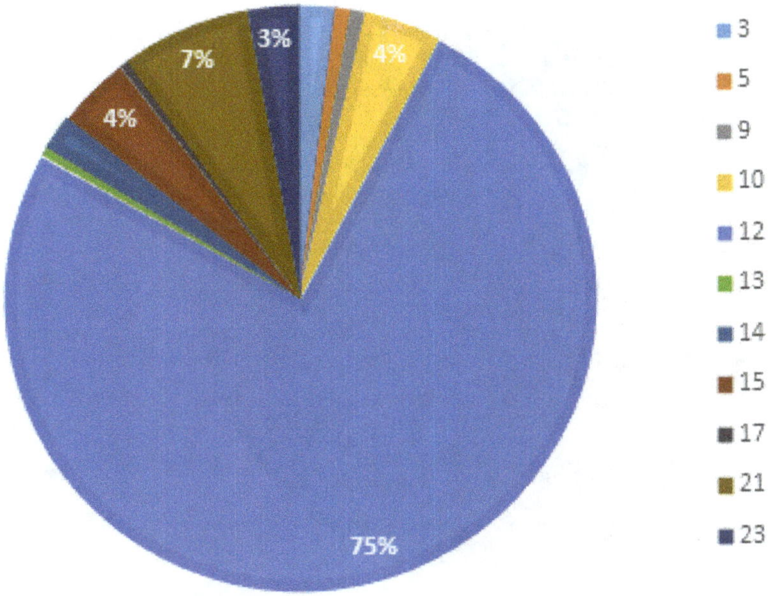

Figure 3: The count of framework categories during the second 24 hour sample of the Kaikoura earthquake (21/11/2016). Framework category numbers are listed on the right hand column.

5 DISCUSSION

As evidence by Figures 2–5, there are significant changes in the framework categories, which indicate changes in the online content of twitter. The first sample shows a dominance of categories 10, 11, 12 and 14, demonstrating that although a large hazard had just occurred there was relatively little online information actually created to help with decision-making processes. This, combined with the low death toll and casualty rate, suggests that time critical information was instead disseminated effectively and across a range of media channels, reducing the confusion often caused by information overload. In the second sample categories focused on smaller, regional hazards became dominant. This was an unexpected result as often this occurs in the preliminary stages of a disaster as opposed to during. This indicates that emergency responders were disseminating information regarding smaller specific regional events and potential hazards at targeted areas, which proved to be an effective way to mitigate the effects of the multi-hazardous nature of high magnitude seismic events.

In the third sample the framework category contents transitions into a well-documented response to natural disasters: large volumes of people begin expressing emotions or stories about the events, which often increases in volume once the larger triggered hazards have subsided. Finally, following this in the final sample a full month after the initial earthquake, the framework category content begins to shift back to the earlier stages of the life cycle phase. This is reflective of geographical regions that experience frequent seismic events as preparedness efforts and public information to effectively tackle these is highly encouraged and supported. The quick transition back to these early stages is arguably one of the main reasons that online emergency responders in New Zealand have reduced risk to the population in areas prone to risk.

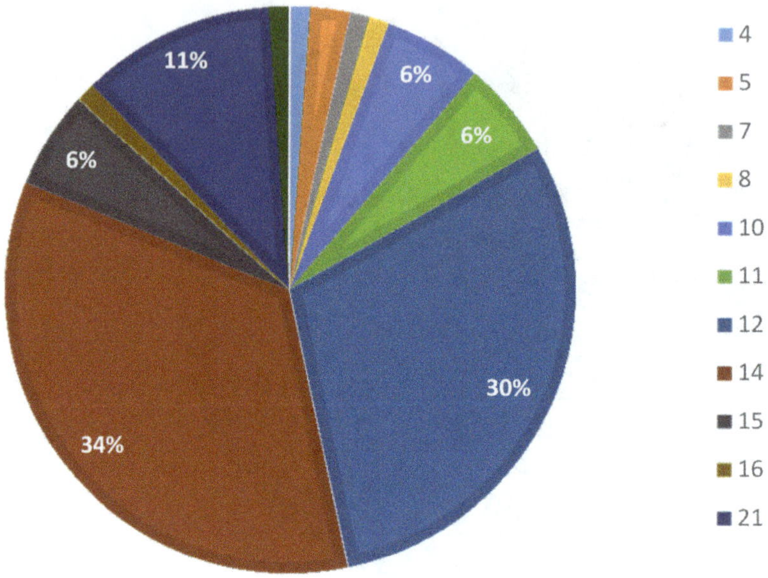

Figure 4: The count of framework categories during the third 24 hour sample of the Kaikoura earthquake (28/11/2016). Framework category numbers are listed on the right hand column.

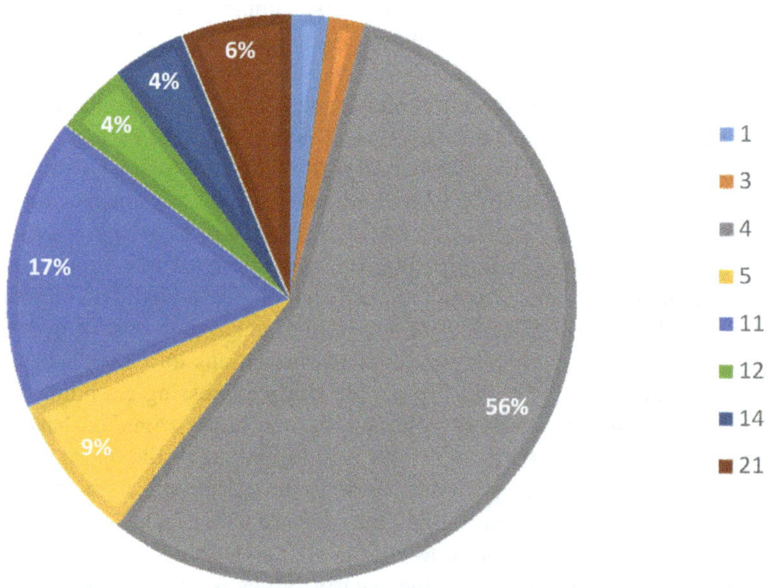

Figure 5: The count of framework categories during the fourth 24 hour sample of the Kaikoura earthquake (05/12/2016). Framework category numbers are listed on the right hand column.

6 CONCLUSION

In conclusion, multi-hazard disasters are becoming more frequent and devastating than previously expected. Social media and other forms of online crisis communication are increasingly being used to help reduce risk and the negative associations of such events. New Zealand, as an area prone to seismic hazards, has invested heavily not only in effective emergency response in the physical world, but has also developed effective online strategies for information dissemination. In these, series of accounts are dedicated to passing on vital information, while avoiding creating new information themselves. This avoids common issues such as information overload. The ability of these channels to move from generalised time-critical information to producing tailored, region-specific information means that smaller areas that are more prone to particular hazards, for instances mud-slides, can be targeted to reduce risk further. Finally, the ability of these channels to rapidly transition back into sharing information associated with the early stages of the disaster lifecycle phases ensures that preparedness and mitigation is strongly supported online, thus making future hazards easier to respond to.

REFERENCES

[1] Ruggiero, A. & Vos, M., Social media monitoring for crisis communication: process, methods and trends in the scientific literature. *Online Journal Communication Media Technologies*, **4**(1), pp. 105–130, 2014.

[2] Haddow, G. & Haddow, K.S., *Disaster Communications in a Changing Media World*, Butterworth-Heinemann, 2013.
https://doi.org/10.1016/c2012-0-06592-1

[3] Alexander, D.E., Social media in disaster risk reduction and crisis management. *Science and Engineering Ethics*, **20**(3), pp. 1–17, 2013.
https://doi.org/10.1007/s11948-013-9502-z

[4] Carver, L. & Turoff, M., Human-computer interaction: the human and computer as a team in emergency management information systems. *Communication of the ACM*, **50**(3), pp. 33–38, 2007.
https://doi.org/10.1145/1226736.1226761

[5] Sobkowicz, P., Kaschesky, M. & Bouchard, G., Opinion mining in social media: modeling, simulating, and forecasting political opinions in the web. *Government Information Quarterly*, **29**(4), pp. 470–479, 2012.
https://doi.org/10.1016/j.giq.2012.06.005

[6] Dashti, S., Palen, L., Heris, M.P., Anderson, K.M., Anderson, T.J. & Anderson, S., Supporting disaster reconnaissance with social media data: a design-oriented case study of the 2013 Colorado floods. *Proceedings of the 11th International ISCRAM Conference*. University Park: Pennsylvania, USA. pp. 632–641, 2014.

[7] Sweta, L.O., Early warning systems and disaster management using mobile crowdsourcing - MDIwMTMxNDI5.pdf. *International Journal of Science Research*, **3**(4), pp. 356–365, 2014.

[8] Crooks, A., Croitoru, A., Stefanidis, A. & Radzikowski, J., #Earthquake: twitter as a distributed sensor system. *Transactions in GIS*, **17**(1), pp. 124–147, 2013.
https://doi.org/10.1111/j.1467-671.2012.01359.x

[9] Lindsay, B., Social media and disasters: current uses, future options, and policy considerations. *Congressional Research Service*, 2011.

[10] Bharosa, N., Lee, J. & Janssen, M., Challenges and obstacles in sharing and coordinating information during multi-agency disaster response: propositions from field exercises. *Information System Frontiers*, **12**(1), pp. 49–65, 2009. https://doi.org/10.1007/s10796-009-9174-z

[11] Lu, Y. & Yang, D., Information exchange in virtual communities under extreme disaster conditions. *Decision Support Systems*, **50**(2), pp. 529–538, 2011. https://doi.org/10.1016/j.dss.2010.11.011

[12] Westerman, D., Spence, P.R. & Van Der Heide, B., Social media as information source: recency of updates and credibility of information. *Journal of Computer Mediated Communication*, **19**(2), pp. 171–183, 2014. https://doi.org/10.1111/jcc4.12041

[13] Shklovski, I., Palen, L. & Sutton, J. Finding community through information and communication technology in disaster response. In *Proceedings of the ACM 2008 Conference on Computer Supported Cooperative Work - CSCW '08*, p. 127, 2008.

[14] Wentz, B., Lazar, J., Stein, M., Gbenro, O., Holandez, E. & Ramsey, A., Danger, danger! evaluating the accessibility of Web-based emergency alert sign-ups in the northeastern United States. *Goverment Information Quarterly*, **31**(3), pp. 488–497, 2014. https://doi.org/10.1016/j.giq.2014.02.010

[15] Fothergill, A., Maestas, E.G.M. & Darlington, J.D., Race, ethnicity and disasters in the United States: a review of the literature. *Disasters*, **23**(2), pp. 156–173, 1999. https://doi.org/10.1111/1467-7717.00111

[16] Laituri, M. & Kodrich, K., On line disaster response community: people as sensors of high magnitude disasters using internet GIS. *Sensors*, **8**(5), pp. 3037–3055, 2008. https://doi.org/10.3390/s8053037

[17] Kaplan, A.M. & Haenlein, M., Users of the world, unite! The challenges and opportunities of social media. *Business Horizons*, **53**(1), pp. 59–68, 2010. https://doi.org/10.1016/j.bushor.2009.09.003

[18] Tapia, A.H. & Moore, K., Good enough is good enough: overcoming disaster response organizations' slow social media data adoption. *Computer Supported Cooperative Work*, **23**(4), pp. 483–512, 2014. https://doi.org/10.1007/s10606-014-9206-1

[19] Goodchild, M.F. & Glennon, J.A., Crowdsourcing geographic information for disaster response: a research frontier. *International Journal of Digital Earth*, **3**(3), pp. 231–241, 2010. https://doi.org/10.1080/17538941003759255

[20] Lu, Y. & Yang, D., Information exchange in virtual communities under extreme disaster conditions. *Decision Support Systems*, **50**(2), pp. 529–538, 2011. https://doi.org/10.1016/j.dss.2010.11.011

[21] Kappes, M.S., Papathoma-Köhle, M. & Keiler, M., Assessing physical vulnerability for multi-hazards using an indicator-based methodology. *Applied Geography*, **32**(2), pp. 577–590, 2012. https://doi.org/10.1016/j.apgeog.2011.07.002

[22] Kousky, C., Informing climate adaptation: a review of the economic costs of natural disasters. *Energy Economics*, **46**, pp. 576–592, 2014. https://doi.org/10.1016/j.eneco.2013.09.029

[23] Lung, T., Dosio, A., Becker, W., Lavalle, C. & Bouwer, L.M., Assessing the influence of climate model uncertainty on EU-wide climate change impact indicators. *Climatic Change*, **120**(1–2), pp. 211–227, 2013.
https://doi.org/10.1007/s10584-013-0825-1

[24] Hoeber, O., Hoeber, L., El Meseery, M., Odoh, K. & Gopi, R., Visual Twitter Analytics (Vista). *Online Information Review*, **40**(1), pp. 25–41, 2016.
https://doi.org/10.1108/oir-02-2015-0067

[25] Gray, B., Weal, M. & Martin, D., Social media and disasters: a new conceptual framework. *Proceedings of the ISCRAM 2016 Conference*, 28 February 2016.

[26] Houston, J., Hawthorne, M.F., Perreault, E.H., Park, M., Goldstein Hode, M.R., Halliwell, S.E., Turner McGowen, R., Davis, S., Vaid, J.A., McElderry, S.A. & Griffith, S.A., Social media and disasters: a functional framework for social media use in disaster planning, response, and research. *Disasters*, **39**(1), pp. 1–22, 2012.
https://doi.org/10.1111/disa.12092

SUPPORTING THE EU RESPONSE TO ENVIRONMENTAL EMERGENCIES: EUROPEAN MULTIPLE ENVIRONMENTAL THREATS EMERGENCY NETWORK

LISBETH HALL[1], JOLANDA ROELOFS[1], SJORS SCHULPEN[1], ARNOUT DE BRUIN[1], SANDER BANUS[1], RAQUEL DUARTE-DAVIDSON[2], EIRIAN THOMAS[2], EMMA-JANE GOODE[2], OWEN LANDEG[2], ANGIE BONE[2], ELISABETH WIGENSTAM[3], BIRGITTA LILJEDAHL[3], ANNICA WALEIJ[3], LOUISE SIMONSSON[3] & ANN GÖRANSSON NYBERG[3]

[1]National Institute for Public Health and the Environment (RIVM), The Netherlands.
[2]Public Health England (PHE), United Kingdom.
[3]Swedish Defence Research Agency (FOI), Sweden.

ABSTRACT

The European Union (EU) has mechanisms in place to support countries when an environmental emergency strikes and it deploys experts to assist the local community. These experts may find themselves in a chaotic situation, in which local and national authorities are overwhelmed. Collating the necessary evidence from scattered sources to conduct a robust risk assessment is a major challenge and deployed experts may not have access to the necessary technical expertise.

Therefore, the European Multiple Environmental Threats Emergency Network (EMETNET) project is building a simple, efficient and robust methodology to enable the rapid risk assessment (RRA) of environmental emergencies. The RRA, which will be delivered to the European Commission (EC), presents an overview of potential and actual health and environmental impacts of the event. A network of risk assessors is being built to carry out this RRA and provide back-office support to deployed experts, enabling them to have rapid access to information and expertise. EMETNET builds on existing methodologies for the RRA of environmental emergencies, in particular, on a methodology developed for serious cross-border chemical health threats.

The EMETNET methodology will support the selection of experts and assets deployed to affected regions and provide a snapshot assessment to experts en route to disaster zones. It will aid defensible decision-making, communication, planning and risk management. Furthermore, it will ensure a harmonised understanding of the associated health, environmental and cross-sectoral impacts of an environmental emergency.

Keywords: DG ECHO, Environmental emergencies, Expert network, Impact, Natural disasters, Public health, Environment, Rapid risk assessment, Union Civil Protection Mechanism, UCPM.

1 INTRODUCTION

Every year the world is hit by numerous natural and manmade disasters, such as the severe mudslides in Columbia and Peru in spring 2017. Many lives are lost, the environment is badly damaged and infrastructure and livelihoods are severely disrupted. In 2015, there were 198 natural catastrophes, the highest ever recorded in one year, according to a 2016 Swiss Re *sigma* study [1].

EU Member States (MSs) differ in their ability to (rapidly) assess both natural and man-made disaster risks. As recognised by EU Decision 1313/2013/EU on a Union Civil Protection Mechanism (UCPM) [2] and EU Decision 1082/2013/EU on serious cross-border threats to health [3], there is a need to strengthen and improve cooperation and coordination in the fields of civil and health protection, in particular to improve the prevention, preparedness and response to natural or man-made disasters that can impact upon one or more MS.

The EU has mechanisms in place to support countries when an environmental emergency strikes and they deploy experts to the scene to assist the local community. On arrival, these experts often find themselves in the midst of a chaotic situation, in which local and perhaps

DOI: 10.2495/SAFE-V7-N3-324-336

even national authorities are overwhelmed. Collating the necessary information from scattered sources to coordinate a robust risk assessment is a major challenge. Moreover, these experts often do not have access to the technical back up and expertise they receive in their own countries.

The EMETNET project has been initiated to provide a simple, robust, inexpensive and authoritative methodology to support the EU and MSs in undertaking rapid risk and impact assessments of emerging environmental threats. This will enable a harmonised understanding of the associated health, environmental and cross-sectoral impacts of such events to aid risk management and the planning and response process. The project is building a pilot network of expert risk assessors to support the Emergency Response Coordination Centre (ERCC) within EC Directorate-General for European Civil Protection and Humanitarian Aid Operations (DG ECHO) in the rapid scientific assessment of risks to inform responders in the field and to contribute to harmonised risk management. Lessons learnt from past disasters, best practise identified in previous EU-funded projects and from stakeholder engagement with other European and MS authorities will help to ensure that the proposed approach is consistent with the needs of EU and MSs.

EMETNET builds on the methodology developed for serious cross-border chemical health threats in the European Chemical Emergency Network (ECHEMNET) project co-funded by EC Directorate-General for Health and Food Safety (DG SANTE) [4].

2 RAPID RISK ASSESSMENT OF ENVIRONMENTAL EMERGENCIES

2.1 Introduction to rapid risk assessment methodology

A rapid risk assessment (RRA) methodology is being developed to address the risks from emerging environmental threats, either natural or manmade, which have the potential to cause significant damage to public health and the environment. It is based on a RRA methodology originally developed to deal with chemical incidents and poisonings that can have significant impacts on the health and wellbeing of people locally and across borders [4]. The original methodology has been adapted to deal with the broader issues relevant to natural and manmade incidents, including impacts on the natural environment. The RRA methodology provides a quick and robust means of assessing the risks associated with these threats and will be performed on request by DG ECHO/ERCC. Table 1 presents a summary of the natural and manmade hazards considered in the RRA methodology.

The methodology has been tested by using case studies of previous disasters corresponding to the hazard categories in Table 1 to ensure that all relevant information needed to estimate the risks to health and the environment has been captured.

2.2 RRA methodology

The RRA methodology being developed in EMETNET follows other methodologies by containing three core components: hazard assessment, exposure assessment and context assessment [5]. The RRA itself is a comprehensive template containing several sections and is designed to act as a checklist to the risk assessors to ensure all the relevant information is captured.

Table 1: Natural and man-made hazards.

Natural Hazards [6, 7]	
Weather and climate-related hazards (includes atmospheric and hydrological processes)	
Drought	An extended period of unusually low precipitation that produces a shortage of water for people, animals and plants and includes meteorological drought, agricultural drought, and hydrological drought.
Extreme Temp.	A climatological hazard relating to heat waves, cold waves or severe winter conditions.
Flood	When water overflows or soaks land that is normally dry. This includes coastal flooding, riverine flooding, flash floods, and ice jam floods.
Storm	A period of disruptive weather most often accompanied by strong, high force winds. There are several types of storms; severe storm (thunderstorm), tornado, sand/dust storm, and severe winter storms.
Storm (Cyclonic)	Tropical storm; a rapidly rotating, cyclonic storm normally originating over tropical waters.
Wildfire	Any uncontrolled and non-prescribed combustion or burning of plants in a natural setting such as a forest, grassland or brush land which spreads based on environmental conditions.
Geological/ Geophysical hazards	
Earthquakes	A sudden release of built-up energy in the earth's crust usually caused by movement along a geological fault or by volcanic activity resulting in the generation of seismic waves which can sometimes be destructive.
Mass movements	Mass movements can be caused by rivers, heavy rain and/ or seismic activity. This includes; rock fall, landslides, lahar (made up of volcanic ash), flows, avalanches, and subsidence.
Tsunami	A series of waves that are generated by a displacement of massive amounts of water through underwater earthquakes, volcanic eruptions, landslides or extra-terrestrial impacts.
Volcanic Activity	Any type of volcanic event near an opening/vent in the Earth's surface including volcanic eruptions of lava, ash, hot vapour, gas, and pyroclastic material.
Biological hazards	
Harmful Organisms	The spread or infestation of an organism and its toxic substances or vector-borne diseases they may carry which can cause damage or ill health. This includes insect infestations, algal blooms, encounters with dangerous or venomous animals or plants, or the intentional or unintentional introduction of an organism into a new environment.

(*Continued*)

Table 1: (*Continued*)

Natural Hazards [6, 7]	
Infectious disease	Infectious diseases are caused by pathogenic microorganisms, such as bacteria, viruses, parasites or fungi; the diseases can be spread, directly or indirectly, from one person to another.
Extra-terrestrial hazards	
Extra-terrestrial Impact	An extra-terrestrial body such as a meteor, comet or asteroid which either impacts with the Earth's surface or explodes upon entering Earth's atmosphere.
Space Weather	A collective term for a number of physical processes which occur in space that can affect the human activities and infrastructure on Earth including energetic particles, geomagnetic storms, and solar flares.
Technical and Man-made hazards	
Anthropogenic; non-intentional	
Chemical	An unexpected uncontrolled release of a chemical from its containment.
Environmental Pollution/ Contamination	The introduction of toxic chemicals or industrial wastes into the natural environment which may cause adverse effects. This includes air pollution, water pollution and soil pollution.
Radiation	The release of radioactive particles leading to exposure to the human population exceeding established safety levels.
Technological disasters and Major Accidents	Danger originating from technological or industrial accidents, dangerous procedures, fires and explosions, infrastructure failures or certain human activities which may cause the loss of life or injury, property damage, social and economic disruption or environmental degradation.
Anthropogenic; intentional	
Wars, Conflict and Civil Strife	Threatens environment, health, safety or wellbeing of a community or large group of people. This includes: mass shootings, civil disobedience, terrorism, weapons of mass destruction (chemical, biological, nuclear/radiological, explosives)

2.2.1 Hazard Assessment

The Hazard Assessment identifies the particular hazard(s) that have caused the event, e.g. an earthquake, and any subsequent incidents that follow as a consequence of the first hazard. For example, the earthquake has caused a landslide and the landslide has blocked a water course which has caused flooding further upstream. The hazard assessment section will identify and summarise the causal incident (i.e. earthquake), the consequential incidents (i.e. landslides, flooding) and the potential for any further incidents (i.e. major accident: a major dam is experiencing structural issues after seismic activity, failure might occur at some point). The hazard assessment reviews key information about the hazards, e.g. location of each of the identified and potential incidents, and ranks the potential hazards [5]. Where possible, consideration should be given to any similar events that have happened in the region, including the effectiveness of previous risk assessments and any lessons learnt.

2.2.2 Exposure Assessment

The Exposure Assessment evaluates the exposure and impact of the incident on individuals and the population at risk [5]. Consideration needs to be given to the number of people that are likely to be exposed to the incident, the duration of the exposure as well as how the geography of the location may impact on exposure. Within the EMETNET methodology, two sections are included which make up the exposure component of the template: Exposure and Impact Assessment, and Health Assessment. The Exposure and Impact Assessment not only captures information on the consideration above, but also identifies any critical infrastructure pertinent to the region at risk and how this could further impact the affected population and the environment. The Health Assessment evaluates the risk to health and public health of the population affected, as well as identifying any particularly vulnerable groups such as elderly, children, gender-specific or race-specific groups.

2.2.3 Context Assessment

A context assessment is the evaluation of the environment in which the incident has taken place and must consider all the factors that can affect the risks [5]. The RRA template contains a number of sections that make up the context assessment; Environment Assessment, Response Capacity Assessment, and Social, Economic and Cross-sectoral factors. Each section is designed to address a specific aspect of the environment in which the event has occurred. The Health Assessment can also be included under this heading as it also addresses public health issues, such as impacts of the food or water supply, impacts on waste, sanitation and the likelihood of communicable disease outbreaks.

The Environment Assessment evaluates the physical environment of the event such as the impacts on air quality, water quality, land and soil issues, i.e. land use changes, topographical changes and soil pollution, vegetation, biodiversity, protected areas and ecosystem services. Both the natural environment and the managed environment need to be considered, i.e. farms, agricultural land and livestock. The RRA methodology also allows for the use of a number of environmental assessment tools such as the Flash Environmental Assessment tool (FEAT) [8] and the Post Disaster Needs Assessment (PDNA) tool [9], and where applicable these tools can be used to help highlight and inform the risks.

The Response Capacity Assessment along with Social, Economic and Cross-sectoral factors considers the risks associated with other sectors (social, technical, ethical, economic, political, legal etc.) that will have an impact on the population and the environment. The WHO refers to this as the STEEEP analysis [5]. The impact of the event on these issues can affect the level of risk by increasing or decreasing the probability of exposure [5].

The next step in the RRA methodology is that of risk characterisation, which will estimate the likelihood and the consequences of the risk. The EMETNET RRA has adapted the risk matrices from the WHO Risk Assessment manual [5] to include more detailed criteria for estimates of consequences of the event. Within this RRA, the estimates of risk are based on expert risk assessor opinion.

Once the RRA is complete, a summary is produced containing information on the key risks identified in the section assessments above. The RRA summary is the only part of the document that is shared with external parties to ensure that they have access to the essential information requested, rather than detailed information of the event that is used to inform the risk assessment within the different components of the RRA. It is important to bear in mind that as the incident progresses the risk estimates may change and therefore the RRA may need to be updated accordingly. The detailed information, however, will be available through 'back-office' support and the more detailed material, sources and advice contained within the RRA document can be shared if requested by DG ECHO, ERCC or supporting experts deployed through the UCPM.

The RRA template gives an auditable and defensible trail to illustrate what the risks are and why specific steps were taken in response to a disaster. It ensures that resources are not wasted and that appropriate response measures are taken.

The RRA has been through several iterations and has been tested at two workshops: an internal workshop with the project team in Oxford UK and an external workshop in Amersfoort, Netherlands. In the external workshop, key stakeholders and experts were asked to work through the RRA using a disaster scenario and to comment on the suitability of the RRA. A command post exercise, which took place in March 2017, resulted in further suggestions for improvement. The comments and suggestions from the workshops and exercise were collated and have been used to inform further iterations of the RRA.

3 NETWORK OF EXPERT RISK ASSESSORS

3.1 Network of expert risk assessors and areas of expertise

Within the EMETNET project, a network of expert risk assessors is being built to provide rapid, dynamic and authoritative risk assessment of serious acute environmental emergencies using the methodology described in Chapter 2. The experts recruited to the network are contributing to the development of the RRA methodology and network mechanism to ensure that these are fit for purpose. They have also been invited to participate in exercises to test the network methodology and mechanism. Once sufficiently established and exercised, the network will function in 'pilot mode' during the remainder of the project duration to provide risk assessment support should a 'live' event take place.

The areas of expertise that the network needs to cover have been discussed with experts who have experience of being sent on missions (e.g. by DG ECHO/ERCC and Joint UNEP/ OCHA Environment Unit (JEU)), and with stakeholders who deal with the deployment of experts in the field (DG ECHO/ERCC and national civil protection focal points). The experts have provided information on areas of expertise they require to support their work in the field and how back-office support can best be delivered to them. Stakeholders have suggested areas of expertise they feel are required, based on the deployments they have coordinated and the terms of reference for deployed experts. The areas of expertise required in the network were further explored during an interactive session at a project workshop in November 2016 using a scenario based on a combination of hazards listed in Table 1 (storms, earthquakes and floods). The areas of expertise which have been recognised so far as those which the network should ideally cover are shown in Table 2, with those areas of expertise covered by the experts already in the network being highlighted in grey.

For recruitment purposes, a questionnaire has been developed for potential members of the network of experts to complete, including an expertise framework with tick boxes based on the areas of expertise defined during the aforementioned process. The questionnaire includes

Table 2: Areas of expertise which the EMETNET network should ideally cover and those currently available among experts within the network (highlighted in grey).

Technical experts: cross-cutting	Technical experts: scenario-specific	Environmental assessment	Public Health & Health Impact Assessment	Water and food	Risk / crisis mitigation	Aftercare & livelihood
Cartography & mapping (GIS)	Ecologists	Environmental science (water, air, land)	Public health science	WASH	Disaster response (CPM)	Clean-up / Recovery
Critical Infrastructure	Limnologists (fresh water biology)	Fate and behaviour modelling (air pollution, aquatic systems)	Health outcome / Severity / Population Assessment	Drinking and surface water	Public health measures (shelter-in-place & evacuation)	Remediation
Structural engineering (e.g. dam stability)	Wildfires	Environmental sampling	Disaster medicine & care	Food safety	Decontamination (mass, personal)	Food security
Climatologists	Flood specialist	Environmental detection, identification & monitoring (field)	Emergency medical response & care	Water management	Disaster waste management	Water security
Meteorologists	Seismologist	Environmental detection, identification & monitoring (laboratory)	Environmental epidemiology			Biomonitoring
Animal health experts	Astronomy (space weather)	Meteorological Modelling	General toxicology			Psychosocial care
Hydrologists	Oceanographers	Eco / Environmental / Aquatic Toxicology	Clinical / Medical toxicology			Sociologists (gender experts, vulnerable groups, livelihood)
	Volcanologist		Signs and symptoms (toxidromes)			Fisheries / Pisciculture
	Oil spill response		Epidemiology			Agriculture
	Marine biologists		Points of entry (IHR)			

inter alia questions to elucidate the level and range of experience of the risk assessors, especially in the acute phase of an incident/disaster. This questionnaire has been sent to experts identified by project partners and their networks as having the expertise necessary for the network. Identifying and inviting experts with areas of expertise needed within the network is an ongoing process throughout the project.

3.2 Current proposed structure of the EMETNET network

Discussions are ongoing with DG ECHO/ERCC to ascertain the structure and mechanism whereby the RRA and back-office support can best be delivered and to ensure the sustainability of the network and the RRA methodology beyond the life of the project. The currently envisaged lines of communication of the network are shown in Figure 1, including the RRA delivery chain.

4 LESSONS LEARNT FROM PAST DISASTERS AND EXERCISES

4.1 Introduction

An analysis of lessons learnt from past disasters is being conducted to test and improve the design of the RRA. This includes the identification and evaluation of knowledge gaps and issues that have lacked attention in the acute phase of environmental emergencies. Due consideration of these knowledge gaps and issues would help improve response and crisis management, and would increase the possibilities of a sustainable recovery. To complement

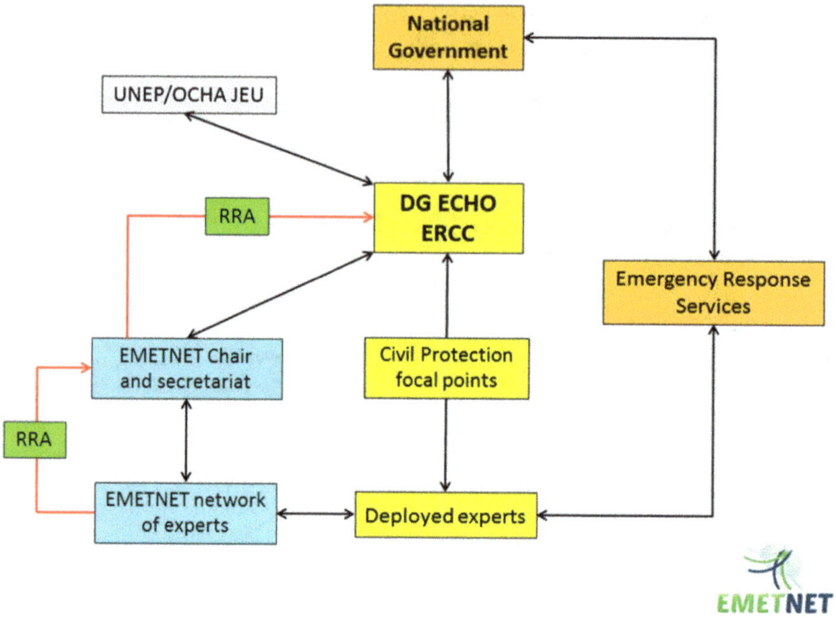

Figure 1: Proposed structure of EMETNET network response and lines of communication (black arrows = proposed lines of communication; red arrows = RRA delivery chain).

this work, exercises are being held in which experts actively test and evaluate the RRA using a scenario that has been formulated based on the work on identifying emerging environmental threats and lessons learnt from past disaster events.

4.2 The European Union Civil Protection Mechanism and environmental emergencies

The Union Civil Protection Mechanism (UCPM) aims to strengthen the cooperation between the EU and the MSs and to facilitate coordination in the field of civil protection in order to improve the effectiveness of systems for preventing, preparing for and responding to natural and man-made disasters [2]. The protection under the UCPM primarily covers people, but also the environment and property against all kinds of natural and man-made disasters, including environmental disasters, marine pollution and acute health emergencies, occurring inside or outside the EU [2]. The UCPM has been activated for a number of such events.

From January 2007 to March 2016, 49% of UCPM activations took place in Europe. Table 3 lists the five most common reasons for activation of the UCPM worldwide and in Europe. Forest fires caused the most activations; both within and outside Europe, with floods and severe weather also frequently triggering the UCPM worldwide. Earthquakes, on the other hand, are a fairly common reason for activation outside Europe but somewhat less common in Europe, whereas activations due to marine pollutions and explosions (e.g. ammunition storage sites) were more common in Europe.

Interesting to note is that activations due to a need to respond to civil unrest related to the refugee crises have increased in the past few years. In 2016 alone, civil unrest, conflict and displaced persons accounted for 17% of the UCPM activations, only exceeded by forest fires (25%) and followed by earthquakes and storms (14% each).

Table 3: Some of the most common reasons for activation of the UCPM from January 2007 to March 2016. [10]

Worldwide (total number: 287)	Europe (total number: 142)
Forest fires (60)	Forest fires (35)
Severe weather* (53)	Floods (27)
Floods (50)	Severe weather* (13)
Conflict/refugees (31)	Marine pollution and explosions (9)
Earthquakes (23)	Refugee crises (9)

* i.e. storms, cyclones, typhoons, etc.

4.3 Selection of cases for lessons learnt

As the purpose of the EMETNET project is to support DG ECHO and ERCC, events that have triggered the activation of the UCPM have been selected from the UCPM database. By examining these events, the UCPM response itself can also be evaluated as well as how and if the EMETNET RRA could support the mechanism.

Selection of event types to study further includes forest fires, floods, earthquakes, extreme weather, conflict/refugee situations and NaTech accidents (Natural Hazards Triggering Technological Accidents). The criteria for selection of cases to study in depth are based on the following factors: type of event, temporal and spatial coverage, effect on human health, the environment and social, economic and political impact. The cases should represent major causes of disasters in Europe and globally and ideally also be emerging, that is, show a future trend of an increase in these types of disasters. Disaster events are, however, often case specific and context-bound, which is why the decision on the number and types of cases to include needs to be made cautiously. Also challenging for the analysis is that extreme events are rare, which means there are few data available to make assessments regarding changes in their frequency or intensity. The more rare the event, the more difficult it is to identify long-term changes. Global-scale trends in a specific extreme event may be either more reliable (e.g. for temperature extremes) or less reliable (e.g. for storms) than some regional-scale trends, depending on the geographical uniformity of the trends in the specific extreme event.

4.4 Initial observations

Although the evaluations of lessons learnt is ongoing, some generic observations can be identified which are described in more detail below.

4.4.1 Lessons on response triggers

It is not necessarily the most common events or the events that result in most human suffering or casualties that generate response activities. For instance, of the response activities performed by JEU in 2004–2014, 38% addressed manmade and/or technological disasters. In addition, preparedness and capacity development for such events accounted for 22%.

There are no records of droughts in the UCPM database, even though the EM-DAT database run by the Belgian-based Centre for Research on the Epidemiology of Disasters (CRED) recorded 136 drought events in Africa alone between 1995 and 2015 [6]. Droughts generally affect a very large number of people and the CRED recommendation is that there needs to be improved data collection on indirect deaths from droughts [11].

4.4.2 Lessons on key determinants of disaster risk and impacts

Disasters are not neutral. Recovery interventions are not neutral either since they can increase, reinforce or reduce existing inequalities. Recovery interventions may hence have a negative impact on gender roles, poverty, the environment and the livelihoods of vulnerable groups [12]. One example is the 2004 tsunami, when recovery aid was concentrated on the fishing sector but not so much on the informal sector, agriculture or business. Within the fishing sector, it was the fishermen who got assistance, rather than women who traded the fish [13].

An understanding of exposure and vulnerability as key determinants of disaster risk and of impacts is critical. For instance, a tropical cyclone can have very different impacts depending on where and when it makes landfall. Similarly, a heat wave can have very different impacts on different groups of populations depending on their vulnerability. Severe impacts can also result from non-extreme events where exposure and vulnerability are high or from a compounding of events or their impacts.

4.4.3 Lessons on secondary impacts

Environmental degradation, deforestation and soil erosion are common vulnerability and sensitivity factors which worsen the impacts of natural disasters. Insufficient attention to environmental considerations can trigger significant degradation and secondary impacts for populations and decreasing socio-ecological resilience. Despite attempts from several organisations to highlight the environmental challenges for upcoming crisis response efforts, very few mechanisms to mainstream this knowledge into the execution of the responses have been implemented [14].

The risks of infectious disease outbreaks following natural disasters are often raised but are generally overstated and lead to misallocation of resources. The overwhelming number of deaths immediately after a natural disaster are directly associated with blunt trauma, crush-related injuries and burn injuries. No certain risk of an infectious disease epidemic occurring in the short-term period after a disaster has been well documented [15]. Flooding is the most common natural disaster described with an increase in cases or outbreaks of infectious diseases, such as diarrhoea, malaria and leptospirosis, whereas a 2006 review showed that only three out of 600 geophysical disasters led to disease epidemics [16]. In fact, the real risk posed by e.g. handling dead bodies is physiological stress caused by grief and shock [17]. Rather, the risk factors for disease outbreaks after disasters are associated primarily with population displacement. The availability of potable water and sanitation, the degree of crowding, the underlying health status of the population, and the availability of healthcare services all interact within the context of the local disease ecology to influence the risk for communicable diseases and death in the affected population [18].

4.4.4 Lessons on the local and cultural context

The inability of international response teams to adapt to the cultural and social contexts in the affected countries can have grave consequences. For instance, the response to the Ebola crisis in West Africa created a significant resistance among the population towards the response since it was too technical and medical in nature and lacked cultural, psychological and sociological understanding [19]. In fact, the majority of life-saving work in any disaster is done by the local populations themselves. Agencies must give good information to communities so they can plan their own recovery from the start.

Another lesson is that recovery is the overriding challenge whereas the need for relief has often been overstated. Recovery also includes social, political, economic and environmental dimensions – not just physical security – and there is a limit to what humanitarian assistance can do. Sustainable livelihoods are key to recovery and it is paramount to listen to affected people about their priorities for livelihood recovery which is often closely linked to the environment and ecosystem services.

4.5 Reflections on 'lessons learnt'

Lessons learnt is an expression that implies that one actually has *learnt* something as well as implemented new knowledge in crisis management procedures. However, the need to produce results in the field and report back that a mission has been successful to ensure future funding may emerge as an unintended obstacle. This impacts both the possibility for efficient coordination and the problem of a lack of 'true' reporting on failures [20]. One such example is the evaluation of the humanitarian response to earthquakes, which has concluded that emergency responses have not effectively applied lessons from previous emergencies [21]. Another example is provided by water and sanitation (WASH) interventions which frequently fail, although the results are not necessarily reported as such [22]. Examples where toilets are left unused because the affected communities did not realise they were built for them are all too common.

4.6 Exercises

Within the EMETNET project, two exercises are being held. As described previously, the main objectives of these exercises are to apply lessons learnt from previous disasters, test the RRA methodology and ensure it is fit for purpose, and test the activation and function of the expert network. Both exercises are run in real-time and are of command-post character, i.e. the players work from their normal workplaces with access to all communication tools via internet and/or telephone lines.

The first exercise was run on 21st March 2017. Fourteen participants from the Netherlands, UK, Sweden and Slovenia joined and were assigned to perform a RRA for a major heat wave in Europe with resulting cascading events. The expert risk assessors managed to deliver answers to critical questions using the RRA template, despite the lack of expertise in some technical areas and reported back to exercise control with a thorough report of the incident(s). A quick wash-up immediately after the exercise showed that participants generally felt it was a successful exercise. They found it challenging with regards to the assignment and the time limit and some found it hard to manage the documents on the collaboration platform used. Nevertheless, they felt the collaboration within the network went well and they were satisfied with the assessments they produced.

5 CONCLUSIONS

Within the EMETNET project, a network of experts is being built to carry out a rapid risk assessment (RRA) and provide back-office support to experts deployed to an environmental emergency. So far, the network covers a number of the defined areas of expertise. Recruitment of network members is ongoing to cover more areas of expertise and to ensure resilience of the network. The RRA methodology being developed has been tested and exercised with promising results. During the first exercise, the network successfully collaborated remotely to carry out a number or RRAs in response to a complex scenario. During the coming year, the network will be further tested and exercised and a structure for a sustainable network response beyond the life of the project will be explored. From the analysis of lessons learnt, it is clear that accountability to the affected people and honest assessment of true lessons learnt is of real importance. Translating these lessons into doable actions is a collective responsibility for all emergency response actors and can help ensure the success of emergency response.

ACKNOWLEDGEMENTS

The project partners, RIVM, PHE and FOI, are grateful to the EU Civil Protection Financial Instrument for the co-funding of this project (Grant Agreement No. ECHO/SUB/2015/718592/ PREV14).

REFERENCES

[1] Bevere, L., Sharan, R. & Vipin, K., Natural catastrophes and man-made disasters in 2015: Asia suffers substantial losses. Swiss Re *sigma* no. 1/2016

[2] European Parliament and Council. Decision No. 1313/2013/EU of the European Parliament and of the Council of 17 December 2013 on a Union Civil Protection Mechanism. *Official Journal of the European Union.* L 347 20.12.2013; pp. 924–927. Available at: http://eur-lex.europa.eu/LexUriServ/LexUriServ.do?uri=OJ:L:2013:347:0924:0947:EN:PDF. (accessed 3 April, 2017).

[3] European Parliament and Council. Decision No. 1082/2013/EU of the European Parliament and of the Council of 22 October 2013 on serious cross-border threats to health and repealing Decision No 2119/98/EC. *Official Journal of the European Union.* L 293 05.11.2013; 1–15. Available at: http://eur-lex.europa.eu/ LexUriServ/LexUriServ.do?uri=OJ:L:2013:293:0001:0015:EN:PDF. (accessed 3 April, 2017).

[4] Orford, R., Hague, C., Schapers, A., Hall, L., Roelofs, J., Morgenstern, P., Göranssen Nyberg, A., Wigenstam, E., Trnka, J., García Cazalilla, M., Ocaña García-Donas, J. & Duarte-Davidson, R., European Chemical Emergency Network (ECHEMNET): dealing with serious cross border chemical health threats. *In preparation.*

[5] World Health Organization. *Rapid Risk Assessment of Acute Public Health Events,* 2012. Available at http://www.who.int/csr/resources/publications/HSE_GAR_ARO_2012_1/en/. (accessed 15 March, 2017).

[6] Centre for Research on the Epidemiology of Disasters. *EM-DAT: The International Disaster Database*. Available at http://www.emdat.be. (accessed 15 March, 2017).

[7] International Federation of Red Cross and Red Crescent Societies. *Types of disasters: Definition of hazard.* Available at http://www.ifrc.org/en/what-we-do/disaster-management/about-disasters/definition-of-hazard/. (accessed 15 March, 2017).

[8] Joint UNEP/OCHA Environment Unit. *The Flash Environmental Assessment Tool (FEAT) v2.0. 2016*. Available at https://docs.unocha.org/sites/dms/ Documents/FEAT_pocket_final.pdf. (accessed 15 March, 2017).

[9] International Recovery Platform. *Post Disaster Needs Assessment*, 2008. Available at http://www.recoveryplatform.org/pdna/. (accessed 15 March, 2017).

[10] European Community Civil Protection Mechanism Activations overview 01.01.2007 – 01.05.2016, on file with authors.

[11] Centre for Research on the Epidemiology of Disasters (CRED) and United Nations Office for Disaster Risk Reduction (UNISDR). *The human cost of weather-related disasters 1995–2015, 2015.*

[12] Neumayer, E. & Plümper, T., The gendered nature of natural disasters: the impact of catastrophic events on the gender gap in life expectancy, 1981–2002. *Annals of the Association of American Geographers*, **97**(3), pp. 551–566, 2007. https://doi.org/10.1111/j.1467-8306.2007.00563.x

[13] ActionAid. *Tsunami response: a human rights assessment.* London: ActionAid International, 2006.

[14] Abrahams, D., The barriers to environmental sustainability in post-disaster settings: a case study of transitional shelter implementation in Haiti. *Disasters,* **38**(s1), pp. S25–S49, 2014.
https://doi.org/10.1111/disa.12054

[15] Kouadio, I., Aljunid, S., Kamigaki, T., Hammad, K. & Oshitani, H., Infectious diseases following natural disasters: Prevention and control measures. *Expert Review of Anti-Infective Therapy*, **10**(1), pp. 95–104, 2012.
https://doi.org/10.1586/eri.11.155

[16] Floret, N., Viel, J-F., Mauny, F., Hoen, B. & Piarroux, R., Negligible risk for epidemics after geophysical disasters. *Emerging Infectious Diseases*, **12**(4), pp. 543–548, 2006. https://doi.org/10.3201/eid1204.051569

[17] de Ville de Goyet, C., Epidemics caused by dead bodies: a disaster myth that does not want to die. *Revista Panamericana de Salud Pública*, **15**(5), pp. 297–299, 2004. https://doi.org/10.1590/s1020-49892004000500002

[18] Watson, J.T., Gayer, M. & Connolly, M.A., Epidemics after natural disasters. *Emerging Infectious Diseases*, **13**(1), pp. 1–5, 2007.
https://doi.org/10.3201/eid1301.060779

[19] Leach, M., Time to put Ebola in context. *Bulletin of the World Health Organization,* **88**(7), 481–560, 2010.

[20] Waleij, A. & Liljedahl, B., *The gap between the buzz words and excellent performance; lessons drawn from the environmental footprint from military and civilian actors in crises and conflict settings*. Prepared for the Environmental Dimensions of Sustainable Recovery: Learning from Post-Conflict and Disaster Response Experience Project. FOI-R-4246-SE. Stockholm, Sweden, 2016.

[21] Cosgrave, J., Responding to earthquakes 2008: Learning from earthquake relief and recovery operations. *ALNAP Lessons Papers*, 2008. Available at http://www.alnap.org/resource/5239. (accessed 3 April, 2017).

[22] James, J., Analysis: Sanitizing the truth - when WASH fails. *IRIN News*, 17 September 2013.

HOMELAND SECURITY AND EMERGENCY MANAGEMENT IN INSTITUTIONS OF HIGHER EDUCATION (IHE): TEXAS CASE STUDY

MAGDALENA ANNA DENHAM & ASHISH KUMAR KHEMKA
Department of Security Studies, Sam Houston State University, U.S.A.

ABSTRACT

This exploratory case study adopted the classical content analysis (CCA), cross case mixed strategy and Qualitative Data Miner (QDA) correspondence analysis of websites across 41 accredited public Institutions of Higher Education (IHE) in the state of Texas. The conceptual framework guiding the study was based on adaptive resiliency and Disaster Resilient University (DRU) paradigms. The goal of this inquiry was to determine (a) the most common organizational arrangements adopted across the IHEs in Texas to integrate emergency management functions on their campuses?; (b) the scope of most common activities and engagement performed by IHEs in Texas with respect to supporting the DRU mission across the cycle of emergency management?; (c) competencies and job definitions of officials specifically involved in supporting the all-hazards approach to emergency management on IHE campuses in Texas?; (d) training, exercise and certification standards as well as emergency notification systems commonly adopted among the IHEs in Texas; and (e) the type of educational, awareness, and outreach programs which can be identified across Texas IHEs in support of the DRU mission? Common and divergent themes, as well as implications for IHE leadership practice will be discussed.
Keywords: emergency management, homeland security, higher education, disaster resiliency, incident management cycle, campus safety and security, preparedness, response, recovery, mitigation.

1 INTRODUCTION

'Emergency management was an afterthought at many IHEs prior to 2001, and if it was a recognized need, it was an ancillary responsibility of a police officer, administrator, or health and safety professional. There simply was no officially designated position that dealt with campus emergencies, and the level of support for such position was extremely low' [1]. American Institutions of Higher Education (IHEs) face a variety of potential threats and hazards, be it human-caused [2–6], technological, or natural within their campus settings [3]. From the Homeland Security and Emergency Management (HSEM) perspective, IHEs have been defined as critical pieces of American infrastructure under the Education Facilities Subsector of the Government Facilities Sector [7, 8]. Brenner [9] explained the national importance of IHEs, underscoring their role in producing cultural, intellectual, and social resources, housing physical buildings on campuses, material resources used or housed in campus facilities, human resources, intellectual resources such as research, and infrastructure like important utilities.

Aligning with one of the core missions of national Homeland Security strategy [10] – strengthening national preparedness and resilience – a multi-threat and all-hazard approach has been proposed when creating plans along the emergency management cycle comprised of preparedness, response, recovery and mitigation [12]. Inclusion of sound practices in HSEM along the emergency management cycle among IHEs would contribute to what Kapucu and Khosa [13] defined as a Disaster Resilient University (DRU); a DRU represents an institution capable of adapting to the aftermath of a disaster and learning through its experience to become better prepared for such instances in the future [13, 14].

DOI: 10.2495/SAFE-V7-N3-337-351

Inarguably, events such as the 2007 Virginia Tech shooting have become a growing concern for campus officials [15]. Concurrently, natural disasters such as 2005 Hurricane Katrina have highlighted the complexity of navigating relationships with emergency response organizations such as FEMA, the National Guard, the American Red Cross (ARC) and other relief personnel by the IHEs. For example, during Hurricane Katrina the Louisiana State University (LSU) assembly center was transformed into an acute care facility, triaging 15,000 evacuees and treating 6,000 patients while the institution worked to effectively communicate essential messages to the public through previously set-up methods and avenues of communication [16]. In 2003, then FBI Director Robert S. Mueller III in a report to Senate Select Committee on Intelligence of the United States Senate identified IHEs' vulnerabilities to a variety of threats and difficulties related to maintaining the balance between allowing for the free exchange of ideas and safeguarding the integrity of the campus and its students [9]. Meanwhile, modifications to the Jean Clery Disclosure of Campus Security Policy and Campus Crime Statistics Act of 1990 (20 U.S.C. §1092 (f) [hereafter referred to as 'the Act']) have expanded responsibilities of campus officials to: distribute immediate emergency notification warnings to the campus community upon confirmation of a dangerous situation on campus involving an immediate threat to the health or safety of students or staff regardless of whether a crime is involved. Examples of emergencies that would warrant such a notification include: an active shooter on campus, a riot, a bomb threat, a tornado, a fire and similar situations involving active and palpable threats. Every institution must conduct yearly tests of emergency response and follow-through activities designed for assessment and evaluation of emergency plans and capabilities [17]. Moreover, in states such as Texas noting the increasing phenomenon of legislative support for the right-to-carry on campus [18], IHEs' administrative leaders have been challenged with defining policies, implementation procedures, and evaluation criteria to ensure compliance with new state laws while providing for safety and security of the entire campus community. Admittedly, the increasing role of HSEM functions among IHEs requires increased commitment to dedicate resources, personnel, and organizational structures in their support. Of increasing importance is thus defining what are the best organizational arrangements, collaborative practices, or supportive mechanisms available to IHEs in their all-hazards mission towards achieving the DRU mission. Some arguments have been raised that IHEs are not as properly secure as they should be highlighting campus officials tasked with such planning and securing lack the training and experience necessary to prevent, plan for, or respond to incidents [19, 20]. Furthermore, collaboration challenges between IHEs' administration and law enforcement have been reported [20, 21]. Kerr [22] who examined public, private not-for-profit and proprietary IHEs in the Midwest for Clery Act compliance reported 'a majority of the institutions did not comply with the requirements of the law'.

Furthermore, Nicoll and Owens [23] noted emergency planning often meets with conflict or resistance as some departments or administrators within the IHE may not want to spend limited resources, such as time and money, on developing response measures that may never be used. Support from those in leadership positions fosters implementation and funding for preparedness and resiliency-building measures [24]. In turn, evidence of administrative support may be reflected in the level of institutionalization of HSEM functions at the IHE. Nonetheless, beyond standardized resources disseminated to guide IHEs through planning and preparedness activities [25, 26], there is lacuna of systematic research conducted among the IHE's regarding common structures and arrangements adopted by IHEs themselves to institutionalize the emergency management functions to fit into the overall concept of IHEs' operations. To administrative leadership in IHE this may be especially troubling given legal, psychological, and societal impacts that characterize high consequence events. Moreover, in spite of increasing legislative mandates both at federal and state levels such as the Clery Act, National Incident Management System (NIMS) [27], or campus carry provisions, absence of frameworks for integration of HSEM functions by administrative leaders may severely impede the capability of those institutions to implement those mandates successfully.

To that end, our exploratory case study aims to examine, using IHEs' data from the state of Texas, the degree and scope of the integration of the HSEM functions (if evident) into IHEs' operations. Specifically, the goal of our study is to elicit answers to the following questions: (a) What if any are the most common organizational arrangements adopted across the IHEs in Texas to integrate emergency management functions on their campuses?; (b) What is the scope of most common activities and engagement performed by IHEs in Texas with respect to supporting the DRU mission across the cycle of emergency management?; (c) What are the competencies and job definitions of officials specifically involved in supporting the all-hazards approach to emergency management on IHE campuses in Texas?; (d) What are the training, exercise, and certification standards as well as emergency notification systems commonly adopted among the IHEs in Texas; and (e) What type of educational, awareness, and outreach programs can be identified across Texas IHEs in support of the DRU mission?

2 CONCEPTUAL FRAMEWORK

Integration of HSEM functions within IHEs will be assessed though the adaptive resilience model [28] represented by Figure 1. Specifically, institutional adaptive capacities related to mitigation and preparedness in support of response and recovery and reflected in organizational arrangements, leveraging of community capital, utilization of social capital, and application of communicative and collaborative processes will be sought. Because of 'diversified and decentralized structure' [29] of the IHE campuses into numerous colleges or schools, with each containing various colleges or departments, challenges could arise stemming from the differences of how each operates, communicates, and is organized. Therefore, assessing organizational arrangements adapted for emergency management across a range of IHEs may illuminate specific solutions to best address unique challenges inherent in such decentralization. A DRU is an IHE that has built 'adaptive capacities through social capital development, community competence, and strong communications and information systems' [13]. By emphasizing resilience, IHEs are focusing more on what they can do for themselves and how they can strengthen their own capabilities [30]. Incidentally, exploring whether emergency management practices across the cycle of emergency management have become institutionalized within the IHEs themselves, is an important step in gauging reliance on internal capabilities.

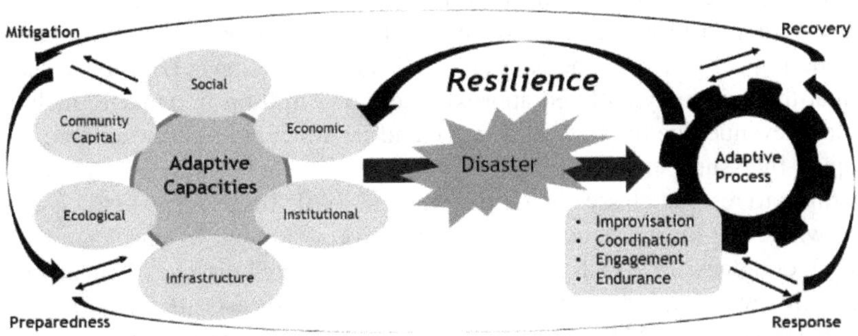

Figure 1: Conceptual model of local disaster resilience. *(Source: Ross, 2013. Copyright)*

In addition, universities become resilient by openness and transparency with their students, staff, faulty, and visitors about emergency management plans [13]. One example of this is how universities are publicizing disaster preparedness and emergency management information on their campus websites, making it possible for more people to become knowledgeable about response and resilience steps of the IHE [9]. Another aspect of resilience relates to training and certification of certain faculty and staff on campuses [13]. Griffin [29] suggested reaching beyond students and integrating faculty and staff should be included in the safety and security education on campuses. Through this education, faculty and staff will be better prepared to assist the IHE in its safety and security efforts, resulting in more people on campus equipped with the knowledge to assist in the preparation for and response to a threat or hazard, as well as more people to spread the knowledge to even more people on campus [29]. By training students, faculty, and staff through such means as presentations and corresponding documents, IHEs become better capable to prepare for and respond to a threat or hazard through the help of the individuals on their campuses. Awareness of what roles they will need to fulfil to ensure the continuity of university operations after a disaster has occurred increases the preparedness and resilience of the IHEs and allows for fostering of partnerships with larger communities [31]. When IHEs cooperate, coordinate and communicate externally, they develop social bonds with groups who will assist them in crisis. Familiarity with the first responders, government entities, and non-profit organizations further enhances preparedness and resilience emergencies [23]. Among the partnerships, coordination with first responders and emergency personnel [29] especially helps to encourage communication, assistance and the sharing of information between the groups involved.

3 METHODOLOGY

Because the goal of our research is exploratory, a qualitative research paradigm will be applied [32]. Specifically, due to the pragmatic lens of the researchers, the case study design has been selected as optimal. The unit of analysis (i.e. IHE) reflects criteria suggested by Yin [33] because it represents an organization. Organizations are bounded both geographically and administratively and as such lend themselves to case study design inquiries. The embedded unit of analysis [33] is represented by IHEs' websites.

3.1 Sampling Scheme and Data Source

Texas has endured numerous disasters and catastrophic incidents over the past century, dating back to the hurricane that made downfall in Galveston in 1900, which claimed over 6,000 lives [34] to more recent events such as Hurricane Rita in 2005 and Hurricane Ike in 2008 which claimed 120 and 135 lives respectively, and resulted in mass scale evacuations [35]. In a study on the impact of Hurricane Rita on a IHE in Texas, Lamar University, the findings indicated that the university suffered approximately $38 million in damage, including $4.4 million in lost revenue due to physical damage and insufficient response and recovery planning prompted by issues such as communications failures, information technology failures and the inability to resume classes in a timely manner [36]. Texas is prone to diverse natural hazards, of which the most frequent are a) inline river flooding; b) hurricane and tropical storms and c) tornadoes. Texas has had more federally declared disasters than any other state [34], in fact, in several instances, it has received multiple disaster declarations in a single year. In 2016, it received FEMA disaster declaration for severe storms and flooding in March, April and June [37].

In addition, the state has also been subject to numerous technological and manmade disasters, several of them impacting IHEs directly. One of the watershed manmade events at an IHE in Texas was the 1966 mass shooting at the University of Texas at Austin (UTA) campus, in which a single individual killed 14 victims. The ineffective response by the police prompted the formation of Special Weapons and Tactics (SWAT) teams around the country [38]. Another notable incident caused by a combination of human error and technology was the 1999 bonfire collapse at Texas A&M University, which killed 11 individuals and sent 28 to hospitals [39]. While these have been some of the focusing IHE disasters in Texas, there continue to be persistent incidents of similar, yet less severe magnitude periodically, such as an explosion in 2010 at a Chemistry laboratory at Texas Tech, the investigation of which highlighted systemic deficiencies in university safety management practices [40].

The IHE's selected for this study were obtained from a dataset of 545 accredited IHEs in Texas which is provided by the US Department of Education (US DOE, 2016). The selection process is outlined in Figure 2. Of the 545 IHEs, a further criterion of the IHE being accredited by the 'Southern Association of Colleges and Schools Commission on Colleges' (SACSCOC) was applied, 170 IHE's fulfilled this criterion. This dataset was further filtered, using the criteria of whether the IHE belonged to one of the 7 state systems in Texas, resulting in a dataset of 41 IHEs. The entire dataset selection was limited to publicly controlled universities because they are subject to various federal and state laws, such as the Homeland Security Presidential Policy Directive 5 (HSPD-5) [27], which established the National Incident Management System (NIMS), and Executive Order 40 by Governor Perry in Texas, which establishes NIMS as the incident management framework for the state [27]. An example of its adoption by IHEs which meet the criteria is IHE's in the University of Texas System, which are guided by policy stating that their emergency management program will incorporate NIMS and the Incident Command System (ICS) as part of their emergency preparedness program [41].

SACSCOC was selected as the accreditor because since its foundation in 1895, it has established itself as the primary regional accreditor for 11 southern states in the nation, including Texas. A regional accreditor, in comparison to a national accreditor, conducts a comprehensive assessment of IHEs in a designated geographic area [42]. The assessment includes all instructional sites, including online, and comprises of multiple standards. Of

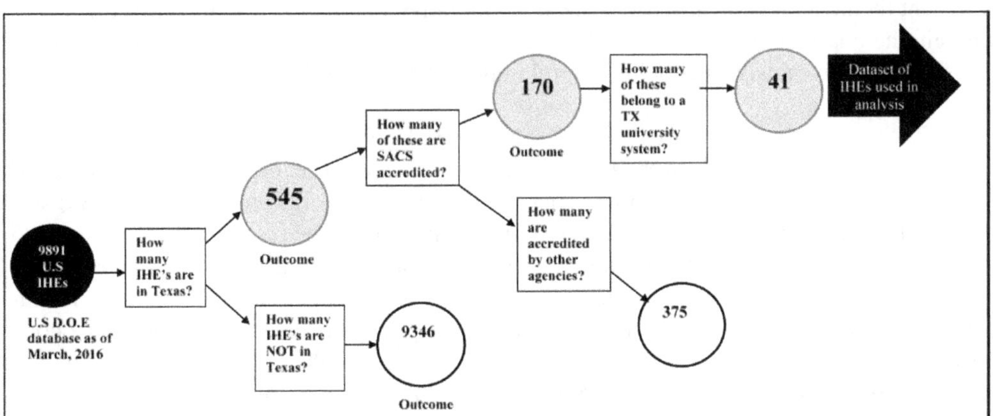

Figure 2: IHE case selection process.

significant relevance to study are SACSCOC standards 3.11.1, which requires the institute to maintain 'control over its physical resources' [43] and standard 3.11.2, which requires the institution to take steps to provide a 'healthy, safe and secure environment for all members of the campus community' [42]. In addition to applying rigorous standards, regional accrediting bodies such as SACSCOC maintain comparative standards across each other and serve as the gatekeepers for Title IV funds, which is a major source of federal aid [44]. In comparison, national accreditors typically cater to private IHEs or those whose missions focus either on career education or religious education [42].

4 DATA ANALYSIS

In order to conduct text and image analysis of IHE websites to search for content related to HSEM functions comprised therein, classical content analysis (CCA), [45, 46] was adopted. CCA 'allows the analyst systematically to examine small or large amounts of textual information and systematically identify the most salient features contained in its communication content by determining, comparing, and contrasting the frequencies of each feature' [47]. Furthermore, considering the [48] call for using multiple qualitative data analysis techniques whenever appropriate and possible, CCA was supplemented with a strategy developed by Miles *et al.* [49] to organize, arrange, and visually represent the data. Miles *et al.* [49] expressed that cross-case mixed strategy involved discerning a set list of elements from literature, applying the elements to chosen cases (i.e. IHEs), evaluate the cases using for each a matrix created from said elements, and then identifying the salient elements by comparing across the matrices.

Finally, after the themes had been extracted via CCA, and cross-case mixed strategy, they were further subjected to correspondence analysis, a graphical technique that allows researchers to conduct a mapping cross-case analysis of emergent themes. Specifically, a correspondence analysis is an exploratory multivariate technique that involves factoring categorical (i.e. nominal level) variables and graphing them (i.e. mapping them) in a property space that displays their associations in multiple (i.e. two or more) dimensions [50]. The QDA Miner 5.0 software program [51] was used to conduct the correspondence analysis. This analysis represented what [52] referred to as a crossover mixed analysis, whereby the analysis types associated with one tradition (i.e. quantitative analysis: correspondence analysis) were used to analyse data associated with a different tradition (i.e. qualitative data: emergent themes). Specifically, we utilized a qualitative-dominant crossover mixed analysis wherein the qualitative analysis was dominant, while, simultaneously, we deemed the addition of quantitative analysis as being helpful in providing richer data and interpretations [52].

5 RESULTS AND DISCUSSION

Overall, the CCA of 41 IHEs' websites yielded 10 supra-codes (i.e. organization, job role and title, competencies, scope of activities, crisis typology, frameworks and DRU features, exercises, emergency notifications, education and outreach). The codebook contained 85 sub-codes that contributed to subsequent cross-case correspondence analysis. Website data were organized in each case by document variables (e.g. organizational charts, safety manuals, organizational notes, checklists). Organizationally, as illustrated by frequency of code distribution in Figure 3, the HSEM function may be nested at the University Police Department (or Office of Public Safety), Environmental Health and Safety, may be shared by various divisions or departments, or may reside under a separate Office of Emergency Management (OEM). The latter organization was evidenced in 25.6% of cases. Our findings partially

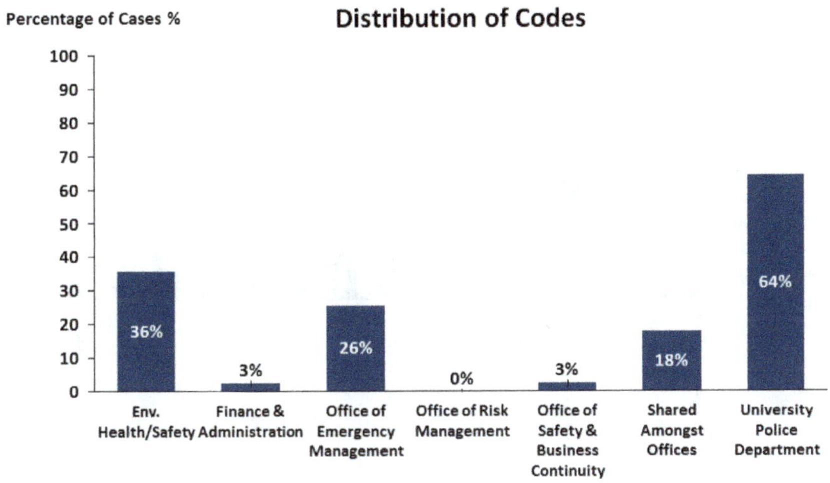

Figure 3: Emergency management function placement. *(Source: Author data using QDA Miner 5.0)*

confirm previous reports [1] that emergency management departments at IHEs in the United States are a division of police or public safety (37.9%) and 24.1% are stand-alone departments. Emerging themes in this study further suggest four general models for integrating HSEM have been adopted by IHEs in our sample: (a) collateral function of unit/division with the most relevant safety and security expertise and engagement on campus; (b) collateral shared function of units/ divisions with the most relevant safety and security expertise and engagement on campus; (c) collateral function of collegiate administrative and academic teams (i.e. executive groups, incident management teams and behavioural response teams) with support of units/divisions with the most relevant safety and security expertise and engagement on campus; and (d) dedicated unit/division established for the purpose to oversee and coordinate HSEM at the University.

Analysis of codes related to job role and title illuminated that HSEM functions have been assumed at various levels of organizations by officers (e.g. Safety Officer, Public Affairs Officer), mid-level managers (e.g. Emergency Manager), teams, and mid to executive level Emergency Management Coordinators (EMCs) and Directors. Specifically, administrative and academic collegiate teams in charge of HSEM were evidenced in 36% of cases and EMC/Director roles in 25.6% of cases. Notably, the only discernible competencies were extracted in relation to the EMC/Director roles theme whereby those positions were characterized by a combination of HSEM academic preparation or HSEM experience preparation or mixture of both. For example, individuals in selected EMC/Director positions possessed degrees in Emergency Management and Disaster Planning, Earth Sciences, Geology, Applied Geography, Public Administration, Political Science or experience in local, state, or federal service through continuity planning, exercise evaluation, incident command experience.

The identification of the most common themes depicting activities related to HSEM (excluding crime/violence and fire prevention functions) is portrayed by Figure 4. Visibly, individuals, teams, and organizational units engaged in HSEM at selected IHEs focused the most on Emergency Response Planning (49%), Preparedness (46%), Evacuations (41%), Mitigation Planning (41%), Recovery Planning (38%), integrating Policies and Laws that

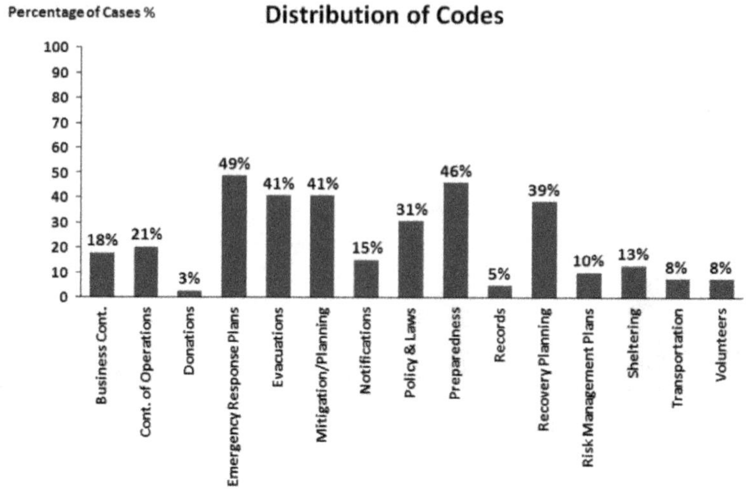

Figure 4: Scope of EM activities evidenced across IHEs in Texas. *(Source: Author data using QDA Miner 5.0)*

regulate HSEM (33%). The typology of HSEM activities underscores IHEs' trend towards conceptualization of HSEM along the crisis management cycle (e.g. preparedness, mitigation, response, and recovery). It is noteworthy that Continuity of Operation Planning (COOP) cited by researchers [53, 54] as critical to achieving resiliency was reported across websites of more than one-fifth of the IHEs in our sample. Themes extracted from taxonomic analysis of crises occurring at IHEs in Texas, shown in Figure 5, underscore strong focus on active shooter (60%) events.

Nearly one fifth of cases addressed crises related to terrorism events on their websites. Importantly, the adoption of a*ll-hazards* approach to HSEM was manifested by 25% of IHEs. Similarly, among the frameworks for best practices in national incident management

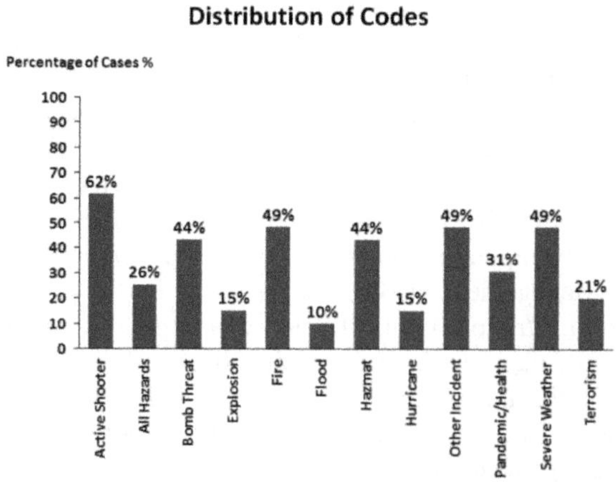

Figure 5: Typology of crises identified across select IHEs in Texas. *(Source: Author data using QDA Miner 5.0)*

(e.g. adoption of the NIMS, Incident Command System [ICS], National Response Framework [NRF], National Disaster Recovery Framework [NDRF], Homeland Security Exercise and Evaluation Planning [HSEEP]) as well as training and certifications, IHEs' specific vulnerability and risk assessments, or state guidelines, Figure 6 represents endorsement distribution that reflects moderate and varying levels of integration for each component. Among the exercises, categorized by the sub-codes into discussion-based (games, seminars, table top and workshops) and operations-based (drills, full-scale and functional), IHEs were reporting the largest involvement in table-tops (28.2%), drills (28.2%), functional exercises (10.3%), workshops (7.7%), seminars (5.1%) or full-scale multi-jurisdictional (5.1%). Relatively low percentage of Universities reported using games such as simulations (2.6%). Website records signalling adoption of emergency notifications systems confirmed that 71.8% of all IHEs in this inquiry possess mass communication methods to deliver crisis messaging, advisories and warnings to its populations. Generally, systems such as Mavalert adopted by the University of Texas Arlington, KatSafe adopted by Sam Houston State University, Miner Alert adopted by University of Texas El Paso, or Code Maroon adopted by the Texas A&M among others represent multi-tiered emergency communication systems to keep campus community informed about emergency situations and rapidly changing conditions from hazardous weather to campus closures, to building emergencies and life-threatening criminal activities. Students, faculty, staff (and in some cases parents) register to receive text messages, voice communications or alerts online on network computers and through wide-range of registered electronic devices. Moreover, emergency communication systems evaluated in this study were amplified by reported web notification applications using social media (33.3%), sirens (15.4%), specially dedicated websites (28.2%), or hotlines (7.7%) and 25.6% of the IHEs indicated capability to reach out external communities, agencies and jurisdictions.

Whereas CCA allowed for theme extraction and selected frequency analyses, the results of subsequent cross-case correspondence analyses in QDA Miner 5.0 [51] facilitated force-based multidimensional scaling to explore deeper connections among codes. Specifically,

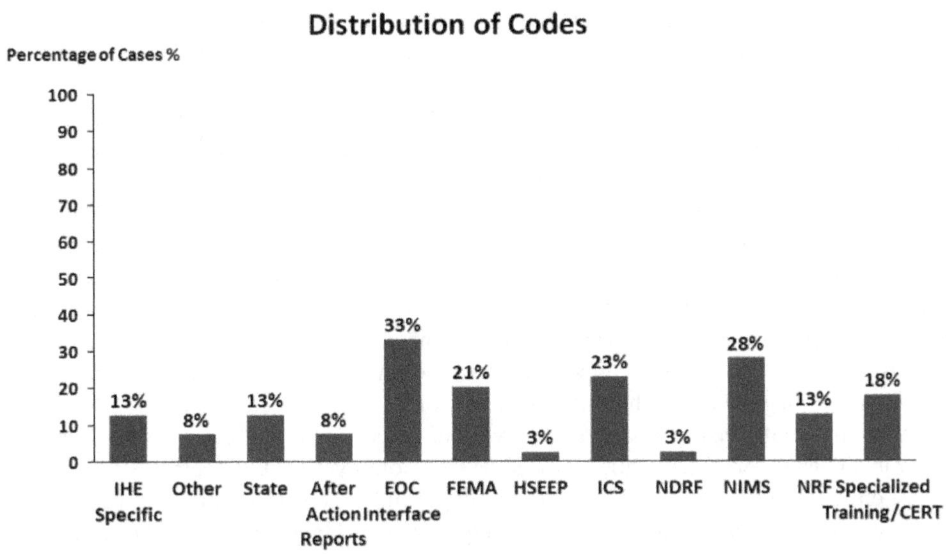

Figure 6: Disaster Resilient University features. *(Source: Author data using QDA Miner 5.0)*

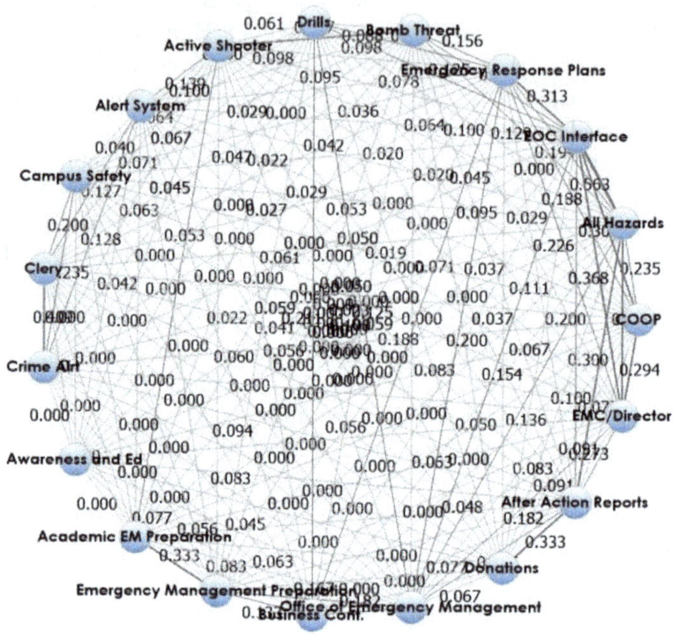

Figure 7: Link analysis. *(Source: Author data using QDA Miner 5.0)*

visualization of link analysis represented by Figure 7 points to clustering nodes and connections around the position of Emergency Management Coordinator/Director.

Specifically, the presence of a EMC/Director position is associated with Continuity of Operation Planning (COOP), adoption of all-hazards approach to HSEM, compilation of After Action Reports (AARs) and interactions with the Emergency Operations Center (EOC) interface. Similarly, the presence of OEM in organizational structures can be associated with emergency management practical and academic expertise, focus on business continuity, emergency response planning and utilization of drills. Meanwhile, the Clery Act node evidences connections to emergency notification systems and crime alerts as well as campus safety. It may be argued that relatively well developed emergency notification systems among IHEs evidenced in our study result from laws and policies related to national Cleary Act reporting requirements. Concurrently, the multidimensional scaling represented by Figure 8 allows discernment of several clusters of codes. Of most pertinence to our discussion and interpretation is the co-occurrence of items associated with DRUs around the function of the EMC/Director whereby planning across all stages of emergency management cycle (response, mitigation, preparedness and recovery) tends to concentrate. It is also around that node that HSEM frameworks and concepts appear to converge. In contrast, presence of administrative and academic collegiate teams that fulfil the role of HSEM on selected campuses show focus on institution-specific risk assessments as well as links to external agencies such as local EOC or local law enforcement entity for HSEM functions even though their link to the EMC/Director cluster can be noted as well. Not as strongly linked, clusters around HSEM function residing within Environmental Health and Safety as well as University Police Department (or Public Safety) structures point to generally hazard-specific approach to managing crises based on type; predominant threat represents active shooter events. Inarguably, amidst a variety

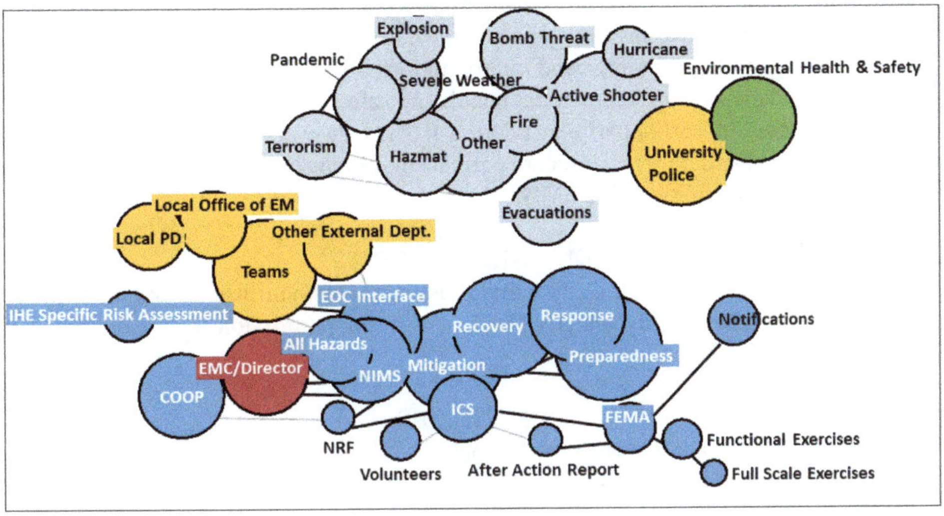

Figure 8: Multidimensional scaling for HSEM integration representing 3 most concentrated theme co-occurrence clusters among 17 yielded by the results. *(Source: Author data adapted from output using QDA Miner 5.0)*

of HSEM activities, and organizational arrangements that focus on them, only the designated position of EMC/Director at an IHE represents the high level of integration. Those findings can be meaningful and impactful in several ways. First, 'When emergency management is a component of university police, public safety, or environmental health and safety, the department competes for scarce resources not only among university departments but also potentially within its own department. What's more, when the mission, vision and effort of emergency management is intertwined with the programs and organizational dynamics of an overarching or parent department, the result is an obfuscation of emergency management objectives and strategies, which can diminish the quality and effectiveness of emergency management programs' [1]. Therefore, consolidation of various HSEM activities under the designated EMC/Director position and/or within the campus OEM may delineate a shift towards a new administrative and organization entity evolving in educational administration. Consequently, HSEM discipline-focused practical and academic expertise characterizing individuals assigned to EMC/Director roles may suggest further future need for HSEM professionalization across IHEs. The new profession as it evolves will require ongoing research on future competencies of HSEM practitioners on campus. Clearly, documented presence of the fourth model of organizational arrangement for HSEM (i.e. OEM and dedicated EMC/Director roles) on campuses reaching 25% representation merits attention. Based on Ross, [28] resiliency model of capability building and adaptation, the OEM model is the only one in our study that mirrors comprehensive emergency management cycle processes.

Markedly, activities of individuals and units/department tasked with emergency management cut across a wide range of hazards and threats, all of which regulated by diverse policies and laws. Indeed, professionals who focus on supporting the DRU mission (DRU, 2015) have called for a *crosswalk* of policies in various safety and security domains (e.g. Clery Act; DOE Action Guide 2009; National Fire Protection Agency [NFPA] 1600; FEMA Community Preparedness Guide 101; Occupational Safety and Health Act [OSHA] 1910, among others).

Such efforts may further consolidate requisite knowledge, abilities, skills and competencies of those entrusted with managing crises on college campuses. Clearly, the goal of our study was not prescriptive and our findings and interpretations do not suggest that traditional policing, crime, fire prevention, or occupational health and safety concerns would be relegated to new organizational units and departments across IHEs. However, it is plausible that distinct organizational HSEM structures may secure permanent presence in the ever more complicated environment of campus safety and security.

ACKNOWLEDGEMENTS

The authors would like to thank Provalis team for assistance with utilization and interpretation of QDA Miner 5.0 functionality and Dr. Ashley Ross for permission to use the resiliency model in our study.

REFERENCES

[1] Farris, D. & McCreight, R., The professionalization of emergency management in Institutions of Higher Education. *Journal of Homeland Security and Emergency Management*, **11(1)**, pp. 73–94, 2014. https://doi.org/10.1515/jhsem-2013-0074.

[2] Hellwig-Olson, B., Jacobsen, M., & Mian, A., Contemporary issues in campus crisis management. *Campus crisis management: A comprehensive guide to planning, response, and recovery,* eds. E. L. Zdziarski, N. W. Dunkel, & M. J. Rollo, John Wiley and Sons, Inc.: San Francisco, pp. 285–328, 2007.

[3] Booker Jr., L. (2014). Crisis management: Changing times for colleges. *Journal of College Admission*, **222**, pp. 19–23, 2014.

[4] Hargis, B., Bird, L.E., & Phillips, B.D., Building resilience to natural disasters across the campus ecosystem. *Managing the unthinkable: Crisis preparation and response for campus leaders,* eds. G.M. Bataille & D.I. Cordova, Stylus Publishing LLC: Sterling, pp. 18–36, 2014.

[5] LaBanc, B.H., Krepel, T.L., Johnson, B.J., & Herrmann, L.V., Managing the whirlwind: Planning for and responding to campus in crisis. *Enough is enough: A student affairs perspective on preparedness and response to campus shooting*, eds. B.O. Hemphill & B.H. LaBanc, Stylus Publishing, LLC: Sterling, pp. 53–82, 2010.

[6] Zdziarski II, E.L., Dunkel, N.W., & Rollo, M.J.(2007). *Campus crisis management: A comprehensive guide to planning, response, and recovery*, John Wiley & Sons, Inc.: San Francisco, 2007.

[7] Department of Homeland Security. *Critical infrastructure sectors* Web Site, available at: www.dhs.gov/critical-infrastructure-sectors. (accessed 24 February, 2017).

[8] The State of Texas. *Texas homeland security strategic plan 2015–2020*, available at: www.txdps.state.tx.us/director_staff/txHomelandSecStratPlan2015-2020.pdf. (accessed 04 February, 2017).

[9] Brenner, S., Protecting America's colleges and universities. *Homeland security: Protecting America's targets,* ed. J. Forest, Praeger Security International: Westport, pp. 133–162, 2006.

[10] Department of Homeland Security. *Quadrennial Homeland Security Review*, available at: www.dhs.gov/sites/default/files/publications/2014-qhsr-final-508.pdf. (accessed 08 February, 2017).

[11] Department of Homeland Security. *Quadrennial Homeland Security Review*, available at: www.dhs.gov/sites/default/files/publications/2014-qhsr-final-508.pdf. (accessed 02 Febraury, 2017).

[12] Haddow, G.D., Bullock, J.A., & Coppola, D.P., *Introduction to emergency management*, Elsevier: Waltham, 2014.

[13] Kapucu, N., & Khosa, S., Disaster resiliency and culture of preparedness for university and college campuses. *Administration & Society*, **47**, pp. 3–37, 2013. https://doi.org/10.1177/0095399712471626.

[14] Federal Emergency Management Agency, Building a disaster-resistant university, available at: www.fema.gov/media-library-data/20130726-1457-20490-1338/dru_report.pdf. (accessed 26 January, 2017).

[15] A *study of active shooter incidents, 2000–2013*. Texas State University and Federal Bureau of Investigation, U.S. Department of Justice, Washington D.C. available at: https://www.fbi.gov/about-us/office-of-partner-engagement/active-shooter-incidents/a-study-of-active-shooter-incidents-in-the-u.s.-2000-2013. (accessed 6 January, 2017).

[16] LSU in the eye of the storm: A university model for disaster response, available at: www.lasfaa.org/docs/news/disaster/SWASFAAPresentation.ppt. (accessed 12 January, 2017).

[17] A President's guide to the Clery Act. American Council on Education, available at: http://www.acenet.edu/news-room/Documents/A-Presidents-Guide-to-the-Clery-Act.pdf. (accessed 08 December, 2016).

[18] Hayter, J.K., Shelley, G.L., & Stevenson, P., Right-to-Carry and campus crime: evidence from the not-so-wild-west. *Libertarian Papers,* **6(1)**, pp. 1–20, 2014.

[19] Gansler, J.S., & Gast, A.P., Academics and national-security experts must work together. *Chronical of Higher Education,* **54**(44), 2008.

[20] Soft target hardening: Protecting people from attack, available at: http://www.crcnetbase.com.ezproxy.shsu.edu/isbn/9781482244229. (accessed 14 November, 2016).

[21] Virginia Tech Review Panel. *Mass shootings at Virginia Tech: Report of the review panel*, available at: http://www.washingtonpost.com/wp-srv/metro/documents/vatechreport.pdf. (accessed 28 Decemeber, 2016).

[22] Kerr, Jr., S.D., *Disclosure of campus security policies and crime statistics among colleges and universities in the Upper Midwest*. Unpublished doctoral dissertation. Vermillion, SD: University of South Dakota.

[23] Nicoll, S.R., & Owens, R.W., Emergency response and business continuity: the next generation in planning. *Professional Safety,* **58(9)**, pp. 50–55, 2013.

[24] Centre for Community Enterprise. *The community resilience manual: A resource for rural recovery and renewal*, available at: http://www.ontario-sea.org/Storage/37/2905_Section_1_-_The_Community_Resilience_Manual.pdf. (accessed 02 December, 2016).

[25] Federal Emergency Management Agency. *Guide for developing high-quality emergency operations plans for institutions of higher education*, available at: https://www.whitehouse.gov/sites/default/files/docs/rems_ihe_guide_508.pdf. (accessed 18 November, 2016).

[26] Perry, R.W., & Lindell, M.K., Preparedness for emergency response: Guidelines for emergency planning process. *Disasters,* **27(4)**, pp. 336–350, 2003. https://doi.org/10.1111/j.0361-3666.2003.00237.x

[27] Homeland Security Presidential Directive 5, available at: https://www.dhs.gov/publication/homeland-security-presidential-directive-5. (accessed 04 October, 2016).

[28] Local disaster resilience: Administrative and political perspectives, available at: http://reader.eblib.com.ezproxy.shsu.edu/(S(3pp1c5hvgo5u1ws1tnd3xbhy))/Reader.aspx?p=1562246&o=3159&u=KMN27TIaEyc%3d&t=1459728400&h=302254CCE1C6BE5387C50DD3661A263250C9F7CE&s=43662860&ut=10382&pg=1&r=img&c=-1&pat=n&cms=-1&sd=2#. (accessed 23 February, 2017).

[29] Griffin, O.R., Constructing a legal and managerial paradigm applicable to the modernday safety and security challenge at colleges and universities. *St. Louis University Law Journal*, **54**, pp. 241–270, 2009.

[30] Twigg, J., *Characteristics of a disaster-resilient community*, 2009. Available at: http://community.eldis.org/.59e907ee/Characteristics2EDITION.pdf

[31] Federal Emergency Management Agency. *Building a disaster-resistant university*, available at: http://www.fema.gov/media-library-data/20130726-1457-20490-1338/dru_report.pdf. (accessed 02 December, 2016).

[32] Creswell, J.W., *Research design: Qualitative, quantitative, and mixed method approaches* (4th ed.), Sage: Thousand Oaks, 2014.

[33] Yin, R.K., *Case study research: Designs and methods* (5th ed.), Sage: Thousand Oaks, 2014.

[34] Texas Department of Public Safety, Texas Division of Emergency Management. *Texas Emergency Management Executive Guide,* pp. 1–108, 2016.

[35] Texas Department of Public Safety, Texas Division of Emergency Management. *State of Texas Hazard Mitigation Plan 2010–2013,* pp. 1–366, 2010.

[36] Beggan, D.M., The impact of Hurricane Rita on an academic institution: Lessons learned. *Disasters*, **34(1)**, pp. 93–111, 2010. https://doi.org/10.1111/j.1467-7717.2009.01116.x

[37] Disaster Declarations for Texas, available at: https://www.fema.gov/disasters/grid/state-tribal-government/24. (accessed 02 December, 2016).

[38] *Patrol response to contemporary problems: enhancing performance of first responders through knowledge and experience*. Charles C. Thomas: Springfield, 2006.

[39] Walton, W.D., Bryner, N.P. & Jason, N.H., *Fire Research Needs Workshop*. Proceedings: Presented at Maryland, October 20, 1999.

[40] U.S Chemical Safety Board. *CSB Releases Investigation into 2010 Texas Tech Laboratory Accident*, available at: http://www.csb.gov/csb-releases-investigation-into-2010-texas-tech-laboratory-accident-case-study-identifies-systemic-deficiencies-in-university-safety-management-practices/. (accessed October 16, 2016).

[41] University of Texas System Board of Regents policy. UT 172 Emergency Management, Board of Regents Policy, available at: https://www.utsystem.edu/board-of-regents/policy-library/policies/uts172-emergency-management. (accessed 04 January, 2017).

[42] Southern Association of Colleges and Schools Commission on Colleges, available at: http://www.sacscoc.org/about.asp. (accessed 08 September, 2016).

[43] Southern Association of Colleges and Schools Commission on Colleges. *Resource Manual for the Principles of Accreditation* (pp. 1–151), 2012.

[44] What are Title IV Programs?, available at: http://federalstudentaid.ed.gov/site/front2back/programs/programs/fb_03_01_0030.htm. (accessed 12 February, 2017).

[45] Berelson, B., *Content analysis in communicative research*. Free Press: New York, 1952.

[46] Wilcox, F.D., The American newspaper: A study in social psychology. *Annals of the American Academy of Political and Social Science,* **16**, pp. 56–92, 1990.

[47] Onwuegbuzie, A.J., & Denham, M.A., Qualitative data analysis techniques. *Oxford Bibliographies* in "Educationß. Oxford University Press: New York, 2014.

[48] Leech, N.L., & Onwuegbuzie, A.J., An array of qualitative data analysis tools: a call for qualitative data analysis triangulation. *School Psychology Quarterly,* **22**, pp. 557–584, 2007.
https://doi.org/10.1037/1045-3830.22.4.557

[49] Miles, M.B., Huberman, A.M., & Saldaña, J., *Qualitative data analysis: A methods sourcebook* (3rd ed.). Sage: Thousand Oaks, 2014.

[50] Michailidis, G. Correspondence analysis. *Encyclopedia of measurement and statistics*, ed. N. J. Salkind, Sage: Thousand Oaks, pp. 191–194, 2007.

[51] Provalis Research. QDA Miner (Version 5.0.5) [Computer software]. Montreal, Quebec, 2016.

[52] Onwuegbuzie, A.J., & Combs, J.P., Emergent data analysis techniques in mixed methods research: A synthesis. In: *Sage handbook of mixed methods in social and behavioral research* (2nd ed.), eds. A. Tashakkori & C. Teddlie, Sage: Thousand Oaks, pp. 397–430, 2010.

[53] Sahebjamnia, N., Torabi, S.A., & Mansouri, A., Integrated business continuity and disaster recovery planning: towards organizational resilience. *European Journal of Operational Research*, **242**, pp. 261–273, 2015.
https://doi.org/10.1016/j.ejor.2014.09.055

[54] Somers, S., Measuring resilience potential: An adaptive strategy for organizational crisis planning. *Journal of Contingencies and Crisis Management,* **17(1)**, pp. 12–23, 2009.
https://doi.org/10.1111/j.1468-5973.2009.00558.x.

[55] Local disaster resilience: Administrative and political perspectives, available at: http://reader.eblib.com.ezproxy.shsu.edu/(S(3pp1c5hvgo5u1ws1tnd3xbhy))/Reader.aspx?p=1562246&o=3159&u=KMN27TIaEyc%3d&t=1459728400&h=302254CCE1C6BE5387C50DD3661A263250C9F7CE&s=43662860&ut=10382&pg=1&r=img&c=-1&pat=n&cms=-1&sd=2#. (accessed 23 February 2017).

CONSTRUCTING RESILIENCE MODEL OF PORT INFRASTRUCTURE BASED ON SYSTEM DYNAMICS

HYUNGMIN CHO & HEEKYUNG PARK
Department of Civil and Environmental Engineering, KAIST, Republic of Korea.

ABSTRACT

The port industry, which plays an important role in Korea's economy, is exposed to various disasters such as earthquakes, tsunamis, and chemical accidents. Therefore, resilience needs to be assessed to evaluate how properly port system can recover its function even after being damaged, and weak points should be complemented by the policy. However, the port infrastructure is too complicated to analyze all the components, so a systemic approach is needed. Therefore, this study evaluates the resilience of the port infrastructure using system dynamics model, which can compare quantitative performance index. This study sets up the cargo process, the most important economic index of the port, as the performance level and constructs a system dynamics model by finding elements corresponding to attributes of resilience. In addition to disruption and recovery actions in the disaster situation, the model also incorporates socioeconomic factors such as changes in cargo demand and financial state, resulting in close proximity to case studies. Simulation of disaster situations with resilience assessment model can express recovery process of the system and accumulated economic damage. By applying various inputs and scenarios, the result of this study can be used as a basis for comparing the resilience of port infrastructure and establishing the reinforcement policy.

Keywords: infrastructure resilience, port disaster, system dynamics model

1 INTRODUCTION

In South Korea, port industry is one of the nation's most important industries because economy of South Korea heavily relies on international trade compared to other developed countries [1] and 99% of trade freights pass through maritime transportation [2]. In September 2003, typhoon Maemi, which recorded the strongest wind speed in South Korea, caused large damage to south-eastern part of Korea including Busan port. As South Korea confronted serious disruption in maritime transportation and rapid recovery was demanded, it reminded importance of resilience against natural disaster in major ports. In near countries, Hanshin earthquake in 1995 and Tianjin explosions in 2015 were also recorded as severe accidents that damaged major national ports and paralyzed logistics [3, 4]. Likewise, in Korea, risk of earthquake and chemical accidents along industrial areas in coastal regions has been pointed out [5, 6]. In other words, it can be said that major ports in South Korea also have been exposed to various hazards. These recent environments increase necessity of making mitigation policies to port infrastructure system in South Korea. As a first step, port infrastructures need to be evaluated for their current resilience capacity. This study looks at the features of port infrastructure in view of resilience and then discusses method to measure the resilience of port functions.

2 BACKGROUND

2.1 Infrastructure resilience

In recent years, series of disasters revealed vulnerability of infrastructure systems and hence improving resilience of infrastructure became important strategy to reduce the risk [7].

DOI: 10.2495/SAFE-V7-N3-352-360

Although concepts of resilience applied to port are various in meaning and range [8], one of the agreed definitions in disaster studies is that 'ability of system to resist, absorb, accommodate and recover from hazard' [9]. With this definition, ways of measuring resilience used in previous studies are categorized to qualitative and quantitative methods [10]. Quantification investigates total impact of disruptive event to performance level of system [11, 12] or total amount of recovery effort of system from given damage [13]. Bruneau *et al.* [11] also suggested attributes of resilience; Robustness, Redundancy, Resourcefulness and Rapidity. As end-attributes, meaning of robustness and rapidity of resilience with performance level is shown in Figure 1. This also implies that two attributes of resilience are distinguished by before or after event [14]. Furthermore, Francis and Bekera [15] proposed adaptive capacity in addition to absorptive and restorative capacity, which refers improvement of performance level after event. However, this improvement in infrastructure system is achieved by getting new facilities from external investment in contrast with self-organizing ecosystem.

2.2 Port-related disasters

In law of Korea, definition of port, or harbour, covers from seaway, anchorage, all facilities related to function to hinterland. So most of accidents in port area are crashes, construction accidents, fire, blackout, and safety accidents and they do not have significant impact on function of the port. However, typhoon Maemi damaged 11 container cranes reducing 24% of unloading capacity of Busan port [16]. These cases imply that measuring performance decline needs to focus on huge disaster rather than the minor operational failure. Labor strike is also reported as major cause of port disability as well as natural disaster [17], but this study excludes those kinds of social problems because affecting factors are too uncertain to be included in infrastructure system model.

One of the well-known cases of port related disaster is Hanshin earthquake and Kobe port, Japan in 1995. Chang [3] investigated loss of Kobe port after earthquake with comparing competing ports in Japan and abroad. Massive earthquake damaged most of berths and container cranes so market share of Kobe port fell sharply after January, 1995. However, more important implication is that Kobe port could not fully recover its previous market

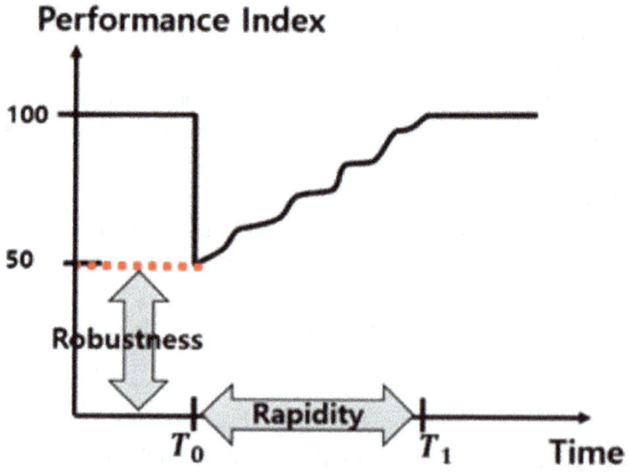

Figure 1: Resilience quantification and attributes (*Source: Tierney & Bruneau, 2007*).

share in line with the facility recovery (Fig. 2). Chang [3] explained this long-term loss with reason that smaller ports of Japan took away market shares without transhipment in Kobe in domestic market. Also, International market share moved to other Asian ports such as Kaohsiung and Busan. This long-term impact is caused by the characteristics of maritime logistics, which has interdependent supply chains and disruption results in shift of customers [18, 19].

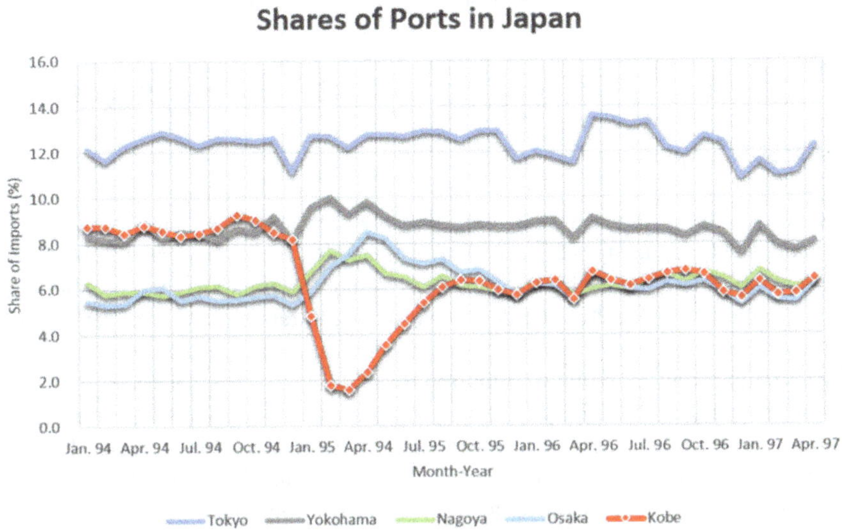

(a)

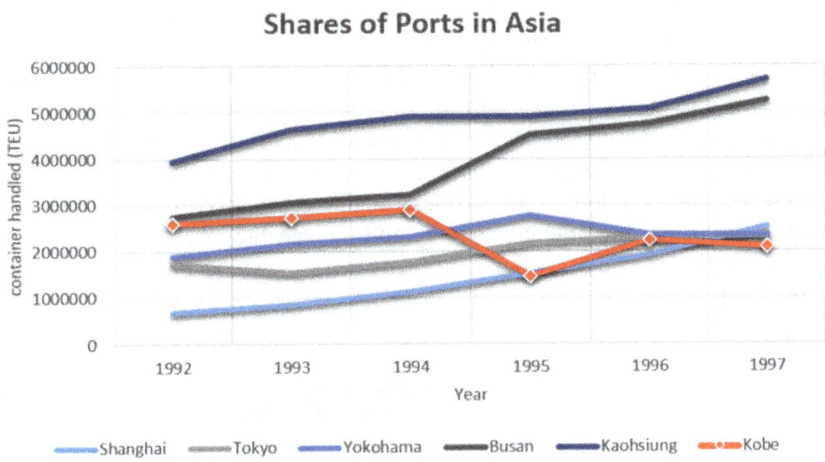

(b)

Figure 2: Changes in market shares of ports before and after the Hanshin earthquake (a) in Japan; (b) in Asia (*Source: Chang, 2000*).

2.3 System dynamics

System dynamics model is a widely used method to represent complex and nonlinear structures and phenomenon [20]. It has been used to investigate the infrastructure system [21] and maritime transportation system [22]. As system dynamics method focuses on how specific variable changes over time [23], it is usable to assess performance of infrastructure system over time after disaster occurs. By previous studies, system dynamics have been applied to maritime transportation as critical infrastructure for measuring post-disaster state and managing it [24–27] but they focused on transportation system not port infrastructures carefully.

3 CONSTRUCT RESILIENCE MODEL OF PORT INFRASTRUCTURE

3.1 Conceptual framework

This study constructs resilience model with two features; assessing performance level and system dynamics analysis using STELLA. Performance level of system should be determined to indicate generally required performance of port system. In previous studies, amount of handled cargo is suggested to measure the port performance in terms of the number of containers [28] or tonnage [29]. Our model sets the performance level as amount of processed cargo per time while assuming a virtual unit has limitation in analysis. A simple feedback loop of system dynamics model which can be applied to any type of disaster is suggested in Figure 3. Positive impact is indicated by blue arrow with label p, and negative impact is indicated by red arrow with label n. Since the key of system dynamics is creating feedback loop which makes nonlinear change [23], it is needed to construct appropriate feedback loop on port infrastructure system in this study. In Figure 3, total loop becomes negative feedback loop which means that disruption is mitigated as decreased performance enhances recovery activity. Consequently, capacity of infrastructure system is restored and performance level returns to normal condition over time as shown in Figure 4.

In model of this study, disaster and disruption are given input value by disaster scenario. Then, other components; capacity of facilities, performance level and recovery activity are able to be affected by external factors. Those variables are derived from literature review and categorized to robustness and rapidity (see Table 1). For variables of robustness, Kim *et al.* [2] studied response to disruption in terms of BCM (Business Continuity Management) which focuses on continuing to provide port service. In this view, damaged service is not same as total physical loss as long as service is available. Therefore, suggested major factors of BCM in port are internal redundant capacity and replaceable level by neighboring ports. This strategy to replace traffic at undamaged facilities is known to be effective in reducing economic loss [30].

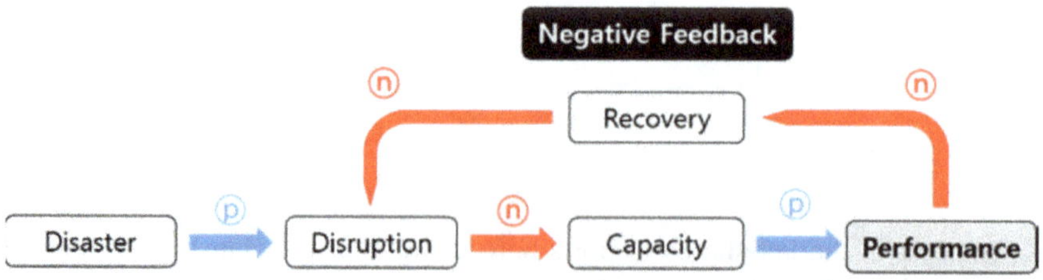

Figure 3: Basic system dynamics model and feedback loop.

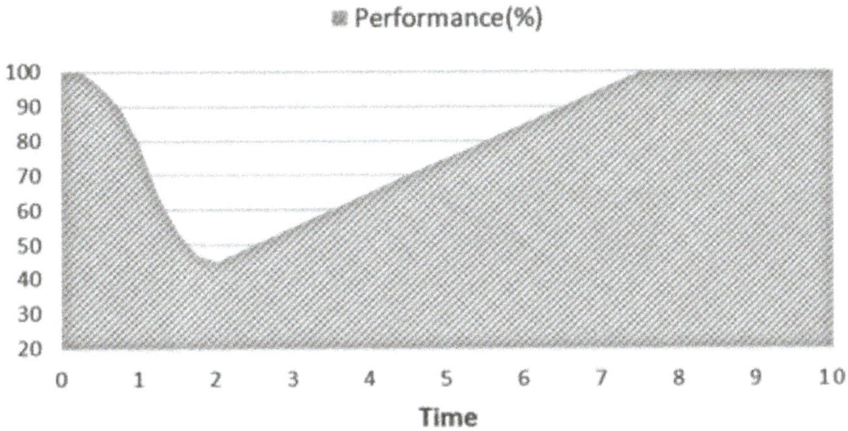

Figure 4: Performance graph of basic model.

Table 1: Attribute and variables of resilience model.

Attributes	Variables	Source
Performance level	Capability of processing cargo	[29]
Robustness	Allowance capacity	[2]
	Replaceable nearby ports	
	Failure ratio	[31]
Rapidity	Repair ratio	[27, 31]
	Time to repair	
	Resource for repair	

Among identified variables in Table 1, allowance capacity, replaceability and resource for repair have positive effect to system components as seen in Figure 5. Figure 6 shows effect of improvement of robustness and rapidity to reducing maximum damage or recovery time. However, system with more rapid recovery, marked in yellow line, got less drop of performance than black line. This is because recovery activity begins as disruption starts to reduce performance level. This feature of system dynamics does not correspond to expected effect of rapidity. Nevertheless, it is still reasonable when disruption in scenario lasts for some time, while some disaster such as earthquake occurs in very short time.

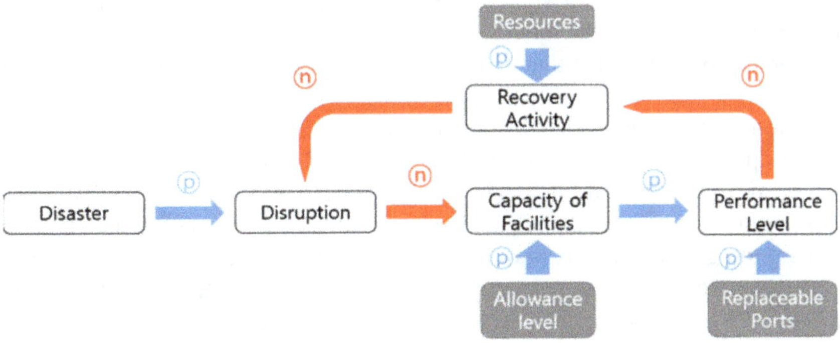

Figure 5: System dynamics model with resilience variables.

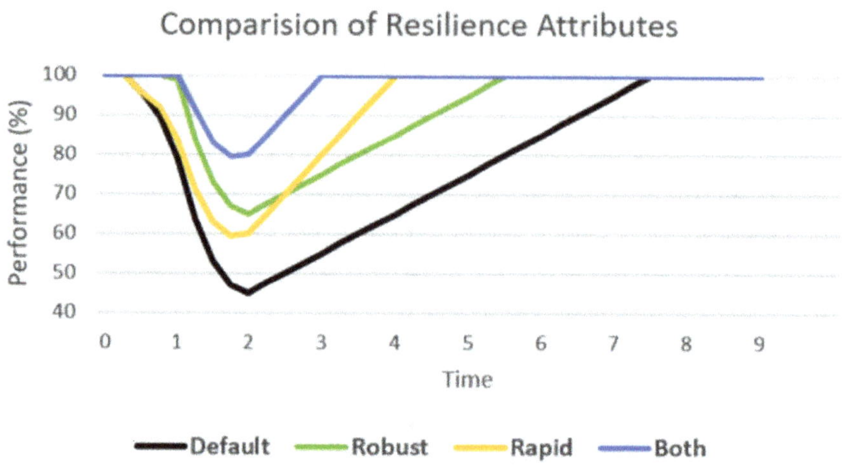

Figure 6: Comparison of performance among different resilience attributes.

3.2 Long-term loss and delayed recovery

As case of Kobe port in Figure 2 shows importance of considering long-term loss, advanced model needs additional long-term elements. Again, components which affect performance level after recovery are related to consumers' choice of port [3]. The major determinants for customers to choose a port are known as the cost and time of using it [28, 32, 33]. On the other hands, Lee [34] pointed out that there is no big difference in cost and facilities among major ports so marketing affects to amount of cargo in meaningful level. When these determinants of port weaken, it can be said that trade volume would be transferred to other ports as seen in Kobe port case [3, 35].

Using STELLA, resilience model including disruption, recovery and long-term effect is constructed (Fig. 7). Looking the bottom loop of Figure 7, cargoes in disabled system are assumed to be moved to other port during disruption and recovery stage. This happens because waiting for recovery increases time and cost for shipping. Moved demands reduce the attractiveness of the port because it takes additional time and cost in returning to original port. However, this long-term loss is not defined as irreparable damage. On the assumption that stabilized financial state of port can create motivating actions such as incentive or advanced facility, attractiveness and performance level recover over time gradually. Financial state of port is assumed to be affected by recovery cost and benefit from cargo process. In long-term loss model, resilience attributes discussed in Figures 5 and 6 are assumed to be constant value that does not affect to result significantly and marked with grey boxes.

Result of long-term loss model is shown in Figure 8. Long-term loss due to reduced attractiveness appears after 20 time units. That is when delayed transportation is processed and facilities are fully repaired but incoming cargoes decrease. As financial state recovers after lowest point, attractiveness and performance of port also recover slowly. Slopes of recovery seem too gentle, but it is reasonable considering disasters such as earthquake and typhoon happens from few hours to days. Furthermore, some factors which can affect financial state and recovery rate are excluded in model since external aids are beyond our model boundary.

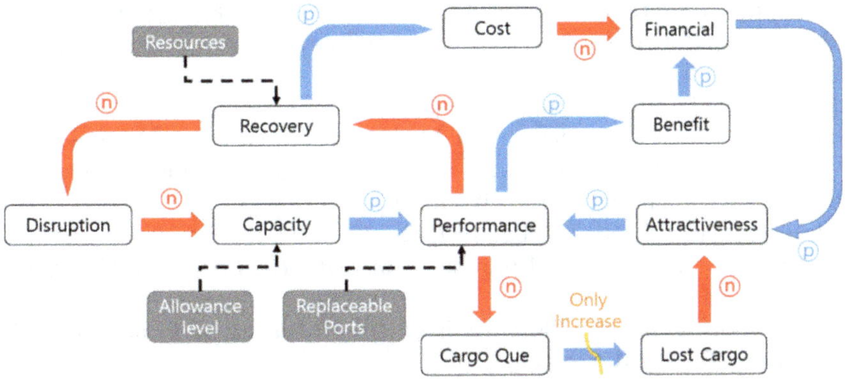

Figure 7: System dynamics model with long-term loss.

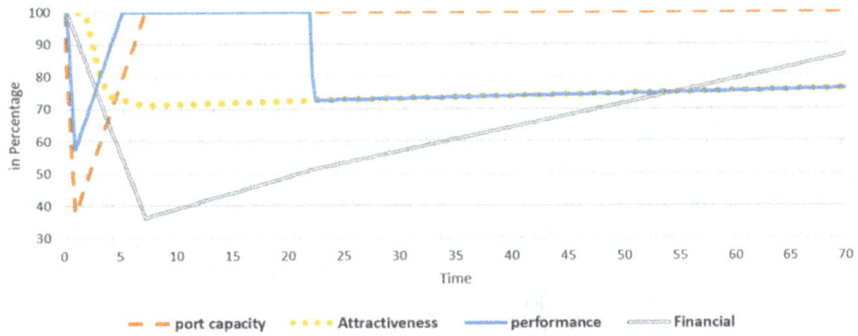

Figure 8: Result of long-term loss model.

4 CONCLUSION

Starting from the need for measurement of resilience to port disasters, we reviewed the related theoretical background and constructed resilience model using system dynamics with STELLA. Although port system is highly complex with numerous number of elements, system dynamics model could show brief conceptual model and approximate measurement of resilience. This study could not apply constructed model to actual site or disaster records due to lack of the detailed settings in model. This study conceptualizes the resilience of port infrastructure system and identifies possibility of using the system dynamics model. It is expected that further studies will be able to construct more detailed model and continue research to measure resilience of port infrastructure system.

ACKNOWLEDGEMENTS

This research was supported by a grant [MPSS-CG-2015-01] through the Disaster and Safety Management Institute funded by Ministry of Public Safety and Security of Korean government.

REFERENCES

[1] OECD, *Factbook 2011: Economic, Environmental and Social Statistics,* 2011.

[2] Kim, H., Lee, J., Ha, T. & Lee, J., *A Study on Impacts of Port Accidents & Disasters and Responsive Measures.* Korea Maritime Institute: Seoul, 2012.

[3] Chang, S.E., Disasters and transport systems: loss, recovery and competition at the Port of Kobe after the 1995 earthquake. *Journal of Transport Geography*, **8**(1), pp. 53–65, 2000.
https://doi.org/10.1016/s0966-6923(99)00023-x
[4] Chung, Y. & Kim, H., On the August 12, 2015 occurrence of explosions and fires in Tianjin, China, and the atmospheric impact observed in central Korea. *Air Quality, Atmosphere & Health*, **8**(6), pp. 521–532, 2015.
https://doi.org/10.1007/s11869-015-0371-2
[5] Kim, M., Seismic damage and prevention of ports. *Facility Safety Magazine*, p. 12, 2003.
[6] Choi, S.-D., Suggestion for Ulsan that is safe from chemical accidents. *Ulsan Development*, **40**, pp. 28–39, 2013.
[7] Chang, S.E., Infrastructure resilience to disasters. *The Bridge*, **44**(3), 2014.
[8] Cao, M., Transportation Resilience: A summative review on Definition and Connotation. *International Conference on Automation, Mechanical Control and Computational Engineering*, 2015.
[9] UNISDR, *UNISDR Terminology on Disaster Risk Reduction*. United Nations: Geneva, 2009.
[10] Hosseini, S., Barker, K. & Ramirez-Marquez, J.E., A review of definitions and measures of system resilience. *Reliability Engineering & System Safety*, **145**, pp. 47–61, 2016.
https://doi.org/10.1016/j.ress.2015.08.006
[11] Bruneau, M., Chang, S.E., Eguchi, R.T., Lee, G.C., O'Rourke, T.D., Reinhorn, A.M., Shinozuka, M., Tierney, K., Wallace, W.A. & von Winterfeldt, D., A framework to quantitatively assess and enhance the seismic resilience of communities. *Earthquake Spectra,* **19**(4), pp. 733–752, 2003.
https://doi.org/10.1193/1.1623497
[12] Tierney, K. & Bruneau, M., Conceptualizing and measuring resilience: A key to disaster loss reduction. *TR news*, **250**, pp. 14–17, 2007.
[13] Vugrin, E.D., Warren, D.E., Ehlen, M.A. & Camphouse, R.C., A framework for assessing the resilience of infrastructure and economic systems. *Sustainable and resilient critical infrastructure systems*, Springer: Berlin, pp. 77–116, 2010.
[14] McDaniels, T., Chang, S., Cole, D., Mikawoz, J. & Longstaff, H., Fostering resilience to extreme events within infrastructure systems: Characterizing decision contexts for mitigation and adaptation. *Global Environmental Change*, **18**(2), pp. 310–318, 2008.
[15] Francis, R. & Bekera, B., A metric and frameworks for resilience analysis of engineered and infrastructure systems. *Reliability Engineering & System Safety*, **121**, pp. 90–103, 2014.
[16] YGPA, *Disaster Cases in Port*. Yeosu Gwangyang Port Authority: Yeosu, 2016.
[17] Lam, J.S.L. & Su, S., Disruption risks and mitigation strategies: an analysis of Asian ports. *Maritime Policy & Management*, **42**(5), pp. 415–435, 2015.
[18] Lam, JSL., Risk management in maritime logistics and supply chains. *Maritime Logistics: Contemporary Issues*, Emerald Group Publishing Limited, pp. 117–132, 2012.
[19] Sheffi, Y. & Rice, Jr, J.B., A supply chain view of the resilient enterprise. *MIT Sloan Management Review*, **47**(1), p. 41, 2005.
[20] Forrester, J.W., System dynamics, systems thinking, and soft OR. *System Dynamics Review*, **10**(2–3), pp. 245–256, 1994.
[21] Ouyang, M., Review on modeling and simulation of interdependent critical infrastructure systems. *Reliability Engineering & System Safety*, **121**, pp. 43–60, 2014.

[22] Oztanriseven, F., Pérez-Lespier, L., Long, S. & Nachtmann, H., A review of system dynamics in maritime transportation. *IIE Annual Conference Proceedings*, p. 2447, 2014.

[23] Yoon, Y.S. & Chae, S.B., *Introduction to complex systems*. SERI: Seoul, 2005.

[24] Lespier, LP. A systems thinking approach to post-disaster restoration of maritime transportation systems. *IIE Annual Conference Proceedings*, 2015.

[25] Mansouri, M., Nilchiani, R. & Mostashari, A., A risk management-based decision analysis framework for resilience in maritime infrastructure and transportation systems. *Systems Conference, 2009 3rd Annual IEEE,* 2009.

[26] Mansouri, M., Sauser, B. & Boardman, J., Applications of systems thinking for resilience study in maritime transportation system of systems. *Systems Conference, 2009 3rd Annual IEEE,* 2009.

[27] Croope, S. & McNeil, S., Improving resilience of critical infrastructure systems postdisaster: recovery and mitigation. *Transportation Research Record: Journal of the Transportation Research Board,* **2234**, pp. 3–13, 2011.

[28] Tongzon, J.L., Determinants of port performance and efficiency. *Transportation Research Part A: Policy and Practice*, **29**(3), pp. 245–252, 1995.

[29] Omer, M., Mostashari, A., Nilchiani, R. & Mansouri, M., A framework for assessing resiliency of maritime transportation systems. *Maritime Policy & Management*, **39**(7), pp. 685–703, 2012.

[30] Pachakis, D. & Kiremidjian, A.S., The use of simulation in disaster response planning and risk management of ports and harbors. *Advancing Mitigation Technologies and Disaster Response for Lifeline Systems*, pp. 425–434, 2004.
https://doi.org/10.1061/40687(2003)44

[31] Laugé, A., Hernantes, J. & Sarriegi, J.M., The role of critical infrastructures' interdependencies on the impacts caused by natural disasters. *International Workshop on Critical Information Infrastructures Security*, pp. 50–61, 2013.

[32] Park, N-K., Moon, D. & Lim, C-K., Port choice model of Transhipment cargo using system dynamics. *Maritime Logistics,* Emerald Group Publishing Limited, pp. 69–90, 2012.

[33] Yeo, G-T., Pak, J-Y. & Yang, Z., Analysis of dynamic effects on seaports adopting port security policy. *Transportation Research Part A: Policy and Practice*, **49**, pp. 285–301, 2013.
https://doi.org/10.1016/j.tra.2013.01.039

[34] Lee, H., Factors affecting port competitiveness. *Korean Public Administration Quarterly*, **24**(1), pp. 1–25, 2012.

[35] Achurra-Gonzalez, P.E., Novati, M., Foulser-Piggott, R., Graham, D.J., Bowman, G., Bell, M.G.H. & Angeloudis, P., Modelling the impact of liner shipping network perturbations on container cargo routing: Southeast Asia to Europe application. *Accident Analysis & Prevention,* 2016.

ROLES AND SCOPE OF SYSTEM INTERFACE IN INTEGRATED CONTROL SYSTEM FOR MULTI DISASTER COUNTERMEASURE

SEUNGHWA PARK & CHANGHEE HONG
Multi Disaster Countermeasure Organization, Korea Institute of Civil Engineering and Building Technology.

ABSTRACT

Damages on human and physical properties have recently become significant due to natural disasters, such as earthquake and hurricanes, as well as social disasters, such as fire in subway and terrorist attacks using explosive. Notably, as buildings are becoming taller and more complex, demands on countermeasure to multi disasters, such as earthquake, fire, and flooding, are increasing for complex skyscraper facilities. Complex skyscraper facilities are equipped with constant management systems, such as facility management, lighting management, CCTV management, and parking management, as well as event-based emergency management system, such as fire control, flooding control, seismic detection, and wind detection. While part of the information for each system is currently being collected by a central monitoring system in the form of SI (system integration), it is difficult to perform SOP (standard operating procedure)-based integrated control because data protocols for sensors and receivers are different. An integrated system for efficient countermeasure/evacuation in complex skyscraper facilities is currently under development, and this study organizes the roles and scope of system interface for linkage of information among the integrated system, disaster information collection system, analysis system, and countermeasure system. The scope of disaster at this point was limited to earthquake, flooding, and fire. The system interface that is designed by this study is one of core modules for implementing an integrated system, and is to be used to present future directions for development of integrated CPS (Cyber Physical System) in countermeasure to disasters.

Keywords: BIM-GIS platform, dashboard, multi disaster, SOP, system interface

1 INTRODUCTION

A convergence research group on disaster response has been launched to develop and practically use an integrated system to protect massive complex facilities such as skyscrapers against earthquakes, fire, flood and other disasters within 3 years. In Korea, there are currently meters such as earthquake accelerometers in operation at around 600 major national facilities, but effective disaster response has been difficult due to the absence of a system and operating body for integrated management of the related information and data. Amid this situation, the Disaster Response Research Group conducted a project to develop an integrated information platform, with the aim of predicting the risk factors associated with disasters and calamities and ensuring a fast response and recovery, in the event of an accident, by promptly collecting and analysing the necessary information. The plan is to develop the key technology for each element of the platform in the first and second years, and to set up a testbed in the third year to perform a wide variety of verification tests. At present, the project is in its first year. Approximately 24 million dollars will be invested into this 3-year convergence research project, which will be undertaken by some 200 researchers at the Korea Institute of Civil Engineering and Building Technology (KICT). The research procedure is as follows (Fig. 1):

The disaster information collection system stage involves implementation of integrated measuring and monitoring technology and establishment of an information collection system to create an information system for responding to disasters such as fire, earthquake, and flood.

DOI: 10.2495/SAFE-V7-N3-361-366

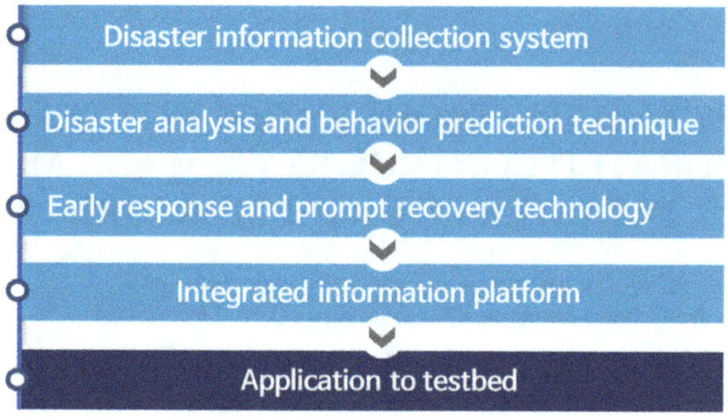

Figure 1: Research procedure.

In the disaster analysis and prediction stage, simulations of a combination of disasters will be performed to predict the risk factors and formulate response and recovery measures, after which the prediction techniques will be set forth.

In the response and recovery stage, an early response and evacuation support system and a prompt response and recovery manual will be established so as to minimize damage and secure the golden hour.

For the integration and control of the aforementioned systems, an integrated information platform will be developed by setting up an integrated CPS based on an open platform server so as to ensure high reliability in information delivery and management.

Lastly, the integrated CPS will be applied to the testbed, and an ROI (return on investment) analysis will be carried out to assess the extent to which the practical use of the system is possible.

This study was conducted on the system interfaces for the BIM/GIS platform[1] [1] server, analysis system, and sensors and measuring systems in the integrated information platform development stage for the purpose of assessing the role and scope of each of the system interfaces comprised of modules.

2 INTEGRATED INFORMATION PLATFORM FOR DISASTER RESPONSE

The integrated information platform for disaster response consists of a BIM/GIS platform server and system interfaces. In this particular chapter, the significance of each system will be examined.

2.1 BIM/GIS Platform Server

One of the limitations of 3D spatial information services such as Google Earth and V-World is that they only present the external appearances of buildings and structures. Since we spend 70% of our time indoors, however, there is a need for technology that can capture and present both the outside and inside of buildings and structures. With respect to this matter, the National Institute of Building Sciences (NIBS) and National BIM Standard (NBIMS) have stated that indoor and outdoor spatial information can be integrated together, centring on the contact points of BIM and GIS [2].

[1] It means the Platform that BIM(Building Information Modelling) data and GIS(Geographic Information System) data can be operated together on one system. It is simply called “BIM on GIS Platform” or “BIM/GIS Platform”.

Figure 2: Main functions of BIM/GIS platform [3].

Accordingly, a platform characterized by BIM/GIS interoperability was developed to create a database of attributes on a BIM model and to enable a seamless service on indoor and outdoor spatial information. Such platform, which was developed by our team, allows seamless use and management of massive amounts of 3D BIM and GIS data on a large number of buildings within a city or a regional sphere simultaneously, rather than a single building.

The BIM/GIS platform can express both the indoor and outdoor spatial information. In the case of attributes related to disaster management and prevention, for instance, it can be used to express the information on the bearing and non-bearing structures, building materials, and evacuation. In addition, it can be used for disaster information analysis, facility management, as well as fire-fighting training such as XVR [4]. For this reason, it serves as the core platform of the integrated system.

2.2 System Interfaces

As shown in Figure 3, the system interfaces collect and accumulate information in connection with the BIM/GIS platform server, described in Section 2.1, as well as various measuring systems, video streamer, access controller, and analysis system. They are synchronized with a support system, and are responsible for primary response and alarm control.

Each interface module plays a set of unique roles depending on the field of response and its composition. The role and scope of the system interface will be described in Chapter 3.

3 THE ROLE AND SCOPE OF THE SYSTEM INTERFACES

3.1 The role of the system interfaces

The system interfaces performs diverse roles, in connection with the platform, as shown in Figure 4.

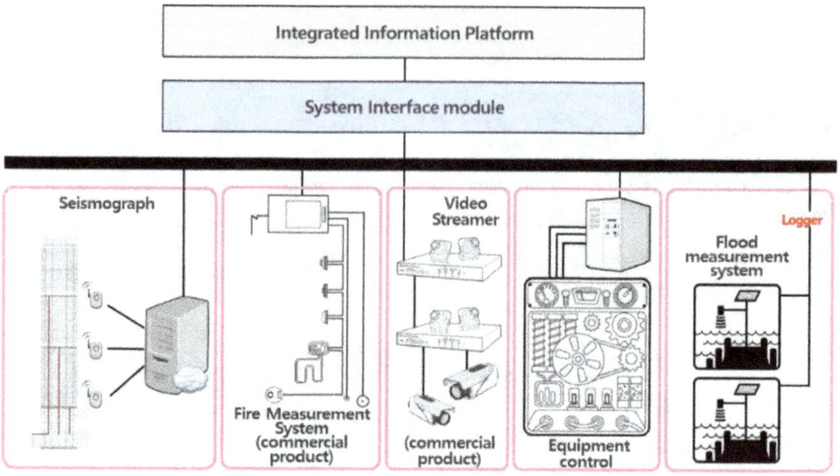

Figure 3: Concept of system interface's architecture.

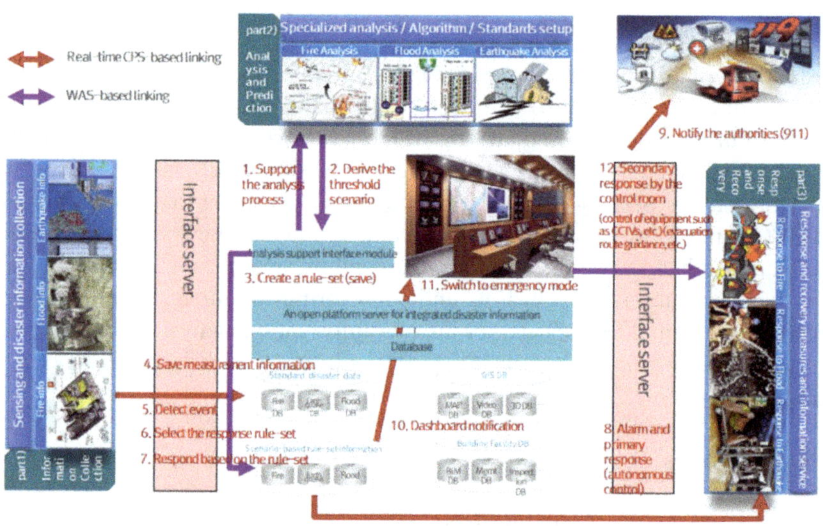

Figure 4: Roles of interface modules.

The disaster information collection interface, for example, stores measuring information, detects events, and selects a response rule-set for optimum response, based on which the instructions for a response are given. In the case of the analysis and prediction interface, it supports the collected disaster information to be communicated to the analysis system, and saves the threshold scenarios derived from the analysis system in the scenario-based rule-set database. As for the response and recovery interface, it supports autonomous control for the primary response and the secondary responses such as control of equipment including CCTVs and evacuation route guidance.

The cases requiring real-time high-reliability data are indicated with a red arrow, whereas data requiring WAS (Web Application Server)-based stability and security are indicated with a purple arrow.

Table 1: Scope of each system interface module.

Module name	Analysis system-linked information module	Measurement information-linked interface module	Autonomous control-linked interface module
Description	• Identifies spatial information to support the analyses of disasters (fire, earthquake, or flood) occurring in super-tall complex facilities • Processes spatial information to support the disaster analysis system • Link with the integrated information platform and the analysis system for disaster (fire, earthquake, or flood) response	• Receive measurement information in real time • Identify the thresholds based on the real-time measurement information (fire, earthquake, or flood) • Select the risk scenarios connected with event signals • Connect to the authorities (emergency call, 911)	• Call the rule-set database for equipment control • Perform automatic equipment control and response based on the rule-set • UX/UI (User Experience / User Interface) for prompt and integrated response • Link information with CCTVs and disaster prevention and management equipment

3.2 The scope of system interfaces

The system interfaces are divided into three interface modules, as shown in Table 1.

The integrated information platform and the interface modules are designed to perform distributed processing and real-time information linking. In order to compensate for lost information, there are plans to develop the system, while taking into consideration equipment protection and data redundancy.

4 CONCLUSION

This paper provides a summary of the roles of the system interfaces in the integrated information platform developed for disaster response. While implementing this research project, the research group came across the following barriers:

1. In the case of fire notification information, which is communicated in the order of a sensor, receiver and alarm, there are limitations to linking it with the interface because the receiver information is kept closed to ensure stable fire alarm and management.
2. There is a need for a standard communication protocol [5] to integrate the information received from each sensor in the interface, but there are limitations in terms of access due to the issues of security and leakage of technology.
3. Even after the integrated information platform and system interface module are set up, there are additional stages such as the testing and certification process that remain.

Once the integrated information platform described in this paper is established, big data analysis of the information obtained from diverse measuring systems can be performed for prompt and accurate decisions to be made during a disaster situation. At present, an integrated system of CPS [6] Level 2 is expected to result from this project, but a complete CPS that can seamlessly link the virtual world and the real world may be developed at a later time by incorporating artificial intelligence into the system.

ACKNOWLEDGEMENTS

This work was supported by the National Research Council of Science & Technology (NST) grant by the Korea government (MSIP) (No. CRC-16-02-KICT)

REFERENCES

[1] Tae-Wook, K., Seung-Hwa, P. & Chang-Hee, H., *BIM/GIS-based data integration framework for facility management*, GEOProcessing 2016 : The Eighth International Conference on Advanced Geographic Information Systems, Applications, and Services, pp. 100–105, 2016.

[2] Dianne, D., *National BIM Standard: BIM GIS Integration*, AEC Infosystems Presentation, 2007.

[3] Seung-Hwa, P., Chang-Hee, H. & Youn-Mi, J., *A study on an application of BIM/GIS interoperable platform for the implementation of SMART city*. Proceeding of the KSCE 2016 Conference & Civil Expo, pp. 127–128, 2016.

[4] Virtual Reality training software for safety and security; available at: http://www.xvr-sim.com/?t=gb. (accessed 26 April, 2017).

[5] Integrated Proprietary Alarm Monitoring System, available at: http://www.digitize-inc.com. (accessed 26 April, 2017).

[6] Grant, C., Hamins, A., Bryner, N., Jones, A. & Koepke, G., *Research roadmap for smart fire fighting*, pp. 6–13, 2015.

SMART MATURE RESILIENCE, SYSTEM DYNAMICS BASED INTERACTIVE LEARNING ENVIRONMENT: A BETA VERSION

MARTA ITURRIZA[1], AHMED A. ABDELGAWAD[2], LEIRE LABAKA[1], JAZIAR RADIANTI[2], JOSE M. SARRIEGI[1] & JOSE J. GONZALEZ[2]
[1] Tecnun, University of Navarre, Spain.
[2] Centre for Integrated Emergency Management, University of Agder, Norway.

ABSTRACT

Natural and man-made disasters are becoming more frequent – unfortunately in most cases inevitably. By applying certain set of policies and guidelines, a city could resist, absorb, adapt to and recover from acute shocks and chronic stresses to keep critical services functioning, i.e. be resilient. In 2016, the Smart Mature Resilience, a Horizon-2020 EU project, defined the Resilience Maturity Model to make cities more resilient. In addition, the project is implementing a System Dynamics simulation model to be used by cities in planning and training for resilience. The System Dynamics model is encapsulated in a web-based Interactive Learning Environment, an easy to use Graphical User Interface (GUI). The beta version of the Interactive Learning Environment was tested by the Smart Mature Resilience project partner cities' representatives and experts in the SMART MATURE RESILIENCE second review workshop in March 2017. The participants formed in groups were supported by moderators; meanwhile recorders took notes and suggestions. This paper introduces the Interactive Learning Environment and its functionalities, in addition to information reported by the recorders for further enhancements.
Keywords: city resilience, disaster resilience, interactive learning environment, resilience maturity model, resilient tool, smart mature resilience project, system dynamics

1 INTRODUCTION

'A disaster (catastrophe) can be interpreted as an event that causes great destruction/damage and human suffering' [1]. Man-made as well as natural disasters are steadily increasing over time [2]. Their effect is enormous as well, in 2014, the United Nations reported that disasters in some way or the other have affected 4.4 billion persons, killed 1.3 million lives and caused economic losses of USD 2 trillion during the two decades prior to the report [3], and according to Swiss Re Sigma Report in 2016, all over the world, 155 man-made events and 198 natural catastrophes were recorded during 2015 causing as twice as the number of deaths than in 2014 [4]. According to the same report, disasters in 2015 caused economic losses of USD 92 billion.

While most natural disasters are mostly inevitable, their physical impact and consequences can be diminished through prevention and mitigation measures [5]. In the aftermath of these disasters, the need to enhance the ability to manage and assess the cities' resilience has gained increasing importance. The United Nations International Strategy for Disaster Reduction (UNISDR) defines resilience as 'the capacity of a system, community or society potentially exposed to hazards to adapt, by resisting or changing in order to reach and maintain an acceptable level of functioning and structure' [6].

In Europe, there is a political interest and pressure to prepare cities to be resilient. As an example, in 2013, the European Commission adopted the European Union Adaptation strategies focusing on cities resilience to climate change [7]. Yet, despite this pressure and the importance of improving cities' resilience, there are currently limited examples of the

DOI: 10.2495/SAFE-V7-N3-367-379

sequential steps that cities should follow in developing resilience [8]. Furthermore, not all cities have the same resilience level, and there is a lack of guidance on which policies should be implemented as a function of the current situation of a city [9]. In addition, there is insufficient understanding of how the city stakeholders should work and collaborate to develop the city's resilience [10]. Smart Mature Resilience is a European Horizon 2020 project started in 2015, and aims at preparing and validating a Resilience Management Guideline through defined set of tools and policies that help city stakeholders enhancing the cities' resilience, and provide the cities with a robust shield against man-made and natural hazards, enabling societies to resist, absorb, accommodate and recover from the effects of a hazard in a timely and efficient manner [11]. The project will be described further in Section 3.

A selected set of these resilience policies were introduced via a mathematical model built using System Dynamics methodology. System Dynamics is a computer-aided modelling approach to capture a complex system and interactions among system's components that will generate the behaviour of the system over time. In the Smart Mature Resilience project, this System Dynamics tool is used to allow city stakeholders simulating different resilience policy options and learn about the consequences of their decisions on the maturity of the city resilience.

The System Dynamics model was encapsulated in an user-friendly Graphical User Interface (GUI) called System Dynamics based Interactive Learning Environment to facilitate better understanding of the Resilience Maturity Model, which is another tool of the project, and accordingly be used as training tool for decision makers and crisis managers to make them well prepared for future crisis. This Interactive Learning Environment is intended as a training tool for city stakeholders. The beta version of the Interactive Learning Environment was tested among approximately 30 city stakeholders of the Smart Mature Resilience project partners in combination with focus group discussions in project workshop done in Donostia/San Sebastian, Spain, March 2017. The evaluation results of the pilot testing are reported in this paper, covering the impression of the Interactive Learning Environment's GUI and the elements included in the model. The objective of this paper is to present this beta version of the Interactive Learning Environment as one of Smart Mature Resilience project tools, and discuss and analyse the results of the pilot testing of this tool during the workshop with the Smart Mature Resilience project stakeholders, in terms of these stakeholders' opinions and perceptions about the Interactive Learning Environment.

This paper is organised as follows: the next section introduces the concept of Interactive Learning Environment. Section 3 familiarises the reader with the Smart Mature Resilience project, which is followed by a brief description of the Smart Mature Resilience System Dynamics model, and the Interactive Learning Environment's GUI and its functionalities. Section 5 describes the workshop held in Donostia and presents the results of this workshop, followed by section 6 analysing these results. Section 7 makes an overall conclusion of the paper.

2 STATE OF ART: INTERACTIVE LEARNING ENVIRONMENT

Generally, an Interactive Learning Environment is defined as a 'software for educational purposes, for supporting the process of learning, where the focus is on learning through the interaction with the computer (human-computer interactivity)' [12]. Interactive Learning Environment is referred to as management flight-simulator, microworld, business simulator, or management simulator by various publications [13, 14]. However, in this paper, we do not differentiate between these terms.

A mental model, in general, could be defined as 'a construct of cognitive psychology. Mental models are internal representations of conceptual and causal interrelations among elements that people use to understand phenomena' [15].

Interactive Learning Environments are developed either for educational or research and validation purposes [16]. For educational purposes, Interactive Learning Environments wish to change the mental models of their users, while for research and validation purposes Interactive Learning Environments aim at identifying their users' mental models.

Interactive Learning Environments have been demanded lately not only because society has evolved into a more technological society [17–19], but also, there has been an evolution in the teaching field. The education has evolved from teacher-centred to learner-centred, from listening model to doing model and from memory learning to search and information use learning [17]. Furthermore, Interactive Learning Environment has the ability to engage user producing a high positive impact on them [17, 18]. For example in Ref. [17], an Interactive Learning Environment is presented with the aim of improving users intercultural awareness, knowledge and communicative competences at the university level. However, in Ref. [20] an Interactive Learning Environment called WeBlog is developed in order to demonstrate the benefits of implementing lean concepts in a company. In both cases users' attention and motivation were high, and the theoretical concepts taught by the Interactive Learning Environments were successfully transferred.

Multiple studies have shown that Interactive Learning Environments can be effective when applying to education [17, 18, 21] providing high benefits in comparison to conventional methods [22]. Ke pointed to content, context and the achieved competences as key factors for success [23]. Sitzman added the necessity of Interactive Learning Environments to be designed as an active learning tool and the importance of having Interactive Learning Environment combined with other tools to ensure its success [24]. Finally, Hamari stated that simulating real world tasks and giving instant feedback help to develop successful Interactive Learning Environments [25]. Related to what these three authors say, Wouters *et al.* [22] described three main aspects to take into account in order to assure effectiveness and success of Interactive Learning Environments: (1) changing the cognitive process (via active learning), (2) affecting motivation (simulating real world tasks for example), and (3) provide the sensation of having learnt (via instant feedback). As a final point in this section, Interactive Learning Environments have proven to be successful educational tools in the context of public policies, both on the national and city levels [26, 27].

3 SMART MATURE RESILIENCE PROJECT

Smart Mature Resilience will deliver a validated Resilience Management Guideline which covers different problem areas such as security sectors in Critical Infrastructures, as well as climate change and social dynamics. The main objective of the project is to provide through the Resilience Management Guideline a robust shield against man-made and natural hazards, enabling society to resist, absorb, accommodate and recover from the effects of a hazard in a timely and efficient manner. The Resilience Management Guideline is based on five tools: The Resilience Maturity Model, the Systemic Risk Assessment Questionnaire, the Engagement tool, the System Dynamics model and the Resilience Building Policies.

In this paper the first version of the System Dynamics model is presented and discussed. The System Dynamics model is encapsulated in an Interactive Learning Environment and based on the policy implementation of each Resilience Maturity Model stage developed on the Smart Mature Resilience project. The System Dynamics model aims to build a tool that facilitates decision makers and crisis managers to better understand the Resilience Maturity Model and therefore train them to be well prepared for future crisis.

The Resilience Maturity Model engages a growing number of stakeholders and multi-level governance in order for cities to become more resilient, defining five maturity stages: *Starting*, *Moderate*, *Advanced*, *Robust*, and *verTebrate* (SMART). Each of these maturity stages includes a description of the objectives of each stage, the agents involved in each maturity stage and a set of resilience building policies to implement in order to reach the objective of each stage and move forward. Therefore, policies are classified depending on their maturity stage and their resilience dimension (see Figure 1). In addition to the five maturity stages, the Resilience Maturity Model is structured in four resilience dimensions: Leadership and Governance, Preparedness, Infrastructure and Robustness and Cooperation. These four dimensions are also divided in sub-dimensions having a total of 10. The presented policies are inter-related between each other with predecessor relationships. Therefore, policy implementation order will determine the efficiency of how the maturity stages are obtained as before applying some policies others should have been implemented before.

4 SMART MATURE RESILIENCE SYSTEM DYNAMICS MODEL

The System Dynamics based Interactive Learning Environment introduced in this paper includes a pilot version as it contains 19 out of the 98 policies the Resilience Maturity Model has. The aim of the Interactive Learning Environment is to provide a tool that enables decision makers and crisis managers to train themselves and better understand the Resilience Maturity Model in order to achieve higher resilience levels for their cities and understand the path toward building resilience. The System Dynamics model and Interactive

RESILIENCE DIMENSIONS		MATURITY STAGES				
		STARTING	MODERATE	ADVANCED	ROBUST	VERTEBRATE
Leadership & Governance	L1	Policy of starting stage L dimension	...	...	...	...
	L2	...	...	...	...	...
	L3	...	...	...	...	...
	L4	...	...	...	...	...
Preparedness	P1	...	...	...	...	...
	P2	...	...	...	...	...
Infrastructures	I1	...	...	...	...	...
	I2	...	...	...	...	...
Cooperation	C1	...	...	...	...	...
	C2	...	...	...	...	Policy of vertebrate stage C dimension

Figure 1: Resilience Maturity Model structure.

Learning Environment were built with the intention of serving users for decision making and mental model enhancement. To develop this Interactive Learning Environment, we were faced with two alternatives; desktop/mobile application and cloud/web-based. Several features including cost, agility, and scalability were provided solely by the cloud/web-based solution [28]. Moreover, it provides the possibility to be accessed anywhere, without the need for particular software installation [28–31]. Accordingly, we went for the cloud/web-based solution.

Several tools could provide the cloud/web-based solution like *Forio Online Simulations*, *iMODELER*, *Insight Maker*, *Sysdea*, *isee Exchange*, and *BROADVIEW*. InsightMaker is distinguished by being totally client-side without any need for a server other than a web-server/web-hotel, in addition to being an open-source tool. Going for the InsightMaker solution saves the cost of having a server dedicated for model solution every time a user wants to simulate her/his scenario, and provides the freedom to distribute the solution among several cities without any additional cost. Furthermore, the InsightMaker solution is integrated, as there is no need for an additional System Dynamics model building software. The following subsection introduces the Smart Mature Resilience System Dynamics Model in a nutshell, followed by a subsection that describes the Smart Mature Resilience Interactive Learning Environment more detailed.

4.1 Smart Mature Resilience System Dynamics Model

The System Dynamics model represents the structure of the Resilience Maturity Model and determines the relationships between policies. As explained in Section 3, the Resilience Maturity Model is composed of five different maturity stages and four resilience dimensions, which are divided into different sub-dimensions. As a consequence of this structure, two types of relationships have been defined: linear relationships and transversal relationships (see Figure 2).

Linear relationships refer to temporal relationships that exist among the different maturity stages. This means that policies in the lower maturity stages should be developed in order to implement the policies in the higher maturity stages. For example, within each sub-dimension, policies in the 'Starting' stage should be developed in order to implement the policies in the 'Moderate' stage, and similarly, policies in the 'Moderate' stage should be developed in order to implement the policies in the 'Advanced' stage.

Transversal relationships refer to relationships among the policies in different dimensions and sub-dimensions. Although the policies have been divided into different sub-dimensions, these sub-dimensions are interrelated with each other. Therefore, within each maturity stage, the relationships among the sub-dimensions have been defined (see Fig. 3). These transversal relationships are maintained from one stage to the next one.

The Smart Mature Resilience System Dynamics model was built taking both linear and transversal relationships into consideration. As a result, when the policies are not implemented in the correct order, they will not be effective, and consequently, the user can inefficiently spend money on implementing policies out of order, without improving the city resilience level in terms of four different indicators computed by the model, namely: 'Leadership and Governance', 'Preparedness', 'Infrastructure and Resources', and 'Cooperation' indicators. The indicators' value is obtained through the average of the sub-dimensions, which are calculated based on the sum of the implementation rate of each policy on that sub-dimension multiplied by its effectiveness.

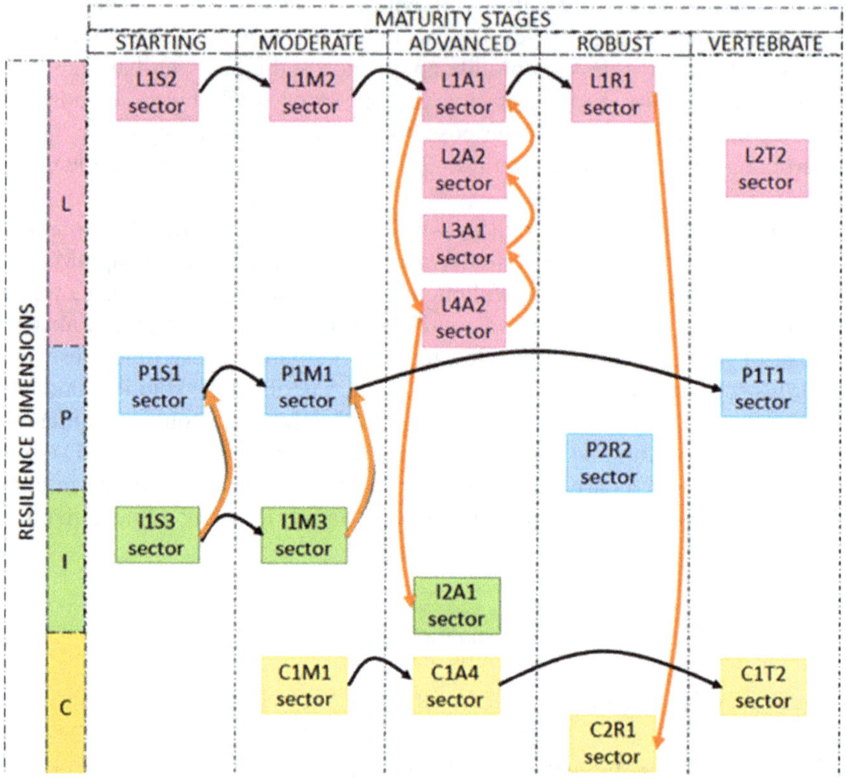

Figure 2: Linear and transversal relationships in the simplified Resilience Maturity Model. Black arrows for linear relationships and orange arrows for transversal relationships.

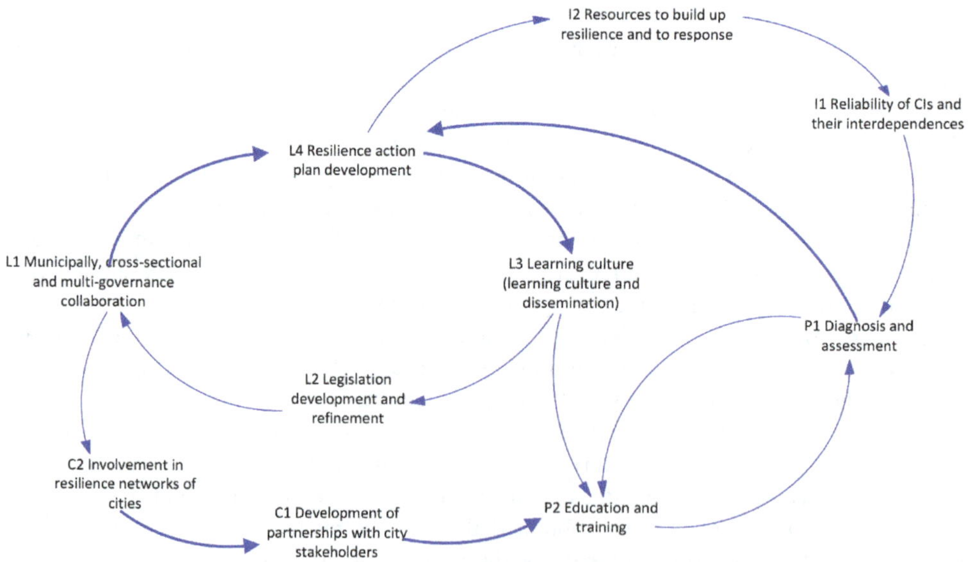

Figure 3: Transversal relationships at sub-dimension level.

4.2 Smart Mature Resilience System Dynamics based Interactive Learning Environment

The Interactive Learning Environment's GUI as mentioned above is web-based. It was developed using web technologies HTML, CSS, and JavaScript. The GUI is structured in three different views; *Initial* view (see Fig. 4), *Decisions* view (see Fig. 5), and *Results* view (see Fig. 6). The 'Initial' view enables users to enter the city's current resilience level, either by directly selecting one of the available SMART stages, or by setting individual policies' current implementation values. In this view, all the applicable policies organized according to their resilience dimension are shown with corresponding sliders to set these initial values. This view has two buttons: one button to recall a dialogue-box which sets the current SMART stage, and another button to change the view to the 'Decisions'. Similar to that 'Initial' view, the 'Decision' view shows all the applicable policies organized according to their resilience dimension. In addition, this view, shows the available budget, the current year of simulation, and 'Simulate' and 'Reset' buttons to advance simulation one step ahead and to stop the current simulation and start a new one respectively. This view also has a button to change the view to the 'Results' view. The 'Results' view summarizes the consequences of the taken

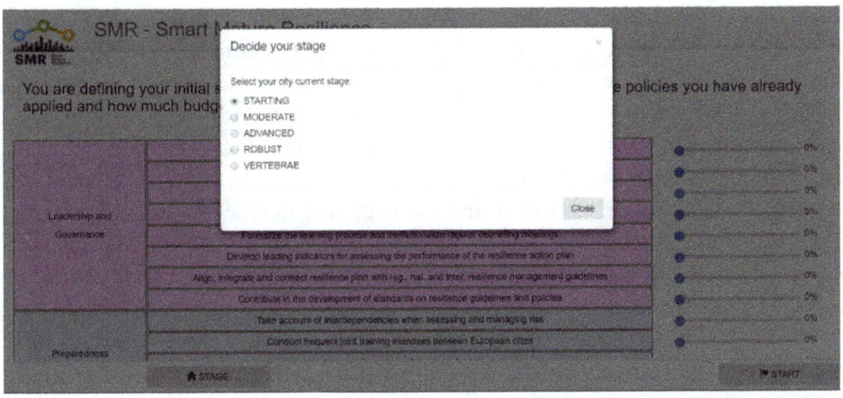

Figure 4: Initial view screen.

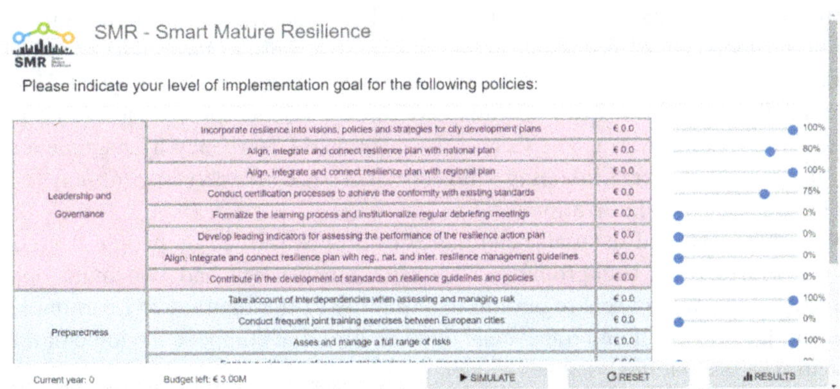

Figure 5: Decision screen.

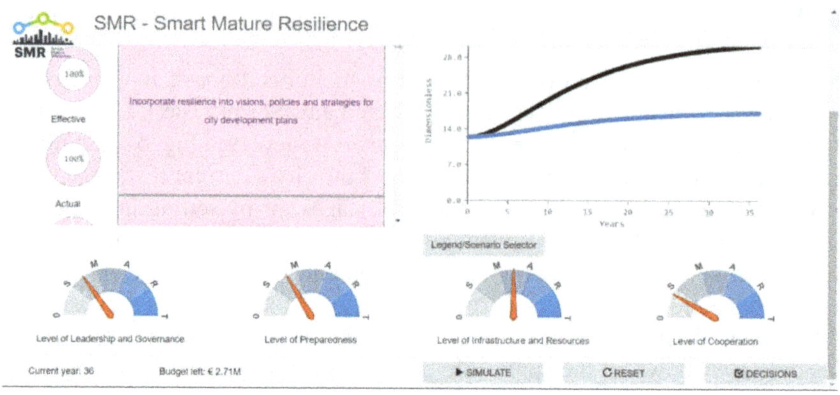

Figure 6: Results screen.

decisions through behaviour-over-time graphs for each of the four indicators mentioned above, implementation percentages for every policy and speedometers showing the resilience SMART maturity stage on the level of each of the four different indicators. Indicators behaviour-over-time graphs were able to show three different scenarios selected by the user, the code behind this behaviour-over-time graphs was adapted from Ref. [26, 27]. In this view the available budget, the current year of simulation, the 'Simulate' and 'Reset' buttons are shown, as well as a button to change the view back to the 'Decision' view.

The user will firstly establish the initial situation defining which maturity level the city is in. Afterwards, the simulation will start and the user will decide which policies implement and up to what percentage. In order to obtain some results and study the consequences of the implemented policies, the user will press the 'Simulate' button stepping forward two years every time. Consequently, the available budget will decrease and the 'Results' view will show the effects of the taken decisions. Apart from that, depending on the taken decisions, every four years the users will receive some pop-up help messages showing the errors they are committing regarding the precedence relationships and suggestions about which policies should be implemented to improve their performance.

5 WORKSHOP

In order to validate and assure the effectiveness of the Smart Mature Resilience pilot Interactive Learning Environment, a one-day workshop was carried out the 6th of March of 2017 with a total of 30 participants from the cities of *Bristol*, *Rome*, *Riga*, *Glasgow*, *Vejle*, *Kristiansand* and *Donostia* in addition to Smart Mature Resilience project academic partners: Tecnun, University of Navarra, The Centre for Integrated Emergency Management (*CIEM*) of the University of Adger, Linköping University (*LIU*), *University of Strathclyde* and the German Institute for Standardization (*DIN*).

The participants worked in small heterogeneous groups formed by people from different origins and with diverse backgrounds. The validation process had two main objectives: receive feedback from the users regarding the usability of the Interactive Learning Environment and validate the linear and transversal relationships that compose the logic of the causal relationships in the System Dynamics model.

Overall, the comments obtained from the participants were positive and they considered that the Interactive Learning Environment was useful for cities to better understand the

Resilience Maturity Model and therefore to train them to be well prepared for future crisis. However, they also made several comments to improve the tool.

6 RESULTS

The received feedback comments were classified depending on whether the comment was related to the System Dynamics model or the GUI.

6.1 The System Dynamics Model

Users found the results given by the model were coherent, yet some could be more realistic. In the following list, the most highlighted comments and suggestions are presented:

1. The Interactive Learning Environment defines the implementation level of a policy through a percentage applied by the user. However, participants pointed it would be more realistic to calculate the implementation level of a policy deciding the amount of resources the user wants to invest in each policy.
2. Regarding policy cost, the pilot version assumed all the policies cost the same. Nevertheless, the participants said it was not realistic to be so, each policy should have different unitary cost and their maintenance should be different too.
3. The Interactive Learning Environment defines the available budget from the beginning, and they pointed out it was not a realistic situation. Therefore, they suggested having an annual budget which could be defined by the user at the beginning and also have the opportunity to change it in the middle of the game in order to represent budget cuts.
4. The Interactive Learning Environment is based on policies' relationships which are defined with the same weight and importance. Participants suggested that in order to be more realistic, both linear and transversal relationships should have stronger or weaker relationships for each case and be more or less important to achieve resilience.

Once the suggestions were discussed, the 1st and 3rd comments were implemented in the new version of the model. For the cases of the 2nd and 4th suggestions, two questionaries' have been carried out in order to parameterize the values of the model.

6.2 The Graphical User Interface

In general, participants found the GUI visually easy to understand, yet not so intuitive. In the following list, the most highlighted comments and suggestions are presented:

1. In the GUI, only the names of the policies are shown, yet they are not defined. The participants suggested a brief explanation of the policy appear when the mouse hover over the policy name. This should help a user to better understand what he/she is implementing.
2. The policies in the GUI were classified only depending on their resilience dimension. However, the participants also asked to sort them depending on the maturity stage to facilitate understanding the implementation order.
3. The resulting graphs and percentages were just defined in the model's user guide. The participants suggested adding a brief explanation of the meaning of these graphs and percentages which appear when the mouse hover over these elements.
4. When defining the initial situation of the city in the Initial page, the participants were confused, as they thought they were already doing the simulation. Therefore, they asked

to differentiate more clearly when the user is defining the initial state and when the simulation starts.

5. Related to the indicators, they commented that the actual indicators were useful, yet the evolution of the spent money should also appear as a result. They suggested adding both a general behaviour-over-time graph with the evolution of the total budget and a specific graph with the budget spent on each resilience dimension as well.
6. In general, the comments related to the feedback messages were positive. However, a group of the participants found the messages to be too long to read comfortably. This group suggested highlighting the policies appearing in the feedback messages in the decisions-screen to easily identify them.
7. Considering that the Interactive Learning Environment the participants were using was a pilot version, they suggested that the final version should give them the opportunity to save the obtained results and be able to share them with colleagues.

After analysing the suggestions, some decisions were taken. For the suggestions 1, 2, 3, 5 and 6 the comments of the participants were implemented in the updated version of the model. In the case of suggestion 4, a third screen with a brief explanation of the model has been designed which appears at the beginning of the simulation. Finally, for the 7th suggestion, the updated version not only enables to save the results but to download them and send them to any other player.

7 CONCLUSION

Disasters are increasing significantly lately leading to a pressing need for developing city resilience. However, there is a lack of practical examples, which provide sequential steps toward building resilience. In this context, the European project 'Smart Mature Resilience' appears with the objective of developing Resilience Management Guideline based on 5 tools. One of these tools is a System Dynamics model, which is encapsulated in an Interactive Learning Environment. Interactive Learning Environments have been recently used for didactical objectives to train and give users better understanding of the issues addressed, as they have shown to provide high benefits compared to traditional methods. Therefore, the main objective of our System Dynamics model is to provide a tool to better understand both the Resilience Maturity Model and the most appropriate path to gain resilience.

This paper presents a beta version of the System Dynamics Interactive Learning Environment developed for the Smart Mature Resilience project. The presented beta version is evaluated and validated through a workshop from which the most highlighted comments are also presented and discussed in this paper. The paper concludes with a set of requirements the Interactive Learning Environment and the System Dynamics model must fulfil in the final version.

ACKNOWLEDGEMENT

Smart Mature Resilience is a European project funded by European Union's Horizon 2020 (H2020-EU.3.7., project ref. 653569) which started the 1st of June of 2015 and will finish the 31st of May of 2018.

REFERENCES

[1] Kundzewicz, Z.W., Disaster aftermath. In *Encyclopedia of public health*, W. Kirch, Ed. Springer Netherlands, pp. 269–284, 2008. Available at: http://link.springer.com/referenceworkentry/10.1007/978-1-4020-5614-7_811. (Accessed 08 May, 2017)

[2] Perrow, C., *Next catastrophe: reducing our vulnerabilities to natural, industrial, and terrorist disasters*. Princeton University Press, 2011. Available at: http://www.jstor.org/stable/j.ctt7t4c1

[3] United Nations Development Programme, *A call for action: disaster risk reduction makes development sustainable*, 2014 [Online]. Available at: http://www.undp.org/content/dam/undp/library/crisis%20prevention/UNDP_CPR_CTA_20140901.pdf. (accessed 08 May, 2017).

[4] Swiss Re, *Natural catastrophes and man-made disasters in 2015: Asia suffers substantial losses*, 2016. Available at: http://media.swissre.com/documents/sigma1_2016_en.pdf. (accessed 08 May, 2017).

[5] de Ville de Goyet, C., Marti, R.Z. & Osorio, C., Natural disaster mitigation and relief. In *Disease control priorities in developing countries*, 2nd ed., D.T. Jamison, J.G. Breman, A.R. Measham, G. Alleyne, M. Claeson, D.B. Evans, P. Jha, A. Mills, & P. Musgrove, Eds. Washington, DC: World Bank, 2006. Available at: http://www.ncbi.nlm.nih.gov/books/NBK11792/. (Accessed 08 May, 2017).

[6] United Nations, *Living with Risk: a global review of disaster reduction initiatives*. New York: UNO-Verl, 2004 [Online]. Available at: http://www.unisdr.org/we/inform/publications/657. (Accessed 08 May, 2017)

[7] "What is the EU doing?," *Climate Action - European Commission*, 23-November-2016. [Online]. Available at: https://ec.europa.eu/clima/policies/adaptation/what_en. (accessed 08 May, 2017).

[8] Valdés, H.M., Amaratunga, D. & Haigh, R., Making cities resilient: from awareness to implementation. *International Journal of Disaster Resilience in the Built Environment*, **4**(1), pp. 5–8, 2013. https://doi.org/10.1108/17595901311299035

[9] Oteng-Ababio, M., Sarfo, K.O. & Owusu-Sekyere, E., Exploring the realities of resilience: case study of Kantamanto market fire in Accra, Ghana. *International Journal of Disaster Risk Reduction*, **12**, pp. 311–318, 2015. https://doi.org/10.1016/j.ijdrr.2015.02.005

[10] Singh-Peterson, L., Salmon, P., Baldwin, C. & Goode, N., Deconstructing the concept of shared responsibility for disaster resilience: a Sunshine Coast case study, Australia. *Natural Hazards*, **79**(2), pp. 755–774, 2015. https://doi.org/10.1007/s11069-015-1871-y

[11] "Smart Mature Resilience Project," *Smart Mature Resilience Project*. [Online]. Available at: http://www.smr-project.eu/home/. (accessed 07 May, 2017)

[12] Sterman, J., *Business dynamics: systems thinking and modelling for a complex world*. McGraw Hill Higher Education, 2000.

[13] Maier, F.H. & Größler, A., What are we talking about?—A taxonomy of computer simulations to support learning. *System Dynamics Review*, **16**(2), pp. 135–148, 2000. https://doi.org/10.1002/1099-1727(200022)16:2<135::aid-sdr193>3.0.co;2-p

[14] Qudrat-Ullah, H., *Better decision making in complex, dynamic tasks: training with human-facilitated interactive learning environments*. Springer, 2014.

[15] Groesser, S.N., Mental model of dynamic systems. In *Encyclopedia of the sciences of learning*, P.D.N.M. Seel, Ed. Springer US, pp. 2195–2200, 2012 [Online]. Available at: http://link.springer.com/referenceworkentry/10.1007/978-1-4419-1428-6_1838. (Accessed 04 February, 2015).

[16] Davidsen, P.I., Issues in the design and use of system-dynamics-based interactive learning environments. *Simulation & Gaming*, **31**(2), pp. 170–177, 2000. https://doi.org/10.1177/104687810003100204

[17] Guillén-Nieto, V. & Aleson-Carbonell, M., Serious games and learning effectiveness: The case of It's a Deal! *Computers & Education*, **58**(1), pp. 435–448, 2012. https://doi.org/10.1016/j.compedu.2011.07.015

[18] Mayer, I., Bekebrede, G., Harteveld, C., Warmelink, H., Zhou, Q., Van Ruijven, T., Lo, J., Kortmann, R. & Wenzler, I., he research and evaluation of serious games: toward a comprehensive methodology. *British Journal of Educational Technology*, **45**(3), pp. 502–527, 2014. https://doi.org/10.1111/bjet.12067

[19] Kiili, K., Digital game-based learning: towards an experiential gaming model. *The Internet and Higher Education*, **8**(1), pp. 13–24, 2005. https://doi.org/10.1016/j.iheduc.2004.12.001

[20] Kuriger, G.W., Wan, H., Mirehei, S.M., Tamma, S. & Chen, F.F., A web-based lean simulation game for office operations: training the other side of a lean enterprise. *Simulation & Gaming*, **41**(4), pp. 487–510, 2010. https://doi.org/10.1177/1046878109334945

[21] Kebritchi, M., Hirumi, A. & Bai, H., The effects of modern mathematics computer games on mathematics achievement and class motivation. *Computers & Education*, **55**(2), pp. 427–443, 2010. https://doi.org/10.1016/j.compedu.2010.02.007

[22] Wouters, P., van Nimwegen, C., van Oostendorp, H. & van der Spek, E.D., A meta-analysis of the cognitive and motivational effects of serious games. *Journal of Educational Psychology*, **105**(2), pp. 249–265, 2013. https://doi.org/10.1037/a0031311

[23] Ke, F., A qualitative meta-analysis of computer games as learning tools. *Gaming and Simulations*, 2009.

[24] Sitzmann, T., A meta-analytic examination of the instructional effectiveness of computer-based simulation games. *Personnel Psychology*, **64**(2), pp. 489–528, 2011. https://doi.org/10.1111/j.1744-6570.2011.01190.x

[25] Hamari, J., Koivisto, J. & Sarsa, H., Does gamification work? - A literature review of empirical studies on gamification. *Proceedings of the Annual Hawaii International Conference on System Sciences*, pp. 3025–3034, 2014.

[26] Abdelgawad, A., Radianti, J., Snaprud, M. & Krogstie, J., Experimentation with a system dynamics based interactive learning environment: a case study of accessibility of Norwegian municipalities websites. *International Journal of Simulation Systems, Science & Technology*, **17**(35), pp. 11, 2016.

[27] Abdelgawad, A.A., Radianti, J., Snaprud, M.H. & Krogstie, J., Disabled people from welfare to jobs: an interactive learning environment experimental investigation. In *Proceedings of the 34rd International Conference of the System Dynamics Society*, Delft, The Netherlands, 2016. Available at: http://www.systemdynamics.org/conferences/2016/proceed/papers/P1158.pdf. (accessed 09 Feburary, 2017).

[28] Nakazawa, M., Koizumi, D. & Hirasawa, S., The influence of QoS on e-learning environment under virtual desktop infrastructure, pp. 174–178, 2012.

[29] Aljenaa, E., Al-Anzi, F.S. & Alshayeji, M., Towards an efficient e-learning system based on cloud computing. In *Proceedings of the Second Kuwait Conference on e-Services and e-Systems*, New York, NY, USA, pp. 13:1–13:7, 2012. Available at: http://doi.acm.org/10.1145/2107556.2107569. (accessed 22 November, 2016).

[30] Kitanov, S. & Davcev, D., *Mobile cloud computing environment as a support for mobile learning*. pp. 99–105, 2012.

[31] Masud, M.A.H. & Huang, X., A novel approach for adopting cloud-based e-learning system. *2012 IEEE/ACIS 11th International Conference on Computer and Information Science*, pp. 37–42, 2012.

INTELLIGENCE ANALYSTS' TRAINING THROUGH SERIOUS GAMES: THE LEILA PROJECT

A. ZANASI, F. RUINI & A. BONZIO
Zanasi & Partners, Italy.

ABSTRACT
Among the various initiatives directed at helping the intelligence community tackling today's security challenges, the EU FP7 research project LEILA (Law Enforcement and Intelligence Learning Applications) focuses on 'serious games' for improving the quality of intelligence analysts' training. LEILA investigates innovative methodologies for the development of core competencies required for performing intelligence analysis. Building upon an in-depth study of intelligence analysts' learning needs and inspired by a multitude of disciplines, LEILA developed a set of serious games, embedded into four comprehensive 'learning experiences' dedicated to intelligence analysis trainees. While the main target of LEILA is law-enforcement intelligence analysts, the competencies addressed by its learning experiences are potentially relevant to analysts operating across various sectors.
Keywords: cognitive biases; inference schemes; intelligence analysis, law enforcement; learning; LEILA; security; serious games; training.

1 INTRODUCTION

Over the last few years, security has become a top priority on the European political agenda. The rise of new threats, together with radical changes in the way people live, work and communicate, have increased the demand for sound intelligence analysis, while creating a set of complex challenges for security services. Improving the quality of intelligence analysis by means of more effective forms of training has been recognised as a key priority for coping with the challenges posed by today's scenarios. Following the multiple initiatives from the US for testing new ways to enhance the quality of training in intelligence analysis, the European Commission (EC) has also started taking steps in this direction. In 2014, the LEILA (Law Enforcement Intelligence Learning Application) project was launched under the EC's Seventh Framework Programme for Research and Technological Development (FP7) with the aim to provide law enforcement agencies with innovative training methodologies for their intelligence analysts, based on serious games.

The aim of this paper is to provide an overview of the work carried out during the LEILA project and to highlight the novelty of its findings. The paper begins by defining some of the key concepts underpinning the LEILA project, from 'law enforcement intelligence' to 'intelligence cycle'. This is followed by a review of some current trends and challenges in intelligence analysis training, with a particular focus on the growing popularity of serious game technology as a learning tool. The paper continues with a description of the main findings of LEILA as well as with an overview of the organisations involved in the project. The final section of the paper sheds some light on the most significant elements of innovation introduced by the LEILA project.

2 LAW ENFORCEMENT INTELLIGENCE AND THE INTELLIGENCE CYCLE

Within the realm of intelligence analysis (IA), the boundaries between governmental and corporate intelligence are progressively blurring. Both disciplines focus on technology, rely on sources such as social media, use text and data mining technologies. Often, the private

DOI: 10.2495/SAFE-V7-N3-380-389

sector takes advantage of governmental intelligence to make business (e.g. McDonnell Douglas vs. Airbus or Raytheon vs. Thomson-Alcatel thanks to ECHELON). At the same time, government authorities often ask the private sector to participate in the development of intelligence (e.g. FBI-Apple encryption dispute, Yahoo! email scandal).

Despite some major overlaps, differences between law enforcement intelligence and other kinds of intelligence activities still exist. According to Carter & Schafer [1], national security intelligence deals with *'social, political, economic, and military issues relating with a nation's stability and safety'*. As the decisions it supports are often political in nature, national security intelligence does not always respond to strict legal requirements. On the contrary, law enforcement intelligence, being ultimately aimed at shaping the decisions and operations of police bodies, is subject to strict legal constraints.

The primary focus of LEILA is on intelligence activities conducted by law-enforcement agencies. Therefore, we borrow from Carter [2] a more comprehensive definition of the concept of law enforcement intelligence as: *'the end product (output) of an analytic process that collects and assesses information about crimes and/or criminal enterprises with the purpose of making judgments and inferences about community conditions, potential problems, and criminal activity with the intent to pursue criminal prosecution or project crime trends to support informed decision making by [law enforcement] management.'*

The concept permeating all intelligence activities is the 'intelligence cycle'. Underlying this concept is the notion that the production and use of intelligence follows a cyclical process consisting of multiple and interdependent steps. Every step of the cycle adds value to the initial input and contributes to its transformation into a completely different product [3]. According to its traditional interpretation, the intelligence cycle comprises five phases [4]: planning and direction, collection, processing, analysis and production, dissemination.

2.1 Planning and direction

This stage refers to the management of the intelligence cycle and the attribution of the single intelligence tasks. During this phase, the intelligence consumer sets out the requirements for the final intelligence product and plans the various intelligence activities needed for its achievement [5]. These activities include establishing the information needs, deciding how information is to be collected as well as defining a timetable for its collection. Requirements may also concern the particular form in which the final intelligence product must be delivered (e.g. a full report, a graph, a single piece of raw information, etc.).

2.2 Collection

A wide range of tools and techniques are employed by intelligence and law enforcement agencies for gathering raw data from all available sources. Collection methods depend on the nature of the information to be collected and the characteristics of the media it is transmitted through. There are five main categories of intelligence sources: Geospatial Intelligence (GEOINT), Human Intelligence (HUMINT), Measurement and Signature Intelligence (MASINT), Open-Source Intelligence (OSINT) and Signals Intelligence (SIGINT) [5]. To these, one may add a set of sub-disciplines, such as Imagery Intelligence (IMINT) and social media intelligence (SOCMINT), which have emerged in parallel to the spread of new technologies.

2.3 Processing

Once collected, the information needs to be converted into a format suitable for analysis. To transform raw and unstructured data into structured and unambiguous information, highly trained personnel and advanced technologies are required [5]. While making unprecedented amounts of data available to intelligence analysts, the rapid growth of IT technologies has created tools allowing to effectively process such data without incurring in the problem of information overload. Processing methods range from decryption, language translations and data reduction [6] to techniques such as data and text mining [7].

2.4 Analysis and production

The main purpose of this phase is to extract meaning from available data, converting raw inputs into a usable intelligence product [8, 9]. Such process requires '*the integration, evaluation, and analysis of all available data and the preparation of various intelligence products*' [9]. An intelligence product may consist either of single pieces of information being disseminated individually (sometimes referred to as 'the dots') – or of finished intelligence reports in which such dots are connected and conclusions about their meaning presented [6].

2.5 Dissemination

During the dissemination phase, the final intelligence product is delivered to the customers, who will then use it to support their decision-making. Dissemination may also induce the intelligence consumers to provide feedback on the received product, thus resulting in a new spin of the intelligence cycle.

3 CHALLENGES AND TRENDS IN INTELLIGENCE ANALYSIS TRAINING

Following the attacks on September 11th 2001 and the military invasion of Iraq in 2003, the US and their allies have devoted a great deal of resources to prevent future intelligence failures by improving the quality of intelligence analysts' training [10]. In the US, new training centres, either independent or belonging to larger organisations, have been set up for this specific purpose. In addition, training courses and curricula with a marked practical component have been introduced by both governmental agencies (e.g. the CIA, with its Career Analyst Programme) and academic institutions (e.g. the Intelligence Studies Department at Mercyhurst College, PA). Similar efforts, though of a lesser scope, have been undertaken by European countries like the UK, as witnessed by the recent flourishing of academic courses in intelligence (e.g. the Intelligence programme offered by the War Studies Department at King's College London [11]). Teaching methods have also undergone important changes over the past few decades. According to Marrin [10], the key to good intelligence analysis is no longer represented by the analyst's expertise on a specific issue but, rather, by the analyst's own reasoning skills. That is why training programmes in intelligence analysis are increasingly focused on structured analytical techniques, such as brainstorming, devil's advocacy, red team analysis, Team A-Team B, Analysis of Competing Hypotheses (ACH), key assumptions check and alternative futures.

Despite such transformations, many problems still undermine the effectiveness of analytic training in the intelligence sector. One of the weaknesses stems from the absence of a common training programme for intelligence analysts at entry-level, with single agencies still maintaining the prerogative to instruct their trainees according to their specific mission and needs [12]. Another challenge has to do with the specific skills being taught during the training. A 2011 study of five different training providers in the UK, including private companies, law enforcement agencies and other public bodies, concluded that none of these institutions offered appropriate training for intelligence analysis [13]. Some of the programmes overestimated the statistical dimension of intelligence analysis at the expense of the investigative one, while others placed undue constraints upon the work of intelligence analysts by granting managers a control function over the analytical process. On a similar note, it has been observed that most currently available training programmes fail to adequately teach critical thinking skills: while training material on critical thinking is often included within courses, the reasoning process behind critical thinking tends to be addressed only minimally [14].

Another skill only marginally dealt with by current training programmes in intelligence analysis is *cognitive biases mitigation*. Richard J. Heuer, author of the first systematic study of cognitive biases applied to the intelligence domain [15], defined cognitive biases (CBs) as '*mental errors caused by our simplified information processing strategies*'. According to Heuer, CBs induce cognitive shortcuts and deviations in judgment based on factors such as '*memory, experience, education, cultural background, political or ideological bias, heuristics or simple rules of thumb*'. As an unconscious phenomenon, cognitive biases cannot be prevented just by being aware of their existence. Their mitigation requires deeper forms of intervention, targeting the very cognitive dynamics underlying people's reasoning.

While CBs are known to have a substantial impact on the work of intelligence analysts, few solutions have been proposed to reduce their negative effects. To fill such gap, the European Commission recently funded the RECOBIA (REduction of COgnitive Biases in Intelligence Analysis, http://www.recobia.eu) project, a three-year research initiative involving researchers from 9 different European organisations. Running from 2012 to 2015, RECOBIA resulted in the identification of a set of mitigation strategies pertaining to the realm of software tools, organisational measures and training. In the latter domain, the project highlighted the benefits of serious games-based training for enhancing analysts' awareness of cognitive biases and helping them to mitigate the associated negative effects.

4 SERIOUS GAMES

Serious games are tools – often digital in nature – exploiting techniques and processes typical of the entertainment sector for purposes other than entertainment [16]. Their popularity as a training tool stems from their capability to simulate real-world situations and foster the acquisition of skills otherwise difficult to develop. Compared to more traditional forms of learning, game-based learning is considered more suitable to provide dynamic, active and implicit learning modes [17]. In particular, gameplay has become highly valued as an educational tool for the possibility of teaching through experiential and declarative learning, by allowing players to fail and to try again [18].

Serious games so far have been applied to several domains, including business, military and healthcare. Over the past decade, however, the US intelligence community has demonstrated a particular interest in their use for improving the training of intelligence analysts. In 2011, the IARPA (Intelligence Advanced Research Projects Activity) launched a 10 million dollar-worth research initiative aimed at developing serious games to '*train participants and measure their proficiency in recognizing and mitigating the cognitive biases that commonly affect all types of intelligence analysis*' [19]. Known as the SIRIUS Programme, the project resulted in the design of a set of computer games addressing various cognitive biases, like 'confirmation bias', 'projection bias', 'bias blind spot', 'anchoring bias', 'representativeness bias' and 'fundamental attribution error'. In one of those serious games, called MACBETH, the player impersonates an intelligence analyst tasked with preventing an impending terrorist attack under increasing time pressure. The game contains a specific section dedicated to mitigating the fundamental attribution error bias, in which the player must review past case files and make threat assessments on profiles of real-life individuals. As a de-biasing mechanism, the game encourages the player to rely on situational - as opposed to dispositional - cues by only allowing correct answers informed by at least 3 situational cues to be rewarded with the possibility of unlocking additional intelligence and progressing with the game [17, 20].

The same set of cognitive biases targeted by the SIRIUS programme has been the focus of another serious game developed in 2013, called 'The Mind's Lie', the game was originally designed as a table-top game, but it has now become available also in the form of an Android app, which can be freely downloaded. The game requires players to examine different real-world scenarios, spot the cognitive biases associated with each scenario and share those findings with the other players. Then, the game foresees that each player has to convince the others about the validity of his/her interpretation [21]. To mitigate cognitive biases, the game uses different pedagogical strategies, including 'peer-learning' and 'retrieval practice' [22]. The former refers to an educational approach in which learning objectives are achieved by students through interactions with other students. The latter relies instead on the idea that constantly recalling details, strategies and patterns from memory facilitates this recall over time.

5 THE LEILA PROJECT

Year 2014 saw the launch of one of the first European initiatives dedicated to the development of serious gaming solutions for improving the training of law enforcement intelligence analysts. Funded by the European Commission (EC) under the Seventh Framework Programme (FP7), the LEILA (Law Enforcement Intelligence Learning Applications, http://www.leila-project.eu) project aims to provide law enforcement agencies with an innovative training methodology based on a set of 'learning experiences' designed to improve cognitive capabilities and reasoning skills both at individual and group level. The ultimate goal of LEILA is the production of effective and innovative serious games to be deployed flexibly to security actors as well as beyond.

The LEILA project involved the following organisations:

- Center for Security Studies (KEMEA, http://www.kemea.gr): research centre of the Greek Ministry of Public Order and Citizen Protection, focused on security policies, acting as project coordinator;
- Alpha Labs (http://www.alpha-simulations.com): French spin-off company of the INSEAD's Centre for Advanced Learning Technologies (CALT), specialised in the development and distribution of advanced learning technologies and Web 2.0 tools;
- FVA (http://www.fvaweb.it): Italian company operating in the field of ICT and new media communications;
- GLOBO Technologies SA (http://www.globogr.com): Greek IT company specialised in the design and implementation of integrated advanced technological services;
- ORT France (http://www.ort.asso.fr): member of the ORT World network, it is an NGO with expertise in innovative strategies related to ICT involved in research, consulting and technology services;

- 'Carol I' National Defence University (NDU, http://www.unap.ro): highest military education institution of the Romanian Ministry of Defence, involved in research and training the area of national security and defence;
- Zanasi & Partners (Z&P, http://www.zanasi-alessandro.eu): Italian SME specialised in research, training and advisory on security, intelligence and their enabling technologies.

The LEILA consortium developed four distinct learning experiences: the 'VUCA (Volatility, Uncertainty, Complexity and Ambiguity) Challenge', the 'WhatATeam! Challenge', 'LabRint' and 'Cyberint'.

5.1 The 'VUCA (Volatility, Uncertainty, Complexity and Ambiguity) Challenge'

The 'VUCA Challenge' learning experience is designed to test the ability of participants to deal with crisis scenarios, within a context of limited time availability, by filtering through large amounts of information and interactions in order to solve a number of complex situations. In doing so, it addresses additional competencies relevant to the work of intelligence analysts, like the ability to mitigate cognitive and behavioural biases such as the 'overconfidence bias' and the 'fundamental attribution error' bias".

The VUCA Challenge begins with a game focusing on overconfidence and estimates quality. The player is asked to provide a series of estimates within 15-second time intervals before being debriefed by the instructor. This game is followed by a learning module based on a security scenario (inspired by the 2014 football World Cup in Brazil), in which players have access to a wide array of information and are tasked with demonstrating their understanding of the situation as well as assessing different types of threats emerging throughout the scenario. The following module consists in a simulation exercise (the 'WhatADay! Challenge') in which the participant has to prove his/her ability to solve three problems in parallel, operating under time pressure and dealing with large amounts of information sources and competing requests that appear in real-time. At the end of the simulation, similarly to what happens at the end of the first module (estimates game), the trainee takes part in a debriefing session focused on the analysis of his/her performance. The VUCA learning experience can also include a follow-up stage in which trainees are asked to implement, in their daily job, an individual action plan agreed with the instructors following the results achieved using the serious games. After a three-month period, the trainees report to the instructor about the effects deriving from following the action plan.

5.2 The 'WhatATeam! Challenge'

The 'WhatATeam! Challenge' learning experience focuses on the development of the key competences needed for successfully operating in diverse and distributed teams, as required by several cross-organisational and international cooperation situations that intelligence analysts face. The critical competencies addressed fall under the umbrella term 'enhanced collaboration skills'. The primary function of these skills is to enable good performance at individual and team level, as well as at organisational and inter-organisational level.

The experience is articulated into progressive learning stages. It starts with a module focused on how to identify barriers to effective collaboration when dealing with complex organisational or inter-organisational scenarios. The reference scenario addresses issues similar to those that affected US intelligence agencies before and during the 9/11 attacks.

This module is followed by a game-like team experience (the 'EagleRacing' phase) in which trainees, over a period of 7 weeks, operate as a virtual team. During this phase, participants are challenged to reach consensus using different collaboration tools, make joint risk assessments and collaborate to determine the best decision in a video-based scenario which evolves based on the teams' decisions. At the end of this simulation, two additional game-based modules help participants to gain awareness of their own natural current strengths and weaknesses as well as to identify the behaviors that may best contribute to enhance the team's effectiveness and cohesiveness.

5.3 'LabRint'

LabRint is a learning experience designed to improve the performance of intelligence analysts by targeting a set of required competencies. The game places particular emphasis on rational thinking, thinking disposition, creative attitude and open mindedness.

LabRint develops along four different stages:

1. An introductory video illustrates the reference scenario, which revolves around a security event taking place in a fictitious Middle Eastern country. At the end of the video, the analyst is presented with three possible hypotheses about the exact nature of the event, and has to choose which one is true:
    - H1. a terrorist attack;
    - H2. an armed robbery;
    - H3: a personal revenge;
2. The analyst receives a sequence of 47 pieces of information ('events'), from which he/she must select the ones that are relevant for evaluating the three hypotheses. Every time an event is presented, the analyst is asked to structure the pieces of information that the event includes into 'evidence'. To that end, the analyst has to look into six different drop-down lists, respectively dedicated to answer the "Who?", "What?", "Where?", "When?", "How?" and "Why?" questions;
3. Once the set of evidences has been created, the analyst is asked to structure it under the form of an inference scheme, the vertices of which are the evidences created and the three hypotheses, while the edges connect vertices that contradict each other. More precisely, given two vertices A and B, there will be an arrow of origin A and extremity B if and only if A being true implies that B is false;
4. The analyst has to use the inference scheme to determine which of the three hypotheses is true. LabRint returns a score based on the correctness of the inference scheme and the hypothesis being selected.

5.4 'Cyberint'

While in 'LabRint', the player is confronted with an armed attack scenario, in 'Cyberint', he/she must deal with a cyber-security scenario instead. Also in this case, the learning experience requires the player to analyse several incoming pieces of information in order to determine which of three hypotheses is correct. Cyberint develops along the same four steps into which LabRint is articulated.

During the LEILA project, all four learning experiences were subjected to test and evaluation by relevant end-users. Such process consisted of a series of pilot workshops giving intelligence analysts, members of law enforcement agencies and other target end-users the opportunity to try out and provide feedback on the LEILA learning experiences. A first round of pilot sessions was carried out in 2015, with dedicated workshops taking place in Romania and Greece. A second pilot round was successfully completed in 2016, with the execution of pilot sessions in Greece, France, Italy and Romania.

6 LEILA'S INNOVATIONS

The LEILA project has brought some important innovations to the domain of intelligence analysis training. One of them pertains to its holistic approach, which combines insights from various disciplines. For its conceptual framework, LEILA relies on insights from the following domains:

- cognitive, psychological and cultural biases that may cause data interpretation errors;
- Bayesian approaches to decision-making under uncertainty;
- preference elicitation and inference schemes through argumentation and dialog games as well as through case-based reasoning;
- Games of Deterrence-based inference schemes as a solution to the problems raised by Bayesian approaches to decision-making [23].

At the beginning of the project, the consortium engaged in a careful study of various issues pertaining to such disciplines in order to ensure the solidity of the project's theoretical foundations. A review of existing training approaches in intelligence analysis was also carried out in an effort to highlight the core skills needed by intelligence analysts as well as existing gaps in current training programmes. This effort resulted in the identification of a set of user requirements and learning needs for intelligence analysts, from which the consortium has drawn a more basic set of abilities and soft skills (defined as 'Competence Development Enablers') to be targeted by the LEILA learning experiences. These represent the preliminary competencies an intelligence analyst must acquire in order to develop the core competencies (e.g. critical thinking) needed to perform at best.

In addition to critical thinking, the array of competencies addressed by LEILA includes rational thinking, creative attitude, collaboration and communication skills, awareness and mitigation of cognitive biases, capabilities in filtering and analysing large datasets as well as decision-making under social and time pressure. Some of these skills are likely to be particularly useful to analysts working in law enforcement and national security intelligence, but may also be beneficial to those operating in industry, economy, society, both in local and global organisations.

An important innovation introduced by LEILA is the flexibility of its proposed learning model. The methodology underlying the design of LEILA learning experiences addresses different categories of end-users by supporting the acquisition of competences at individual but also at team level and, indirectly, at organisational level. The LEILA learning experiences offer various deployment options: they may be delivered to trainees either as part of online/offline sessions or in a blended format mixing virtual and in presence interactions, with or without the support of a facilitator as well as both in real time and asynchronously. Furthermore, they can be easily tailored in terms of sequence and selection of modules, introduction of new content as well as adaptation to specific schedules and curricula. Such element of flexibility is particularly emphasised in the LabRint and Cyberint learning experiences, in which players have the opportunity to re-play the same session in order to spot possible failures and traps, decide whether to play individually or as a team as well as customise scenarios, missions and tasks.

A further peculiarity of the LEILA learning experiences is their ability to successfully combine multiple pedagogical approaches together. Such characteristic is shared by all LEILA learning experiences, as each of them consists of separate modules addressing a specific set of competencies through different forms of learning. Some of the learning modules proposed by LEILA rely on highly interactive computer simulations, presenting trainees with complex situations and requesting them to either answer a series of verification questions or operate autonomously within the reference environment in order to complete a certain mission. Other modules are centred upon 'conceptual model reconstruction games' in which, starting from a validated conceptual model (for example, a model describing the actions to undertake in a certain situation), participants are induced to gradually 'discover' and 'build' the model themselves. To that end, the trainees have to assess a number of situations/scenarios narratives/questions, also relying on computer-based 'dialog games'. Simple game dynamics, such as challenging trainees to accomplish a mission within a clearly defined playful context, are also exploited as a way to help the participants grasping new concepts or expose them to their own level of competence in a given area.

7 CONCLUSION

Traditional training methodologies are no longer sufficient for acquiring the core competencies needed to perform good intelligence analysis in an increasingly complex, IT-dominated and rapidly changing world. With the LEILA project, the European Commission has demonstrated its willingness to explore alternative pathways for empowering law enforcement intelligence analysts to tackle the challenges associated with their modern operational environment. The learning methodology developed by the LEILA consortium, based on a set of flexible serious games integrating insights from various disciplines and extensively validated by European end-users, represents a promising solution in this respect. The maturity level of the solutions built by LEILA makes it possible for European organisations to adopt them immediately. In addition to intelligence analysts operating in the law enforcement sector, domains such as industry, economy and society could also benefit from the findings of LEILA, opening a potentially large diffusion space for its proposed learning methodologies. Further research into the educational potential of serious game technology, building upon the results of LEILA, could lead to new opportunities in the domain of law enforcement intelligence as well as in many other fields where quick and accurate information analysis is critical.

ACKNOWLEDGEMENTS

The research leading to these results has received funding from the European Union Seventh Framework Programme (FP7/2007-2013) under grant agreement n° 608303.

REFERENCES

[1] Carter, D.L. & Schafer, J.A. The future of law-enforcement intelligence. *Policing 2020: exploring the future of crime, communities, and policing*, ed. Schafer, J.A., U.S. Federal Bureau of Investigation: Washington DC., pp. 226–256, 2007.

[2] Carter, D.L., *Law enforcement intelligence: a guide for state, local, and tribal law enforcement agencies.* CreateSpace Independent Publishing Platform: Lavergne, TN, 2012.

[3] Krizan, L., *Intelligence essentials for everyone*, Books for Business: Washington DC, 2003.

[4] The Intelligence Cycle; *Central intelligence agency*, available at: https://www.cia.gov/kids-page/6-12th-grade/who-we-are-what-we-do/the-intelligence-cycle.html

[5] The six steps in the intelligence cycle; U.S. Department of Homeland Security, Fusion Center, available at: http://fusioncenter.golearnportal.org/rawmedia_repository/b4e1b-56dcf572f53b00ee43a31b34223?/document.pdf

[6] Intelligence Branch; Federal Bureau of Investigation, available at: https://www.fbi.gov/about/leadership-and-structure/intelligence-branch

[7] Zanasi, A., (ed). *Text mining and its applications to intelligence, CRM and knowledge management*, WIT Press: Southampton, UK, 2007.

[8] Gill, P. & Phytian, M., *Intelligence in an insecure world*, Polity Press, Cambridge, UK, 2012.

[9] Richelson, J.T., *The US intelligence community*, Westview Press: Boulder, CO, 2011.

[10] Marrin, S., Training and educating U.S. intelligence analysts. *International Journal of Intelligence and Counter Intelligence*, **22(1)**, pp. 131–146, 2009. https://doi.org/10.1080/08850600802486986

[11] Goodman, M. & Omand, D., What analysts need to understand: the king's intelligence studies programme. *Studies in Intelligence*, **52(4)**, 2008.

[12] Johnston, R., *Analytic culture in the U.S. intelligence community. an ethnographic study*, Government Printing Office: Pittsburgh, PA, 2005.

[13] Training; IntelligenceAnalysis.net, available at: http://www.intelligenceanalysis.net/Training.htm

[14] Moore, D.T., *Critical thinking and intelligence analysis*, National Defense Intelligence College Press: Washington D.C., 2006.

[15] Heuer, R.J., *Psychology of intelligence analysis*, Books Express Publishing: Saffron Walden, UK, 2010.

[16] Libes, D. & O'Connell, T., Applying serious games to intelligence analysis. *Proceedings of SEA '07, the 11th IASTED International Conference on Software Engineering and Applications*, ACTA Press: Anaheim, CA, pp. 311–317, 2007.

[17] Dunbar, N.E., *et al.*, Implicit and explicit training in the mitigation of cognitive bias through the use of a serious game. *Computers in Human Behavior*, **37**, pp. 307–318. https://doi.org/10.1016/j.chb.2014.04.053

[18] Gee, J.P., *What video games have to teach us about learning and literacy*, St. Martin's Griffin: New York, NY, 2007.

[19] Sirius; IARPA, available at: https://www.iarpa.gov/index.php/research-programs/Sirius

[20] Dunbar, N.E., *et al.*, MACBETH: Development of a training game for the mitigation of cognitive bias. *International Journal of Game-Based Learning*, **3(4)**, pp. 7–26, 2013.

[21] Richey, M.K., The mind's lie: games-based learning for critical thinking. *The International Journal of Humanities Education*, **12(1)**, 2014.

[22] Game-Based Learning and Intelligence Analysis: Current Trends and Future Prospects; E-International Relations, available at: http://www.e-ir.info/2013/08/07/game-based-learning-and-intelligence-analysis-current-trends-and-future-prospects/

[23] Rudnianski, M., Sadana, U. & Bestougeff, H., Bayesian networks and games of deterrence. *Recent advances in game theory and applications*, eds. Petrosyan, L. & Mazalov, V., Springer Verlag, pp. 201–224, 2016.

MANAGEMENT OF ZIKA OUTBREAK IN A TROPICAL URBANISED COUNTRY: SINGAPORE

DOREEN TAN SOEK CHIN & CHONG SI JACK
Department of Emergency Preparedness, Sengkang Health, Singapore.

ABSTRACT
Zika is a vector borne viral disease that has been associated with significant congenital abnormalities, including microcephaly. Singapore, being a densely populated tropical country, is extremely concerned with this particular disease as the vector Aedes Egypti is endemic in the country. In Aug 2016, a local Zika outbreak occurred in Singapore, which resulted in more than 400 locally transmitted cases. The authors would like to share our experiences and lessons learnt in our acute care hospital's response to this outbreak. We will include the preparation of infrastructure, facilities, logistics, communications, screening and management procedures that allowed us to manage the outbreak expeditiously.
Keywords: aedes Aegypti, outbreak management, Singapore, vector-borne disease, zika, zika virus infection.

1 INTRODUCTION

Sengkang Health (SKH) is a relatively new Public Healthcare Institution (PHI), which operates from Alexandra Hospital (AH), in the south-western part of Singapore. Although SKH is considered an acute PHI [1], the hospital's bed occupancy rate stands at an average of 56% daily, because of (a) the presence of two other major acute hospitals within the vicinity, and (b) AH being only the holding grounds for SKH, while the new SKH building is being built in the north-eastern part of Singapore.

From as early as January 2016, Singapore's Ministry of Health (MOH) and Infectious Diseases experts in the country had alerted that it was a matter of time that Singapore would face the first case of Zika virus, which was already spreading rapidly throughout Latin America [2, 3], due to the presence of the vector Aedes Aegypti in the country.

On 27 August 2016 (Saturday), MOH and the National Environment Agency (NEA) issued a joint press release [4, 5] to announce the first locally transmitted case of Zika virus infection. The patient, a 47-year-old female Malaysian who resides at Aljunied Crescent (south-eastern part of the island) and works in Singapore, had not travelled to Zika-affected areas, and both agencies opined that she was likely to have been infected in Singapore.

The next day, following the press announcement, SKH was tasked by the MOH to ring-fence beds for Singapore's Infectious Diseases Hospital (Tan Tock Seng Hospital (TTSH)), as the latter's isolation ward was nearing full capacity, and would not be able to cope with the influx of the new Zika infection cases.

2 PREPARATIONS

SKH@AH has two closed, naturally ventilated general wards (i.e. no air-conditioning, only fans and open windows). As Zika virus is vector borne (as opposed to airborne or droplet borne, which would require a higher level of isolation facilities for patients), the Chairman, Medical Board (CMB) decided to open one of these wards to take in the Zika-positive cases from TTSH. An urgent meeting was convened on the morning of Sunday, 28 August 2016, and preparations of the ward for patients' occupation commenced that afternoon, and were completed by 13:00 hrs (GMT+8) the next morning (Monday, 29 August 2016).

DOI: 10.2495/SAFE-V7-N3-390-398

Preparations comprised infrastructure, facilities, manpower, equipment, logistics, & supplies, communications, and screening and management procedures.

2.1 Infrastructure

AH is surrounded by ample vegetation and vector activity is abundant (refer to Fig. 1). As Zika virus is vector borne, the first order of business was to ensure that all windows were closed, air coolers were put in place, and any gaps sealed. The Environmental Services Department was also activated to clean and disinfect the ward, supply beds and mattresses, as well as linen (bed sheets and patient pajamas) for the entire ward.

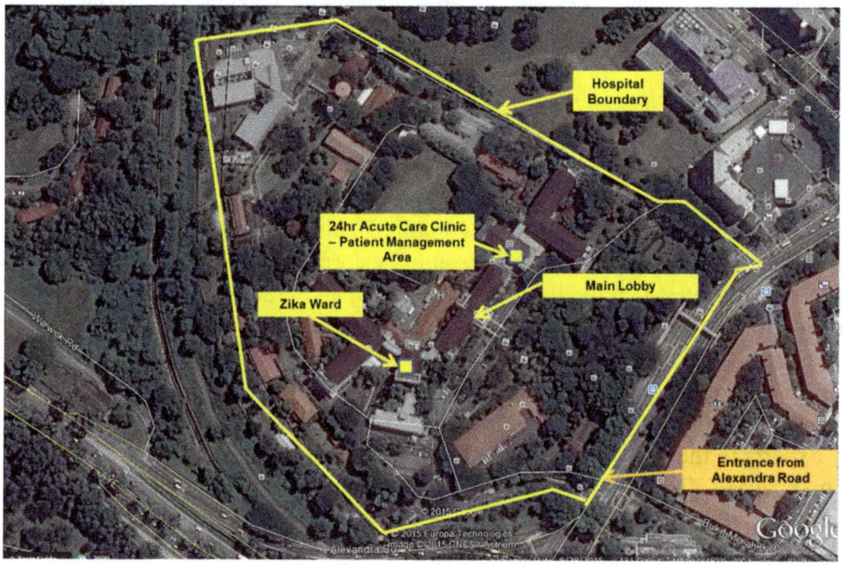

Figure 1: Satellite Image [6] of Alexandra Hospital.

These first order tasks were completed at 21:00 hrs (GMT+8) on 28 August 2016 (within nine hours of activation).

2.2 Facilities

The Zika ward had 35 beds, in cubicles of 5 to 6, as shown in Fig. 2. An additional 2 beds were not operationalized (bottom left of Fig. 2). There is also a patient lounge (just above the 2 unused beds) where patients could gather to watch television programmes, or play board games, which were specially provided for them, or read newspapers, or simply just mingle.

Patients were assigned beds according to sex (females nearer to the front of the ward, and males further to the back, as seen in Fig. 2). The cubicles were segregated by low partition walls (i.e. not full height walls). There are also toilets and shower facilities (also segregated by sex) within the ward.

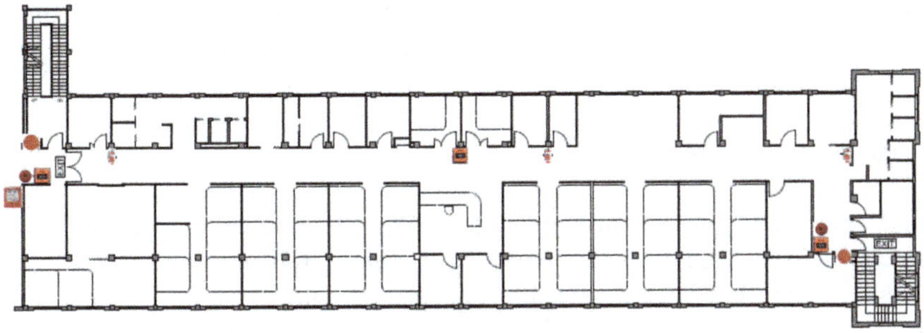

Figure 2: Layout of AH Zika ward.

2.3 Manpower

Nursing manpower to staff the Zika ward was drawn from the non-Zika wards. The Chief Nurse, in consultation with CMB, decided to reduce the number of in-service beds in these non-Zika wards, so that the manpower could be deployed to man the Zika ward 24/7. Bed Management Unit, therefore, was instructed to 'fence off' these closed beds in the Bed Management System, so that patients would not be admitted to these beds accidentally.

Medical manpower (i.e. doctors) came primarily from the Division of Medicine (as opposed to Division of Surgery). The doctors were thoroughly briefed by CMB regarding the disease, and rosters were quickly arranged so that these doctors could include the Zika ward in their morning ward rounds.

The SKH Command Centre, initially manned by the Emergency Preparedness Department, was augmented by the Clinical Governance Department from the second day of operations.

2.4 Equipment, logistics and supplies

A separate Cost Centre was also set up within the day for staff to charge any Zika-related expenditure, in order to expedite operations, and efficiently track expenses.

Medical equipment was drawn from the store, and redeployed from other wards so that there was sufficient equipment for patients who required them.

Additionally, Information Technology (IT) equipment was deployed to the Zika ward within 24 hours of activation. These included Computers-on-Wheels (CoWs), printers, photocopier machines, etc. IT applications and software were also installed and set up within the day.

The Materials Management and Pharmacy Departments worked closely with the nursing and medical staff to provide the required drugs, pharmaceutical and medical supplies for the Zika ward. In addition, the pharmacy also procured, within 24 hours, 2000 bottles of insect repellent sprays, and an additional 1000 tubes of insect repellent cream in the next two days that followed. It is also worthy to note that the country's insect repellent suppliers and retailers ran out of insect repellents within a few days – SKH@AH was therefore one of the only hospitals to receive a substantial supply.

Environmental Services staff were rostered to collect and change the linen, and to clean the ward regularly. The pest control unit was also activated to mist the hospital premises with increased frequency (from once a week to thrice weekly) to limit vector activity.

An insect repellent application station was set up right outside the ward, and any staff or visitors who wished to enter the ward were strongly encouraged to spray or apply the repellent cream liberally.

2.5 Internal communications

Things happened really quickly during the first week – the situation was fluid on the ground, and instructions from MOH were changing by the day. SKH management was cognizant of the need to have timely communications with the staff, and provide clear and accurate information and instructions as the situation changed. Daily bulletins to staff were instituted from the first day of operations of the Zika ward (refer to Fig. 3 for a bulletin sample).

This email was sent in BCC to all Sengkang Health staff. Please help disseminate to colleagues with no email access.

Dear Colleagues

You might already be aware that The Ministry of Health (MOH) confirmed yesterday 41 cases of locally transmitted Zika virus infection in Singapore.

We are gearing up to receive Zika virus patients at Alexandra Hospital as part of our national effort to manage and contain the Zika virus. Wards 2 and 3 are being prepared to receive patients should we be activated by MOH, or if patients present themselves directly at Alexandra Hospital.

Detailed operational instructions will be sent from the Emergency Preparedness team to HODs shortly. Please abide with all the protocols to ensure we maintain a safe environment at the hospital.

The Zika virus is transmitted to people primarily through the bite of an infected Aedes species mosquito. Most people fully recover from the Zika virus without severe complications. Severe illness or deaths from Zika virus are extremely rare, according to the Communicable Disease Centre (CDC) at Tan Tock Hospital. For staff, especially those who are pregnant, should be vigilant, apply insect repellent and monitor yourself for signs of fever, rashes, joint pains, red eyes and headaches. If these symptoms present themselves, please see medical attention immediately.

Around AH premises, our Environmental Services team is stepping up on vector control, i.e., misting to prevent mosquito breeding and ensure we keep mosquitoes at bay.

More information and FAQs on Zika virus can be found at www.moh.gov.sg/zika

We are monitoriing the Zika virus situation closely and will provide you with daily updates.

In the meantime, stay vigilant and as always, please exercise confidentiality with all internal memos and emails.

Yours sincerely,

Ong Biauw Chi
Chairman Medical Board

Christopher Cheng
Chief Executive Officer

Figure 3: Snapshot of First Internal Staff Memo from the CEO and CMB.

Additionally, Operations Instructions were sent out to provide clear instructions on the workflows and situation management (refer to Fig. 4 for a sample of the first set of Ops Instructions sent out).

To : All SKH HODs & Managers
Cc : SKH Mass Casualty & Outbreak Management Committee Members

Dear All,

1. You would already be aware of the Zika Virus Infection (ZVI) situation in Singapore, with the cluster of ZVI positive cases identified at the Sims Avenue / Aljunied Crescent area. **Please take careful note of the revised Ops Instructions for the screening and management of ZVI cases in SKH@AH detailed below.**

Revised Workflow

2. Attached is the revised "**ZVI-1: Workflow for Screening and Management of ZVI Suspects**", updated 29 August 2016, for your immediate implementation and dissemination to your staff.

Workflow: Discharge Advisory:

Updated Screening Form & List of Affected Countries / Areas

3. The Screening Form & List of Affected Countries / Areas have also been updated correspondingly. Please use the updated version, and ensure that all old versions are discarded.

4. Please take special note of the **additional areas in SINGAPORE** indicated in the List of Affected Countries / Areas under the "Zika Virus" section.

Figure 4: Snapshot of first set of operations instructions sent to the staff.

2.6 External communications

In a national crisis situation, communications between healthcare institutions, and with the MOH, is critical, as coordination is key. As the situation developed, there were various avenues of information that were sometimes conflicting. Patients and their loved ones were concerned about how the virus affected them, the financial implications (tests, hospitalisation costs, etc), and generally how they could cope with the uncertainties. The press had many reports on the situation, and social media was abuzz with information and opinions. A calm and trusted voice, therefore, was critical in such a situation.

As per the standard operating procedure for such situations, all queries and correspondences with the media was fronted by the MOH; media queries relating to specific PHIs were then directed by the MOH to the Communications team in such PHI respectively.

Each PHI drafted their own drawer media scripts and prepared their spokespersons for potential media queries.

3 RECEIVING PATIENTS

The first Zika-positive patient was transferred from TTSH to SKH@AH at approximately 21:00 hrs (GMT+8) the day the Zika ward was in operation. Such transfer patients had to pass through the 24hr Acute Care Clinic (ACC), where the Emergency Department doctors and nurses would assess them to ensure that their condition was stable, before allowing the ambulance staff to transfer them to the Zika ward.

While patients arrived from TTSH and also from the Singapore General Hospital (SGH) – Singapore's 1700-bedded flagship hospital – the ACC was also receiving walk-in patients who were concerned that they were infected with the virus.

SKH@AH admitted a total of 34 patients (either transferred from TTSH or SGH, or directly through the ACC) from Day 1 (29 August 2016) to Day 15 (12 September 2016), when the last patient was admitted. Table 1 below shows the breakdown of the number of patients admitted.

Table 1: Breakdown of Number of Patients Admitted to SKH@AH's Zika Ward.

Admission Date	**Number of patients**
29 August 2016	9
30 August 2016	3
01 September 2016	3
02 September 2016	2
03 September 2016	5
04 September 2016	8
05 September 2016	2
08 September 2016	1
12 September 2016	1
Total	34

In comparison, a total of 333 cases had tested positive for the Zika virus in Singapore as of 12 September 2016. The number of confirmed cases stood at 459 as of 21 January 2017 (458 cases in 2016, and 1 case in 2017).

3.1 Screening and management procedures

Every patient entering the hospital system either through the ACC or the Specialist Outpatient Clinics (SOC) was screened by the frontline staff using a standard screening form (refer to Fig. 5). Those who met the case definition for Zika virus infection were then held in isolation at the ACC (as a precaution, because as mentioned earlier, SKH@ AH is surrounded by greenery and vector activity is abound), while their blood and urine samples were drawn and sent to the laboratory to be tested for the presence of the virus.

Patients found to be positive were then admitted to the Zika ward. Those that were found to be negative were then discharged with the necessary medications and documentations.

On 05 September 2016, the MOH issued a press release [7] that the Ministry no longer required suspect cases to be isolated. This signalled the transition from what the Ministry terms the 'Containment Phase' to the 'Mitigation Phase', and was an acknowledgement that the virus was now widespread in the community.

Laboratory testing was now no longer mandatory for all patients presenting with symptoms, and patients could wait for their test results at home. Patients who tested positive for the virus need not be hospitalised unless clinically required, and those who were admitted did not require isolation.

Sengkang Health
SingHealth

Affix Patient Sticky Label if available

NAME: ____________________

NRIC / PASSPORT NO.: ____________________

LOCATION: ____________________

Date: ____________________ **Patient's Mobile No:** ____________

Screening Questions	Yes	No
1a. Travel history to ***affected countries* or areas***** in the past ***2 weeks*** ____________________(state Country / Area)		
1b. Fever (Current Temp: ____________________)		
2. **MERS-CoV**[1]: Flu-Like Symptoms (e.g. Cough, Runny Nose, Sore Throat) ____________________ (state symptoms)		
3. **ZIKA**[2]: Fever **AND** Maculopapular rash (big red spots) **AND** any of the following (**TICK accordingly**): ☐ Arthralgia (Joint Pain) ☐ Myalgia (Muscle Pain) ☐ Headache ☐ Non-Purulent Conjunctivitis (Redness of Eye without Pus)		
4. **AVIAN INFLUENZA**[3]: Respiratory Symptoms **AND** any of the following (**TICK accordingly**): ☐ Contact with confirmed avian influenza cases ☐ Exposure to poultry		

* *Please refer to the latest Affected Countries List.*
** *Please refer to the latest Affected Areas List.*

Note to screening staff:
(1) Refer to List of Affected Countries / Areas for each disease
(2) If "Yes" to Q1 **AND** either Q2, OR Q3, OR Q4, patient should be directed to ACC Isolation Room / Clinic identified Isolation Room / Area Isolation Room / Area.

Accompanied by: ____________________ **(Name)**
____________________ **(Contact Number)**

Screening done by: ____________________

Screening Form Revised on 29 August 2016 *Page 1 of 1*

Figure 5: Snapshot of SKH@AH Travel History & Symptoms Screening Form.

With this announcement, from 06 September 2016 onwards, Singapore stepped down its heightened alert, but operationally, vector control measures continued to be the cornerstone of the National Environment Agency's and MOH's strategy to limit the spread of the disease. This was also translated down to the PHIs.

3.2 Types of patients presenting at the Hospital's 24 hr Acute Care Clinic

Patients presenting at the ACC (either transferred cases from SGH and TTSH, or individuals coming on their own to SKH@AH) were generally well. No pregnant patients presented at the hospital. Some of the patients presented only with fever, some with rashes, and many were actually asymptomatic. At the beginning of the episode, these patients were judiciously tested for the virus if they stayed within the identified clusters' areas (Aljunied at first, and other areas subsequently). With the stepping down of the alert level, tests were only done if the physician felt that it was clinically required.

3.3 Stepping down of measures and closure of the zika ward

There were 11 patients remaining in the Zika ward on the morning of 04 September 2016, and the laboratory results for 10 of them came back negative. The sole remaining patient who could not be discharged would be transferred to a General Ward. Instructions were therefore issued from the SKH Command Centre (after discussion with CMB) to close the Zika ward, and for all future Zika positive cases to be admitted to the Isolation Ward, more for ease of management than for the requirement to isolate them, as the Isolation ward was situated just next to the ACC. The MOH was also informed of this decision.

4 AFTER ACTION REVIEW

An After Action Review (AAR) for the management of the episode was conducted on 16 September 2016. This helped identify key best practices, challenges faced, and areas for improvement.

4.1 Key challenges and learning points

The two main challenges presented by the management of the Zika virus were: (1) the disease was relatively unknown, and there appeared inconclusive and insufficient research on the subject. Patients and staff were therefore concerned about how the disease would affect them, and (2) as the situation was fluid on the ground, guidelines and instructions from the MOH were constantly changing, and sometimes could not be absolutely clear. This sometimes resulted in difficulties in disseminating accurate and timely information to the ground.

With close and constant communications between the senior management and the staff on the ground, as well as an appointed single source of instructions and information (SKH Command Centre), however, the staff were kept abreast of new developments, and were thus able to manage the patients' and next-of-kins' expectations.

4.2 Areas for improvement

As with all crises, there are areas that could be improved upon. These included: (1) the requirement for a Hospital Operations Centre or Command Centre to coordinate and manage crises, (2) the need to leverage on information technology for speedy collation and collection of data, and timely dissemination of critical information to key stakeholders, (3) the need for clarity on financial guidelines as early as possible, because patients were clearly concerned about this, and (4) the opportunity for Standard Operating Procedures to be set in place, so that future similar situations could be better managed.

5 CONCLUSION

In conclusion, communications is critical in the management of any crisis, and a single source of information and instructions is important to ensure that there is no confusion or conflicting directions to the ground. SKH was able to manage this episode relatively effortlessly because the senior management understood this.

REFERENCES

[1] Hospital Services. Ministry of Health Website, Singapore, available at www.moh.gov.sg/content/moh_web/home/our_healthcare_system/Healthcare_Services/Hospitals.html (accessed 03 January 2017)

[2] It's matter of time before Zika virus hits Singapore: Potential threat for pregnant women. The Asian Parent Website, Singapore, available at https://sg.theasianparent.com/zika-virus/ (accessed 20 January 2017)

[3] 'Matter of time' before Zika cases are seen in Singapore: Experts. Channel News Asia Website, Singapore, available at http://www.channelnewsasia.com/news/singapore/matter-of-time-before/2481498.html (accessed 20 January 2017)

[4] First Case of Locally Transmitted Zika Virus Infection. Ministry of Health Website, Singapore, available at www.moh.gov.sg/content/moh_web/home/pressRoom/pressRoomItemRelease/2016/first-case-of-locally-transmitted-zika-virus-infection.html (accessed 03 January 2017)

[5] First case of locally-transmitted Zika virus infection reported in Singapore: MOH, NEA. Channel News Asia Website, Singapore, available at www.channelnewsasia.com/news/singapore/first-case-of-locally/3078974.html (accessed 03 January 2017)

[6] Satellite Image of Alexandra Hospital. Google Maps, available at https://www.google.com.sg/maps/place/Alexandra+Hospital/@1.2866021,103.7988575,687m/data=!3m2!1e3!4b1!4m5!3m4!1s0x31da1bcb19f367f5:0x8f256e299bf58c73!8m2!3d1.2866021!4d103.8010462!6m1!1e1 (accessed 13 January 2017)

[7] Cessation of Isolation, MOH to Subsidize Zika Testing for all Singaporeans with Symptoms. MOH Ministry of Health Website, Singapore, available at https://www.moh.gov.sg/content/moh_web/home/pressRoom/pressRoomItemRelease/2016/cessation-of-isolation--moh-to-subsidise-zika-testing-for-all-si.html (accessed 02 February 2017)

LOCAL SUPPLY CHAINS: THE DISASTER MANAGEMENT PERSPECTIVE

KYLE B. PFEIFFER, CARMELLA BURDI & SCOTT SCHLUETER
Risk and Infrastructure Sciences Center, Argonne National Laboratory, United States of America.

ABSTRACT

Situational awareness of the operational status of specific, critical supply and demand nodes following a major disaster may inform response and recovery activities based on the ability of an infrastructure asset or system to support core facility operations. Near-real-time analysis of infrastructure dependency information is a computationally intensive process that has generally been observed informally by public safety officials. While system-level information may be desired, it has been beyond the capabilities of most local public safety and emergency management agencies. To address this problem, a Grassroots Infrastructure Dependency Model (GRID-M) was developed to enable near-real-time analysis of physical infrastructure dependencies of specific supply and demand nodes within four lifeline sectors: electricity, natural gas, water, and wastewater. The operational status of each node can be characterized as operational, partially operational, or not operational. These statuses are obtained by matching real-time outage or disruption data from utility providers with predetermined specific coping strategies based on a preincident limited infrastructure survey for specific assets within a network. This information can also be paired with a limited damage assessment to provide awareness of the accessibility to, and physical state of, each node within supply chains of interest. GRID-M displays all outputs within a Geographic Information Systems environment with additional prepopulated layers such as real-time traffic and demographic information of the affected communities. As such, GRID-M may be used following a major disaster to support the identification of priority response and recovery objectives based on the disruptions of critical local supply chains and their relationship with affected communities.

Keywords: critical infrastructure, dependency, disaster, emergency management, preparedness, resilience, supply chain.

1 INTRODUCTION

In the context of disaster management, supply chains may be defined as activities associated with the flow and transformation of goods from raw materials to end-users. [1] While other definitions may exist, they are often influenced by industry- or sector-specific attributes that mire the concept for public safety officials. As indicated by Fig. 1, the supply chain concept for this research is rather effort simplistic: materials flow from suppliers to manufacturers who transform the raw materials into finished products. These products are then moved to distributors and eventually received by consumers. This representation of the logistics workflow is useful to scale often complex processes to a manageable paradigm through which local officials can interpret relationships and ultimately affect change regarding the resilience posture of the supply chain.

Managing risk, even for a simplistic supply chain, has become a professional responsibility. In the United States, both small and large private sector organizations employ logisticians, continuity experts, and risk analysts to help identify and mitigate risks to their enterprise. Their focus can be everything from coordination of supply and demand to minimizing disruption of normal activities. For the purpose of this research, disruption risks are examined, as they are more likely to intersect with the roles and responsibilities of public safety officials. [2] Furthermore, those in the public sector may be able to effect the most change around them. Disruption risks can be described in four broad categories: operational contingencies, natural hazards, terrorism, and political instability. Operational contingencies are systematic failures,

DOI: 10.2495/SAFE-V7-N3-399-405

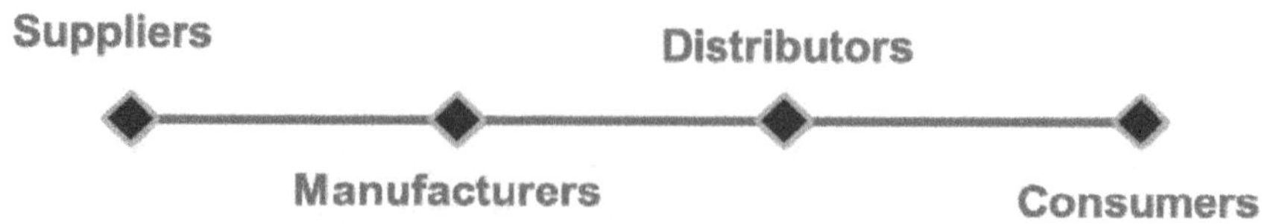

Figure 1: Supply Chain Concept.

such as the August 2003 Northeastern United States grid blackout, which caused curtailment of operations for many supply and demand nodes. [2, 3]. Natural hazards are events such as hurricanes, wildfires, earthquakes, or flooding which can disrupt critical components of a supply chain – such as the fuel shortages in the New York–New Jersey area following Superstorm Sandy in 2012. [4] Terrorism, the next category of disruption risks, may include indiscriminate destruction of critical supply chain components or targeted sabotage of known critical assets and systems. [3] Lastly, political instability may disrupt supply chains, such as oil production curtailment in Iraq and Syria, given the regional strife since 2012.

In the United States, operational contingencies and natural hazards are among the most likely disruption risks to significantly affect supply chains that are critical to communities following a disaster. The effects of these disruptions may directly impact the ability of a community to obtain food, fuel, prescription medication, or other critical goods and services in the minutes, days, or weeks following a disaster. Understanding the behavior of these supply chains, particularly in near real-time after a disaster, may help officials to make more informed decisions regarding response and recovery activities.

2 FRAMING LOCAL SUPPLY CHAIN RESILIENCE ACTIVITIES

Public safety officials are often at a disadvantage as compared to their private sector partners when considering risks to their supply chains of concern. First and foremost, most public safety organizations, such as emergency management agencies, have often not identified specific supply chains that provide critical goods and services to their communities nor do they have a precise understanding of the cascading effects of a degradation to one or more of these critical supply chains. This is often due to a combination of a lack of staffing, mission assignment or tasking, and internal expertise. Furthermore, the characteristics of many supply chains make it difficult for public safety officials, particularly at the local level, to effect change; or even build awareness of their risk landscape. This is a result of the global interdependence of supply chains, the untold number of potential disruptors on the worldwide scale, and the lack of public control or influence of private sector logistics (particularly in the United States). These factors make actions led by public sector entities more difficult.

While public safety officials, particularly at the local level, may be hamstrung by these impediments they can still gain awareness of their local supply chain risk landscape and even build near real-time situational awareness of the behavior of their supply chains of concern before, during, and after a disaster. This can be accomplished by scaling the focus area to the relationship between local and nearby distributors, supply nodes (e.g. a grocery store or pharmacy), disaster commodity stockpiles and points-of-distribution, and the affected community. Once the boundary conditions of the supply chains of interest are set, public safety officials can execute a short research and planning project to gain understanding of the likely disruption risks to the Critical Infrastructure (CI) needed to support a given supply chain.

CI consists of systems and assets, whether physical of virtual, that if disrupted, degraded, or destroyed would have a serious impact on the health, safety, security, and well-being

Figure 2: Infrastructure Dependency Concept as Described by Rinaldi *et al.*

of society. [5, 6] Rinaldi *et al.* defined a critical infrastructure dependency as a 'linkage or connection between two infrastructures, by which the state of one infrastructure influences or is reliant upon the state of the other'. [7] For example, a cold storage warehouse may be dependent upon an external source of electric power for refrigeration. A disruption to primary electric service, in the absence of a backup generator or battery, may degrade critical operations at the warehouse. In this example, a physical dependency has caused the notional disruption of operations at a cold storage warehouse. This concept is illustrated in Fig. 2.

3 GRASSROOTS INFRASTRUCTURE DEPENDENCY MODEL

GRID-M was developed within an Esri environment including ArcGIS for Desktop, ArcGIS for Server, ArcGIS for Portal and/or ArcGIS Online for Organizations (AGOL). GRID-M is located within Esri's model builder application. It is published as a geoprocessing (GP) service to an ArcGIS server in order to be accessible as a defined workflow within a web-based mapping application. This cloud-based access prevents the need for all users to be GIS experts and instead allows for easy access by public safety officials.

The model creation occurs within the Esri ArcGIS Desktop environment. The model builder is 'a flowchart schematic that guides the user to visually lay out the task'. [8] Any ArcGIS for Desktop tools, as well as customized python scripts, are accessible through the model builder and can be customized into a workflow.

The overall model is based on several assumptions regarding the likely supply chains of concern and level of sophistication that local public safety officials have within the logistics domain. First, it is assumed that local public safety officials may cite food, water, fuel, and pharmaceutical supplies among the most critical commodity needs following a major disaster. Furthermore, it is assumed that planners have broad awareness of some of the disruption risks to their local supply chains – particularly in the context of major disasters. While a risk-basis for this framework, combined with a stronger methodology to prioritize essential goods is needed, that was not the focus of this research effort. With time and resources, it is assumed that most planners can make these basic risk-informed judgments to identify their essential supply chains and, in partnership with their private sector partners, identify the likely disruptors in the context of operationally contingent and natural hazard risks. Once that is completed, the initial dataset for the GRID-M would be available.

This initial data set should include a library of georeferenced observation points for each critical supply and demand node and denote the relationship between them (e.g. which distribution warehouses serve which grocery stores). Once this basic dataset is established within the GRID-M, a series of short infrastructure surveys may be completed. These infrastructure surveys are meant to identify reliance on a specific lifeline infrastructure – to include the ability of an entity to cope with a degradation or disruption of service based on coping and continuity strategies. For instance, a grocery store may have a backup generator that can support full operations for a 24-hour period. However, without resupplying on-site fuel, this node will become classified as 'not operational'. An example of the infrastructure survey question sets can be found in Fig. 3. Of note, similar question sets are available to assess the dependency of an asset based on reliance on water, wastewater, and natural gas.

These limited infrastructure surveys must be completed for each node. In many instances, corporations will have similar coping strategies (e.g. a generator) so the data entry may be somewhat streamlined. Of note, research is underway to allow for validation or suggested survey data for common supply chains and nodes rather than force public safety officials to conduct extensive preliminary research. However, that is not available at this time.

Once the limited infrastructure surveys are completed they should be validated by the respective private sector entities and then housed in the native GRID-M database.

The next step in the planning process is for public safety officials to establish informal infrastructure reporting requirements with their utility owners and operators. For instance, an emergency management agency may request that the electric utility provider(s) for their community provide outage data when a certain threshold is met. These reporting thresholds may be based on the percentage of a population affected by an outage or when certain critical geographic areas are without power. Other thresholds may be established based on the information needs of the emergency management entity for each lifeline utility. To be used within the GRID-M, the outage data may be provided in several formats, including.CSV, KML, and. SHP files. Outage areas may also be inserted as polygons using the Esri drawing tool.

Once data is received from a utility provider, it may be imported into the GRID-M. This process usually only takes a few seconds if the data is well structured and in a readable format.

Once imported, an analyst will be able to see how supply and demand nodes are affected within the outage area. The results are generated within the GRID-M by comparing the coping strategies defined within the limited infrastructure surveys and generating results

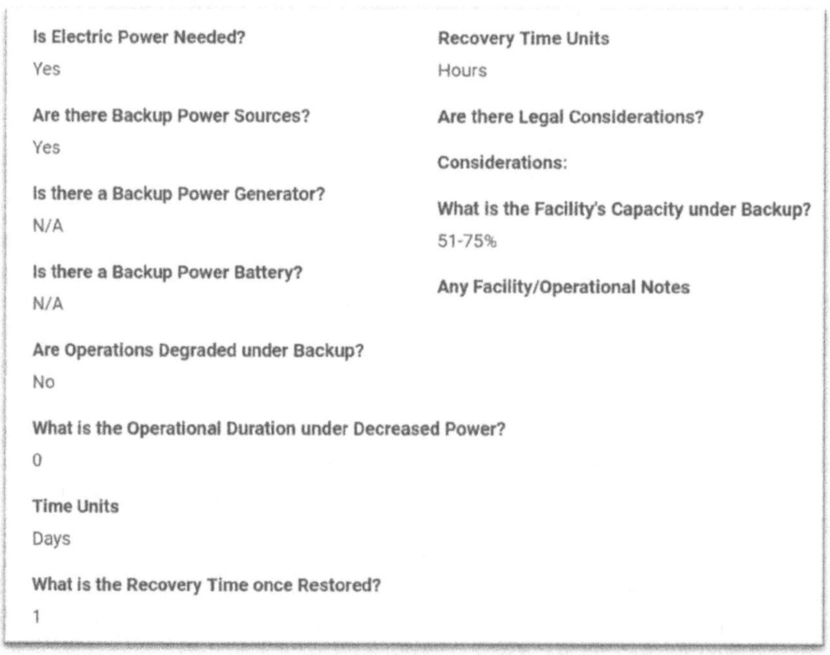

Is Electric Power Needed?
Yes

Are there Backup Power Sources?
Yes

Is there a Backup Power Generator?
N/A

Is there a Backup Power Battery?
N/A

Are Operations Degraded under Backup?
No

What is the Operational Duration under Decreased Power?
0

Time Units
Days

What is the Recovery Time once Restored?
1

Recovery Time Units
Hours

Are there Legal Considerations?

Considerations:

What is the Facility's Capacity under Backup?
51-75%

Any Facility/Operational Notes

Figure 3: Sample Completed Electric Infrastructure Survey.

regarding the operational status of each node based on the disruption areas (represented as polygons). The resulting output of the model builder is published as a GP service to an ArcGIS Server. That GP service is available to users of either a Portal or an AGOL account. The creation of a web application serves as a container or UI for the GP service. Configurable widgets house the GP service and provide a direct link for users of the web application to run the GRID-M without the need for an ArcGIS desktop environment.

The use of GRID-M within a cloud-based environment requires Internet Information Services (IIS) to store or host web applications and node.js as a runtime environment.

Figure 4 provides a notional representation of the near-real-time operational status of grocery stores in coastal Maine in the United States based on a disruption of electric service. The white diamonds are fully operational grocery stores, the gray diamonds are partially operational, and the black diamond signifies that the node is non-operational. The squares with bullseyes represent distribution centers. The same color logic applies in this gray scale example.

In another notional example, Fig. 5 depicts a disruption to the regional pharmaceutical supply chain in coastal Maine based on an electric outage.

Of note, in the web application version of GRID-M, the operational status of supply and demand nodes is displayed in a choropleth fashion using a stoplight color scheme.

GRID-M also makes use of the Esri collector application to support limited damage assessments for nodes affected by disruption risks. Using the standard damage assessment form issued by the Federal Emergency Management Agency, survey and assessment teams can report information to Emergency Operations Center staff regarding the physical state of a node. This reporting can be done through a mobile application as well as the desktop and web applications. GRID-M can also support integration with real-time traffic reporting services, such as Waze™. Therefore, by using this model, public safety officials can receive

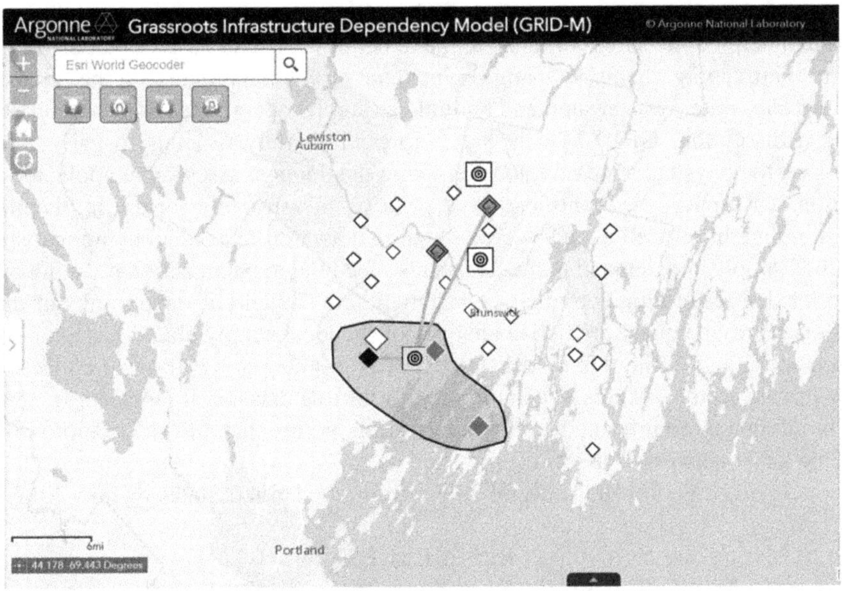

Figure 4: GRID-M Representation of a Notional Disruption to a Local Grocery Supply Chain in Coastal Maine.

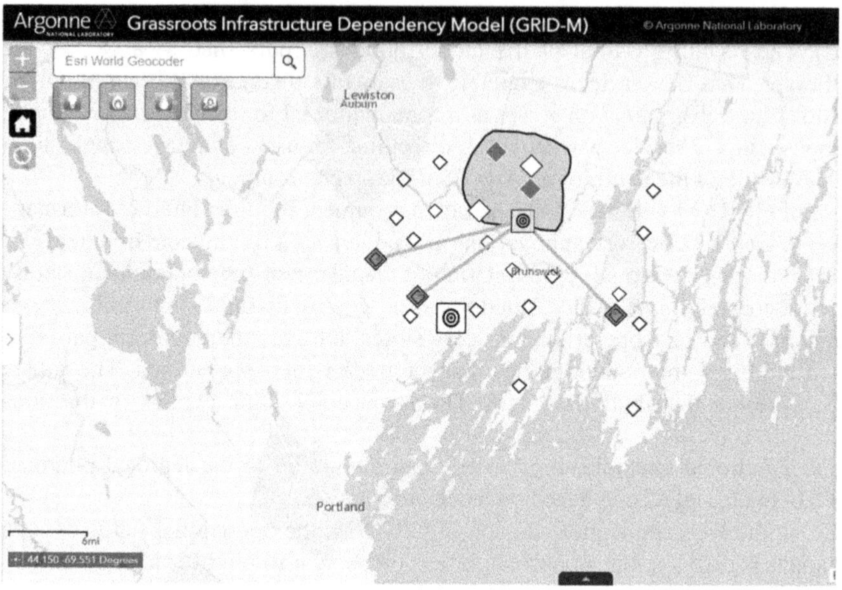

Figure 5: GRID-M Representation of a Notional Disruption to a Local Pharmaceutical Supply Chain in Costal Maine.

moderately vetted information regarding the operational status of an infrastructure based on disruption to primary utility service, its organic coping mechanisms, the physical state of the node, and egress based on surface transportation infrastructure.

4 CONCLUSIONS

GRID-M can help public safety officials to make better approximations about disruptions of concern to their supply chains by using preincident survey data regarding the reliance of a node on electric, water, wastewater, and natural gas and its operational capacity and duration following a disruption. GRID-M can also help public safety officials to gain situational awareness on the physical state of a node by using the damage assessment application. This information can support the identification of priority infrastructure for use in planning and exercise simulations as well as real-world situational awareness of the operational status of preidentified supply and demand nodes. It can also feed the incident action planning process, help to prioritize restoration activities, identify suitable shelter sites and points-of-distribution, and even provide quasi-predictive analysis of the local supply chains based on planned restoration activities. While this model has a barrier to entry in the form of collecting data from key private sector partners in a community prior to a disaster, it represents a new paradigm in building a more informed common operating picture that allows for approximations of the behavior of networked assets using vetted data.

The model will be deployed to several cities within the United States by mid-2017.

REFERENCES

[1] Gaspar, J., Kolari, J., Hise, R., Bierman, L. & Smith, L.M., *Introduction to Global Business: Understanding the International Environment & Global Business Functions*, Cengage Learning: Boston, p. 329, 2017.

[2] Vakharia, A.J. & Yenipazarli, A., Managing supply chain disruptions. *Foundations and Trends in Technology, Information and Operations Management*, **2**(4), pp. 245–255, 2009.

[3] Kleindorfer, P.R. & Saad, G.H., Managing disruption risks in supply chains. *Production and Operations Management*, **14**(1), pp. 53–68, 2005. https://doi.org/10.1111/j.1937-5956.2005.tb00009.x

[4] Hurricane Sandy After-Action Report, United States Department of Homeland Security, Federal Emergency Management Agency, available at https://www.fema.gov/media-library-data/20130726-1923-25045-7442/sandy_fema_aar.pdf (accessed 30 January 2017)

[5] National Infrastructure Protection Plan 2013: Partnering for Critical Infrastructure Security and Resilience, United States Department of Homeland Security, available at https://www.dhs.gov/publication/nipp-2013-partnering-critical-infrastructure-security-and-resilience(accessed 30 January 2017)

[6] Presidential Policy Directive—Critical Infrastructure Security and Resilience, The White House, available at https://obamawhitehouse.archives.gov/the-press-office/2013/02/12/presidential-policy-directive-critical-infrastructure-security-and-resil. (accessed 30 January 2017)

[7] Rinaldi, S., Peerenboom, J. & Kelly, T., Identifying, understanding and analyzing critical infrastructure interdependencies. *IEEE Control Systems*, **21**(6), pp. 11–25, 2001. https://doi.org/10.1109/37.969131

[8] Allen, D.W., *Getting to Know ArcGIS Modelbuilder,* Esri Press, Redlands, California: 2011.

A TAXONOMICAL FRAMEWORK OF SOCIO-CULTURAL HAZARDS IN TRANSPORT HUBS

SIMONE ROZZI[1], ALESSIA GOLFETTI[1], MICHAEL MINKOV[2], MARK ROBINSON[3], FERHAN ŞENGÜR[4] & CARLO DAMBRA[5]
[1]Deep Blue s.r.l., Rome, Italy.
[2]Varna University of Management, Varna, Bulgaria.
[3]University of Leeds, Leeds, UK.
[4]Anadolu University, Eskisehir, Turkey.
[5]Proprs, Ltd, Oxford, UK.

ABSTRACT
This article presents a taxonomical framework that supports the considerations of socio-cultural hazards that may affect crowd management in transport hubs, i.e. airports, ports, underground and train stations, both in normal and emergency situations. Such hazards include communication breakdowns with passengers due, for instance, to language barriers; increased potential for revolts, as in stranded passenger situations; misreporting of security threats; and uncooperative behaviour in case of emergencies. Such socio-cultural hazards are not normally considered from the integrated perspective of transport hub operators, e.g. security staff, first responders and service assistants as well as safety and security managers. The present study provides an integrated perspective of these hazards as a means to increase the performance of transport staff members that interact with the public and with passengers on a daily basis. The methodology used to develop the framework comprises: (i) a focus group with relevant experts, (ii) semi-structured interviews at operational facilities with front-end practitioners, and (iii) a review of academic literature and media reports. The framework has also been qualitatively corroborated with transport operators in dedicated interviews and a focus group session. The study identified 10 socio-cultural hazards that were combined into a single framework comprising three high-level sub-categories: (i) crowd–staff interactions, (ii) crowd–crowd interactions, and (iii) crowd–environment interactions. The framework of socio-cultural factors can increase staff's awareness of relevant socio-cultural hazards, their potential consequences in both normal and emergency situations, and the associated mitigation strategies. In turn, this can increase the quality and continuity of service, safety and security in the management of members of the public and passengers in transport hubs.
Keywords: crowd management, disruptive passenger behaviour, emergency management, emergency preparedness and training, risk and security, socio-cultural hazards, transport hubs.

1 INTRODUCTION

The effective and safe management of the travelling public is an important consideration for front-end operators of airports, ports, railways and underground stations. The operators of such transport hubs include service personnel (e.g. desk information officers, ticket collectors), security staff (e.g. police, security guards and screeners), and first responders (e.g. fire fighters, medical personnel). Collectively, these operators are responsible for monitoring, informing, assisting and instructing an enormous flow of passengers of different ages, nationalities and cultures, both in normal and emergency situations (e.g. immediate or controlled evacuations, stranded passenger scenarios, etc.).

However, ensuring the smooth, secure and safe movement of passengers is not easy. Operators need to be able to address a plethora of needs and expectations of different passengers and crowds, who may also have linguistic difficulties, and, depending on the situation, may exhibit strong distress, fear or anger. In particular, transport hub operators need to be able to address successfully the variety of socio-cultural hazards (SCHs) that may arise in the transport hub.

DOI: 10.2495/SAFE-V7-N3-406-418

SCHs correspond to behaviours of transport hub users and staff that are dysfunctional to the effective, secure and safe transit of the public in the hub. Such hazards may include communication breakdowns among passengers and staff due to, for instance, language barriers; increased potential for revolts, as in stranded passenger situations; misreporting of security threats; as well as uncooperative behaviour in case of emergencies. As their name suggests, SCHs have both social and cultural determinants, such as age, gender, education level, social status and cultural background. Managing these kinds of behavioural hazards is part of the operational demand of front-end operators of transport hubs. Such hazards may be a source of stress and discomfort for both staff and passengers, potentially reducing passenger satisfaction at a given hub. Furthermore, if not adequately mitigated, SCHs may lead to interruptions/slowing down of passenger throughput, potential for injury, (safety) equipment damage, reputational damage for relevant transport organisations, and potential for legal litigation.

SCHs may occur both in normal situations and emergencies. In the former case, SCHs may be the triggers themselves of the emergency. This is the situation that may arise, for instance, when angry passengers stuck in a transport terminal, exasperated by a long delay following a service disruption, behave aggressively with staff, or when passengers of two different cultural groups located in the same terminal start to argue with each other. Although not necessarily triggering an immediate health and safety risk, these situations require an adequate intervention by staff in order to ensure/restore continuity of service (i.e., flow of passengers in transit) and the security of the involved staff and the other passengers.

In the case of ongoing emergencies—e.g., evacuations, stranded passengers following transport service disruptions—SCHs may aggravate emergency consequences. SCHs can in fact create the potential for more serious consequences than would have occurred in the absence of the SCHs concerned. For instance, communication breakdowns between staff members and the public due to language differences may significantly slow down evacuation time, and may result in inadvertent discrimination towards some sociocultural groups of passengers (e.g., less educated, non-English speaking foreigners), which in turn may be more likely to suffer higher injuries and casualties. Similarly, the failure of passengers to comply with a request to remain in a confined environment during an ongoing emergency (e.g., technical failure) may result in passengers leaving the assigned area and exposing themselves to health and safety hazards. Both situations may result in higher potential for injury and casualties than if the specific SCH not materialised.

The notion of SCHs overlaps with, but is not limited to, the notion of disruptive passenger behaviour. This definition captures instances in which passengers may fail to respect good rules of conduct (for instance based on staff's instructions), may disrupt good order and discipline of operation, and may create a potential safety or security threat [1–3]. Examples of disruptive passenger behaviour include, for instance, consumption of narcotics, refusal to comply with instructions, sexual abuse, and making threats of any kind towards the crew, other passengers and staff [3]. Often associated with alcohol consumption, disruptive passenger behaviour has been especially highlighted in relation to in-flight service, during which cabin staff cannot count on external resources (e.g. help from police forces), but have to deal with the passenger(s) concerned, at least until the aircraft is on the ground. Cheng-Hua and Hsin-Li [4] have however observed that ground service staff should also be prepared to identify and mitigate such disruptive behaviours, as their occurrence on the ground (e.g. airport terminal) is a predictor of their occurrence in the air. Also, incidents linked to disruptive behaviour incidents may involve verbal or physical violence. For instance, Hella [5] provides a diagrammatic representation showing how passenger behaviour may move from the initial triggering phase (e.g. luggage dispute, mistreatment by staff), to the following phases of escalation, aggression (verbal and/or physical), and return to normal, baseline behaviour. The ICAO Doc 9811 [6] adopts a four-level scheme to classify the seriousness of unruly/disruptive passenger behaviour, which ranges from verbally disruptive behaviour (level 1) and physical abusive behaviour (level 2), to life-threatening behaviour (level 3) and breach of flight crew compartment (level 4).

Implicit in the definition of disruptive passenger behaviour is a focus on misconduct, potential for violence and criminal offence, and the need for staff to respond with adequate communication and, especially, physical containment actions (e.g. handcuffing). While being highly functional to the safety and security of the transport industry, this characterisation fails, however, to account for (minor) incidents in which the passenger involved does not necessarily intend to create a problem, and in which socio-cultural factors should be considered – such as in situations involving first-time fliers, or passengers of other nations or culture failing to comply with local security regulations.

Indeed, the importance of socio-cultural factors in crisis management has been highlighted in studies of natural disasters such as tsunami, earthquakes, hurricanes and floods [7]. In particular, such studies highlighted that culture can act both as an enabler of and a barrier to effective disaster response. For instance, the positive role of culture towards disaster was observed in relation to the Indian Ocean Tsunami. During that event, some communities had indigenous knowledge about the manifestation of an imminent tsunami, which enabled them to take preventive action (i.e. move away from the sea) and survive, as opposed to migrants and tourists who did not have local knowledge [8]. In particular, the communities were aware of the signs of an imminent Tsunami: unusual behaviours of birds and animals, and low tide. At the same time, culture may play a negative role in effective disaster response. Various cases have been reported in the literature regarding communities that refuse to evacuate a dangerous area – dangerous, for instance, because of the closeness of a volcano – until advised to do so by their local leaders [9]. These instances highlight that cultural background influences the behavioural responses to hazards. These considerations are highly relevant for transport hubs, which usually accommodates large groups of transit passengers of different social and cultural backgrounds, who may react differently to the same instruction, hazard or threat.

With these considerations in mind, the present study aims to investigate the potential SCHs that may occur in transport hubs. In particular, the study provides a preliminary, qualitative, taxonomical framework of SCHs in transport hubs. The goal is to capture the range of SCHs that operators of transport hubs may have to confront in normal and emergency situations. This knowledge base will help transport hub stakeholders in developing appropriate mitigations, such as socio-culturally aware risk management models and training programmes. The study is organised as follows: the Section 2 will describe the methodology; Section 3, the framework; Section 4 will address discussion and conclusions.

2 METHODOLOGY

The framework of SCHs presented in this study was developed by adopting an incremental, qualitative, multiple-data collection approach, which is illustrated in the reminder of this section. The data collection comprised the following activities:

- *1-day workshop with experts*. This was attended by 21 safety and security experts from the aviation, maritime, railway and metro domains, and coming from the European Union, Turkey and Israel. The workshop took place in November 2015 in Amsterdam. It was organised around two main parts. Part one, familiarisation, briefed experts about the research context, aims and objectives, and preliminary results. Part two, data gathering, aimed at collecting expert feedback from participants by means of three group sessions. In group sessions 1 and 2, feedback was collected by means of individual questionnaires and structured group discussions. The questionnaire items covered the areas of security training, risk management practices, communication with passengers, and relevant regulations. The group discussions allowed participating experts to provide further feedback on issues not included in the questionnaire (see below) and to report potential socio-cultural hazards;

- *17 Semi-structured interviews with front-end staff.* The interviews were conducted over the period July 2015–April 2016 during site visits to an airport (n = 1), a high-speed train company (n = 2), and railway police (n = 2) facilities in Italy. Personnel interviewed included front-end staff such as stewards of high-speed trains, passengers' assistants, and security screeners and police officers. Each interview lasted between 40 min and 1 h. They were supported by a 24-item questionnaire organised in four parts: biographical data (n of items = 4), problematic passenger groups (n = 8), current security training (n = 6), and management of emergency situations (n = 6). The items were open ended to allow for the collection of plenty of feedback.
- *Literature and media reviews.* Data collection included also a literature review of relevant academic articles and media reports. The review focused on the identification of incidents between service personnel and passengers in transit at transport hubs.

The collected data was analysed qualitatively in a bottom-up fashion, building on the procedures described in [10, 11]. The first step of the analysis consisted of familiarising the researchers with the data, which consisted of notes, visits reports, audio recordings, and intermediate research reports. Then, this data corpus was reviewed in order to identify initial, preliminary categories of SCHs that emerged from the interviews, the focus group, and the literature review respectively. This process involved numerous judgements about the meaning and the significance of the data examined, and the potential categories of SCHs discussed. Subsequently, the preliminary categories of SCHs were systematically compared with each other. This comparison led to some of the SCHs being either refined and/or aggregated according to similarity into more abstract, generalisable categories of SCHs, or discarded because they were insufficiently substantiated by the available evidence. Subsequently, the SCHs identified were subjected to two internal revisions in which the research team qualitatively checked them for overlaps, completeness, accuracy, and relevance. This process led to an initial definition of a total of 17 SCHs.

The initial version of the ECCS framework was subjected to a corroboration cycle in which feedback was collected by three operational experts and one safety manager of two Turkish airports. While this revision indicated that the framework captured relevant categories of SCHs, it was also instrumental in spotting similar categories that could be further aggregated. This triggered a further revision of the framework by the research team, which led to the final form of the framework: a total of 10 categories of SCHs grouped in three macro categories (as will be described in the next section).

3 RESULTS: THE ECCS FRAMEWORK

The SCHs are presented in the context of the Environment, Crowd, other Crowd members and Staff (ECCS) framework. ECCS assumes that to understand the SCHs that can arise from a multicultural crowd, it is necessary to consider the surrounding context in which the crowd is located. SCHs in transport hubs lie at the interaction between the crowd (C) and the components of the surrounding context. Such components include:

- *Staff* (S), i.e. the operators that can be found at the front end of large transport hubs. These include roles involved in normal day-to-day operations (e.g. screeners, information desk assistants, police officers) and first responders (medics, fire-fighters, etc.);
- *Other crowd members* (C), i.e. the other crowd members that a person or a group can interact with in a transport hub. Also, interaction with other crowd members may trigger some types of cultural-dependent hazardous behaviours, hence why the C component appears twice in the model;
- *The environment* (E), i.e. the information and the physical space that can be found in the transport hub. The public's experience of transport hub includes the gathering of information from various sources, such as aural messages (audio alarms; public messages); digital and physical signage; SMS; social-media (Facebook, etc.). Furthermore, passengers have to move across the hub layout and may interact with various types of equipment.

It is in the interactions between the pivotal components of the framework – i.e. crowd members, and the surrounding contextual components – that one should look at to see the occurrence of socio-cultural hazards. SCHs can be found in the staff–crowd interaction; crowd–crowd interaction (also called within-crowd interaction); and crowd–environment interaction (see Fig. 1). Thus, the ECCS model identifies the interactions between the crowd and the different contextual components that may lead to incidents and service interruptions in a transport hub. The model and its components are shown in Fig. 1. The pivotal element of the model is the crowd, around which the other contextual elements are found. The next sections describe the SCHs that can be found in these interactions.

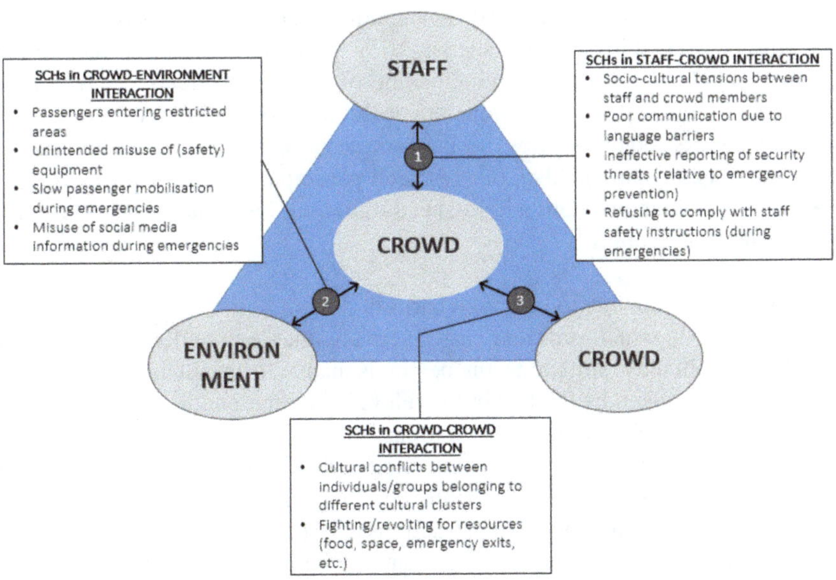

Figure 1: The ECCS model and its components (numbers identify the interactions of the model).

3.1 Crowd–staff interactions

3.1.1 Socio-cultural tensions between staff and crowd members

At least three types of socio-cultural tensions may occur between staff and passengers in transport hubs. First, staff may fail to account for the cultural diversity of the passengers they have to deal with. In particular, they may fail to provide equal (health care/emergency) assistance/treatment to passengers of different cultures (especially ethnical or racial minorities) because of stereotypes and prejudice. Stereotyping can be defined as 'the process by which people use social categories (e.g. race, sex) in acquiring, processing, and recalling information about others' [12]. In general, individuals are frequently not aware of the activation or the impact of stereotyping on their perceptions, emotions and behaviour. Stereotype-linked bias is an automatic and unconscious process, and can occur even among persons who are not outwardly prejudiced [13, 14]. To Jones [15] stereotyped responses of staff can occur especially in situations characterised by intense time pressure, resource constraints, uncertainty and high/cognitive demand, such as emergencies. Ultimately, stereotyping may result in staff discriminating against some cultural groups when providing assistance (for instance because of religious differences).

Second, staff may inadvertently engage in behaviours considered unacceptable/offensive from the perspective of a passenger of another culture [16]. For instance, tensions may arise due to staff improperly handling sacred/personal items at security checks; staff using inadequate body language (e.g. a male security officer talking directly to a Muslim woman while ignoring her husband); staff making requests considered unacceptable by other cultures (e.g. asking a Sikh to remove his hat in public; asking a Muslim woman to remove her head scarf in public; having a Muslim male body searched by a woman officer). In general it is important for staff members to realise how their behaviour may inadvertently spark incidents with passengers from nationalities and cultures different than theirs.

Finally, the staff interviewed reported that another category of tensions concerned the interaction with privileged individuals, such as VIPs, politicians and high-ranking individuals. Some incidents have occurred in which some of these individuals behaved arrogantly towards staff, held unrealistic expectations about the level of service they could expect, or refused security checks and procedures (for instance because they did not want to be inspected by virtue of their social status). Also, our research participants reported at least one occurrence in which a politician on board a high-speed train used his authority to interfere with an on-going emergency operation.

3.1.2 Poor communication due to language barriers

Situations may occur in which staff and passengers are unable to understand each other due to language barriers. In these situations, the person is unable to articulate own thoughts either in the local language spoken by staff or in the relevant international language (e.g. English). This SCH may include different kinds of situations, such as, for instance, injured/sick passengers who are unable to report/communicate their own (deteriorating) health conditions; passengers unable to report a problem to staff (e.g. threat or hazard observed; missing relative); passengers misunderstanding staff instructions (e.g. not following the direction assigned by staff). Note that this hazard may be exacerbated when staff adopt poor communication strategies, such as repeating instructions or questions louder and louder, but without slowly articulating words – something that may trigger a defensive response of the concerned passenger, because of fear and/or embarrassment.

3.1.3 Ineffective reporting of security threats (relative to emergency prevention)
Usually, passengers in transport hubs receive communications about the importance of noticing and reporting to authorities' items that may pose an immediate security threat, such as unattended packages, objects and baggage. However, our research participants reported cases in which passengers failed to report such threats effectively. For instance, passengers may fail to report a security threat, such as an unattended item, because they either are not aware of the threat, or may not perceive reporting it as an obligation. Furthermore, passengers may engage in an improper reporting procedure: they may themselves collect a potential threat so as to bring it to the police (instead of just calling the police). This latter case was reported relative to infrequent travellers, who, while attempting to do something useful, did not realise the potential danger they exposed themselves to. Finally, passengers may err on the cautious side, and over-report potentially security threats, which in fact turn out to be not dangerous. This latter behaviour may increase following the occurrence of a major disaster/terrorist attack, which usually enhances the level of alertness of the general public.

3.1.4 Refusing to comply with staff safety instructions (during emergencies)
During emergencies, passengers may fail to comply with staff instructions. For instance, passengers may act based on their own initiative. Unauthorised, self-initiated responses may be triggered by the increasing sense of danger, especially following a prolonged stay in a confined environment. For instance, in the case of the Norman Atlantic ferry disaster in 2015, some passengers abandoned the ship due to increased fear of the growing fire, and eventually died in deep waters. Again, in an incident mentioned during our interviews, a passenger on board a broken-down train, temporary stuck at a station, broke a window to get off without staff authorisation, and hence exposed himself to various hazards such as electrocution. These behaviours can arise as a consequence of a heightened sense of danger, combined with at least two culture-dependent factors such as low trust in authorities and a tendency for self-organisation (see e.g., [17, 18]).

Also, familiarity with a particular transport hub may limit compliance with staff instructions, especially in emergencies. People have mental scripts for navigating through familiar environments and use these automatically without conscious thought [19, 20]. Thus, regular users of a transport hub, such as commuters, may pay little attention to staff instructions, especially if these involve taking unfamiliar paths – such as taking elevators, stairs, or a tunnel to escape from a threat (e.g., fire) – and even if these are the quickest and safest ways to safety. This was exemplified in the case of the King's Cross underground station disaster in London in 1987, where the majority of the victims had taken the familiar route instead of the one instructed by the staff [19]. Overall, these considerations suggest that in transport terminals, familiar users such as commuters may require clear and targeted information to override their routine automatic routes, particularly in emergencies.

3.2 Crowd–crowd interactions

3.2.1 Cultural conflicts between individuals/groups belonging to different cultural groups
SCHs may include tensions between passengers belonging to different cultural groups. For instance, in an incident reported on the news, an 81-year old woman on an El Al flight from New York to Tel Aviv was invited to change seat, after an ultra-orthodox Jew refused to sit next to a woman [21]. The incident resulted in the woman concerned suing the airline. It seemed that the incident was not a single instance and others were reported to have occurred due to sexual discrimination by religious ultra-orthodox Jewish men, eventually causing

flight delays [21, 22]. Other examples mentioned by the staff interviewed concerned potential conflicts between supporters of different football teams in the same terminal, e.g. train station airport. Although cultural difference, per se, is not necessarily the primary cause of conflict between groups, culture influences significantly the perceptions according to which cross-cultural conflict is formed [23].

3.2.2 Fighting/revolting for resources (food, space, emergency exists, etc.)

The literature on evacuation and crowd disasters has acknowledged the tension existing between the competitive and collaborative behaviour of passengers in emergencies [24, 25], and the fact that passengers may positively engage in collaborative behaviour in emergencies, for instance by keeping calm and supporting others. However, although less desirable, competitive behaviour may still occur. Passengers may fight for safety equipment during emergencies, and may push or even beat other passengers while attempting to evacuate quicker [26, 27]. In one of our interviews, a high-speed train steward reported an incident in which angry and hungry passengers on board a broken-down train revolted against him and other on-board staff in order to find food in the galley.

3.2.3 Crowd–environment interactionPassengers entering restricted areas

This category of SCHs involves passengers entering restricted areas, either inadvertently or intentionally. Inadvertent entrance into restricted areas may occur as wayfinding in transport hubs can be very demanding, especially for unfamiliar users or infrequent travellers. A serious wayfinding failure involves passengers inadvertently entering restricted security areas, something that may trigger an alarm and result in the evacuation of the hub (or part of it). For instance, on May 2016, at Cologne airport, a 62-year-old Spanish passenger rushing to his aircraft inadvertently entered a restricted area of Terminal One [28]. The event triggered a preventive evacuation of the terminal, while police officers searched for explosives, and several aircrafts were delayed. A similar security breach in the same airport with similar consequences involved a woman who was able to walk through the security screening area without being checked [29]. Besides age, familiarity and travel experience, another factor that may lead to this SCH is the lack of understanding of symbols across cultures: as Symonds succinctly puts it: "people of China does not [necessarily] understand the same symbols as American tourists" [30]. Corroborating this point is the observation of one of our research participants, a railway police director, who noted that asylum seekers in railways may not necessarily be familiar with basic symbols, such as the yellow line next to the rail.

Finally, sometimes passengers intentionally enter into restricted areas, such as rail tracks. These may be commuters in a hurry, or immigrants who do not perceive the danger of crossing tracks at railways station.

3.2.4 Unintended misuse of (safety) equipment

This behaviour involves the unauthorised and unjustified use of equipment by passengers who are unaware of the consequences or gravity of their actions. For instance, the railways operators we interviewed reported cases of improper emergency brake activation occurring when people accompanying passengers on board a high-speed train realised that the train was leaving the station while they were still on board. When approached by the train steward, the people concerned were surprised by the expensive penalty they had to pay. Analogous cases have been reported in the airport domain, in which the evacuation slide was unnecessarily deployed following the improper opening of a plane door by a first-time flyer, because of reasons such as mistaking an emergency exit for the toilet [31], or attempting to get fresh air before take-off [32]. Overall, the common factor across these types of behaviours is the lack of awareness of the consequences, such as delays and potential for injuries, arguably due to poor education and lack of familiarity with modern transport systems or a combination of both.

3.2.5 Slow passenger mobilisation during emergencies

While a quick mobilisation of passengers of a transport hub may be needed in certain situations, e.g. evacuation, there are factors that may hamper this process. First, passengers may react differently to public alarms in transport terminals, depending on the level of risk they perceive. This may vary according to different socio-demographic parameters, such as gender and age. Men and younger people are generally less risk-adverse than women and older people as documented by [33, 34]. Thus, front-end operators have to consider that passengers might be more or less willing to evacuate according to their level of risk perception. Second, different passenger crowds may move at different speeds. Crowd walking speed might vary depending on several socio-demographic parameters, such as gender, age and group formation [35]. For instance, men walk faster than women, and adults and children walk faster than the elderly. People also walk more slowly when they are with others than when they are alone. Furthermore, as reported by [36], dress length also has a direct influence on people walking speed. These findings suggest that transport hub operators should be aware that crowds should not be treated as a collection of similar individuals.

3.2.6 Misuse of social media information during emergencies

Our research participants stressed that during an emergency, users of the transport hub may publish incorrect, partial, or incomplete information on social media, which may generate unjustified panic. Furthermore, young people sometimes record the emergency situation on their mobile phones instead of escaping from the danger and evacuating safely. This finding is indeed supported by a video analysis of Robinson and Cichomska (available in [37]) about the terrorist attack at Brussels airport.

4 DISCUSSION AND CONCLUSIONS

SUMMARY

This study has identified 10 socio-cultural hazards (SCHs) that may occur in transport hubs. SCHs correspond to behaviours of members of the public in transit (and staff) that are dysfunctional to the efficiency, safety and quality of operations, and that therefore need to be addressed by front-end operators. SCHs may occur both during normal situations and during emergencies. In the former case, they may trigger incidents; in the latter case, they may add to the chaos and uncertainty of emergency operations, thus increasing their negative consequences. The framework of SCHs provides a means for front-end operators to identify and discuss issues they encounter in their interactions with the public.

The SCHs identified were formed into a framework comprising three high-level areas – crowd–staff, crowd–crowd, and crowd–environment interactions. These three areas capture the fact that SCHS in transport hubs can be found in the interaction between individuals or groups in the crowd and the other elements of the surrounding context: staff members, other crowd members, and the information and physical environment.

4.1 Practical value of the framework

The ECCS framework has potential pragmatic value. It can enhance current transport hub safety and security risk assessment by prompting analysts to consider relevant categories of behavioural hazards that can impact substantially the risk profile of a given hub – either because they can trigger a crisis, or because they can greatly influence the consequences associated with an existing one. Also, the framework can become part of training packages for front-end operators of transport hubs. The framework can be used to sensitise operators regarding potential hazardous behaviours that may arise from the public in transport hubs, and their potential consequences in both normal and emergency situations. Indeed, the effective mitigation of SCH requires awareness and anticipation of them by front-end operators, as well as the use of adequate socio-cultural mitigations. To this end, the research team is currently developing a training package that aims to increase operators' intercultural skills, cross-cultural sensitivity, ability to cope with cultural differences, and awareness of cross-cultural differences in crowd behaviours as well as verbal and non-verbal communication skills, both one-to-one and one-to-many. Such skills are not necessarily covered by existing mandatory training programmes. Usually, these tend to cover the technical, procedural and regulatory aspects of the work, while providing limited coverage of the potential issues that may arise when interacting with members of the public. Yet, knowledge of these SCHs is part of the background of the more senior staff members [4], and is essential to contribute to safe and effective task performance both during normal and emergency situations.

Overall, the methodological approach taken ensured that the SCHs identified by the study were heavily based on the accounts of relevant research participants. As discussed in Section 2, these included both back-end as well as front-end roles from the airport, railways, underground and maritime domains. This provided a relevant knowledge base which was triangulated with relevant academic literature and incidents reported in the media. Furthermore, an initial corroboration was obtained from a small-sample of four airport staff. Despite this, the framework can benefit from at least two types of future studies: (i) studies that will seek further corroboratory feedback from experts of other transport domains about dimensions such as the relevance, completeness, and correctness of the ECCS framework; and (ii) studies that will deepen the understanding of viewpoints of members of the public involved in SCHs, for instance through in-depth interviews and ethnographic observations of passengers in transport hubs. In turn, these efforts are expected to increase the validity and the generalisability of the framework.

Finally, note that, while the framework has essentially stressed the undesirable behaviours that may arise from the interaction of the public and the transport hub, previous research indicates that members of the public may also play a positive role in emergency management. Therefore, it is anticipated that the public may be considered a resource in the development of the socio-cultural mitigations. Thus, the authors advocate that the ECCS framework be extended or complemented with an analogous taxonomy of the positive behaviours that can be played by the public in transport hubs, in normal as well as in emergency situations.

ACKNOWLEDGEMENTS

This work is supported by funding from the European Commission through the IMPACT project (H2020 GA no. 653383). *The authors wish to express their gratitude to the project coordinator, Dr Alessandra Tedeschi, for her support to this study, and to the members of the IMPACT External Stakeholder Group and the front-end staff members involved in the study.*

REFERENCES

[1] Skybrary, 'Unruly Passengers - SKYbrary Aviation Safety', 2016, available at http://www.skybrary.aero/index.php/Unruly_Passengers#The_Problem (accessed 30 January 2017).

[2] ICAO, *Annex 17: Security. Safeguarding International Civil Aviation Against Acts of Unlawful Interference,* 9th edn., International Civil Aviation Organisation, 2011.
[3] IATA, Guidance on unruly passenger behaviour and management. *International Air Transport Association*, 2012.
[4] Cheng-Hua, Y. & Hsin-Li, C., Exploring the perceived competence of airport ground staff in dealing with unruly passenger behaviors. *Tourism Management*, **33**(3), pp. 611–621, 2012.
https://doi.org/10.1016/j.tourman.2011.07.001
[5] Hella, J., Cabin violence: assessment and management. *Cabin Safety Update*, **7**(9), pp. 28–31, 2001.
[6] ICAO, Doc 9811 Manual on the Implementation of the Security Provisions of Annex 6. *International Civil Aviation Organisation*, 2000.
[7] Kulatunga, U., Impact of culture towards disaster risk reduction. *International Journal of Strategic Property Management*, **4**, pp. 304–313, 2010.
https://doi.org/10.3846/ijspm.2010.23
[8] Arunotai, N., Saved by an old legend and a keen observation: The case of Moken sea nomads in Thailand. *Indigenous Knowledge for Disaster Risk Reduction*, **73**, 2008.
[9] Lavigne, F., De Coster, B., Juvin, N., Flohic, F., Gaillard, J.C., Texier, P., Morin, J. & Sartohadi, J., People's behaviour in the face of volcanic hazards: Perspectives from Javanese communities, Indonesia. *Journal of Volcanology and Geothermal Research*, **172**(3), pp. 273–287, 2008.
https://doi.org/10.1016/j.jvolgeores.2007.12.013
[10] Schreier, M., *Qualitative Content Analysis in Practice*, Thousand Oaks, California: SAGE Publications, 2012.
[11] Ritchie, J. & Spencer, L., Qualitative data analysis for applied policy research. In *The Qualitative Researcher's Companion*, eds A.M. Huberman & M.B. Miles, Thousand Oaks, California, USA: SAGE Publications, pp. 305–329, 2002.
[12] Smedley, B.D., Stith, A.Y. & Nelson, A.R., *Unequal treatment: Confronting racial and ethnic disparities in health care (full printed version)*, National Academies Press, 2002.
[13] Devine, P.G., Stereotypes and prejudice: Their automatic and controlled components. *Journal of Personality and Social Psychology*, **56**(1), pp. 5–18, 1989.
https://doi.org/10.1037/0022-3514.56.1.5
[14] Van Ryn, M. & Fu, S.S., Paved with good intentions: do public health and human service providers contribute to racial/ethnic disparities in health?. *American Journal of Public Health*, **93**(2), pp. 248–255, 2003.
https://doi.org/10.2105/AJPH.93.2.248
[15] Jones, A., *Concept paper: Cultural competence: Solutions and strategies for emergency medical services*, 2007.
[16] Sutherland, D., The role of homeland security and training. In *The Arab-American Handbook: A Guide to the Arab, Arab-American, and Muslim Worlds*, 2nd edn., ed. N. Shora, Cune Press, 2009.
[17] Hofstede, G.H., *Culture's Consequences: Comparing Values, Behaviors, Institutions and Organizations Across Nations*, Sage, 2001.
[18] Welzel, C., How selfish are self-expression values? A civicness test. *Journal of Cross-Cultural Psychology*, **41**(2), pp. 152–174, 2010.
https://doi.org/10.1177/0022022109354378

[19] Donald, I. & Canter, D., Intentionality and fatality during the King's Cross underground fire. *European Journal of Social Psychology*, **22**(3), pp. 203–218, 1992. https://doi.org/10.1002/ejsp.2420220302

[20] Charlton, S.G. & Starkey, N.J., *Driving without awareness: The effects of practice and automaticity on attention and driving*, 2011.

[21] Sherwood, H., Retired lawyer sues Israeli airline after she was asked to move seat. *The Guardian*, 01-April-2016.

[22] Paulson, M., When a plane seat next to a woman is against orthodox faith. *The New York Times*, 09-April-2015.

[23] Avruch, K., Cross-cultural conflict. *Conflict Resolution*, **1**, pp. 45–57, 2009.

[24] Chertkoff, J.M. & Kushigian, R.H., *Don't Panic: The Psychology of Emergency Egress and Ingress*, Praeger Publishers, 1999.

[25] Rogsch, C., Schreckenberg, M., Tribble, E., Klingsch, W. & Kretz, T., Was it panic? An overview about mass-emergencies and their origins all over the world for recent years. In *Pedestrian and Evacuation Dynamics 2008*, Springer, pp. 743–755, 2010. https://doi.org/10.1007/978-3-642-04504-2_72

[26] Sabin, L., Italy ferry fire: Rescued passengers describe violence and panic on board stricken vessel, available at http://www.independent.co.uk. [Online], available at http://www.independent.co.uk/news/world/europe/norman-atlantic-ferry-fire-two-british-men-still-not-rescued-from-burning-italian-ship-9947674.html (accessed 07 Febuary 2017).

[27] Rossington, B., Men HIT women in fight to rescue helicopters during Italy ferry fire. *Mirror*, 30 December 2014. Available at http://www.mirror.co.uk/news/world-news/italy-ferry-fire-men-hit-4893979. (accessed 16 September 2016).

[28] Confused Spaniard to blame for alarm at Cologne airport. *The Local*, 30-May-2016. Available at: http://www.thelocal.de/20160530/breaking-flights-stopped-over-security-alert-at-cologne-airport. (accessed 31 August 2016).

[29] Withnall, A., Cologne airport Terminal 2 evacuated due to security breach | The Independent, available at http://www.independent.co.uk, 01-Mar-2016. [Online], available at http://www.independent.co.uk/news/world/europe/cologne-airport-evacuated-due-to-security-breach-a6904766.html (accessed 07 Febuary 2017).

[30] Symonds, P., The Complexities of Human Wayfinding in Airports, 2014.

[31] Branigan, T., Chinese passenger opens plane door for fresh air. *The Guardian*, 18 Dec 2014, [Online], available at https://www.theguardian.com/world/2014/dec/18/chinese-passenger-opens-plane-door-for-fresh-air (accessed 12 July 2016).

[32] Kitching, C., Flyer mistakes emergency exit for loo and triggers inflatable slide, *Mail Online*, 27 March 2016. Available at http://www.dailymail.co.uk/travel/travel_news/article-3511283/China-Southern-Airlines-passenger-opens-emergency-exit.html (accessed 12 July 2016).

[33] Byrnes, J.P., Miller, D.C. & Schafer, W.D., *Gender Differences in Risk Taking: A Meta-Analysis,* American Psychological Association, 1999.

[34] Deakin, J., Aitken, M., Robbins, T. & Sahakian, B.J., Risk taking during decision-making in normal volunteers changes with age. *Journal of the International Neuropsychological Society*, **10**(4), pp. 590–598, 2004. https://doi.org/10.1017/S1355617704104104

[35] Willis, A., Gjersoe, N., Havard, C., Kerridge, J. & Kukla, R., Human movement behaviour in urban spaces: Implications for the design and modelling of effective pedestrian environments. *Environment and Planning B: Planningand Design*, **31**(6), pp. 805–828, 2004
https://doi.org/10.1068/b3060.

[36] Almejmaj, M., Meacham, B. & Skorinko, J., The effects of cultural differences between the west and Saudi Arabia on emergency evacuation—clothing effects on walking speed. *Fire and Materials*, **39**(4), pp. 353–370, 2015.
https://doi.org/10.1002/fam.2227

[37] Cichomska, K., *et al.*, D1.3: Theoretical Framework of Cross-Cultural and Psycho-Social Crowd Behaviour and Management in Transport Terminals. IMPACT Deliverable D1.3, January 2016.

MANAGING EXTREME AGRICULTURE RISKS IN BRAZIL

DIEGO ARIAS, PEDRO A. VIEIRA & PAULO M. MENDES
World Bank, Empresa Brasileira de Pesquisa Agropecuaria & Ministry of Agriculture of Brazil.

ABSTRACT
The agricultural sector is strategic for Brazil's economy and society. Nevertheless, the fluctuation in national agriculture GDP, which reaches 1% per year on average, and can reach more than 3% in some cases, indicates the weakness of disaster risk management in the sector. This study identified that Brazil has some successful public policies and programs dedicated for disaster risk management in the agriculture sector that serve as reference for other developing countries. However, they are mainly focused on risk mitigation. Disaster risk response and transfer mechanisms for agriculture, especially insurance, are still underdeveloped in relation to other similar emerging economies such as Argentina and Mexico. Furthermore, this study points out to the importance of managing extreme agriculture risks in an integrated fashion, as there are clear interrelationships among the main risks facing the agriculture sector of Brazil. Finally, the results of this study also strongly suggest that there is a lack of integration of existing federal-level public policies and programs dedicated to agricultural disaster risk management in Brazil, which constraints the reduction of systemic losses at the national level.
Keywords: agriculture policies, agriculture risks, brazil agriculture, disaster risk, extreme risks, risk management, risk strategies.

1 INTRODUCTION

The agriculture sector of Brazil faces a large number of risks linked to the productive process, which has led to substantial losses to the country in the past years. An adequate and integrated management of those risks can leave farmer incomes less exposed to losses, benefiting the sector and the country as a whole. Therefore, given the always-present resource limitation, it is important to maximize the economic returns of agriculture risk management actions. Brazil built important agriculture risk management policies and programs, but there are several signs that it is possible to improve their efficiency and effectiveness with more coordination and a prioritization in the treatment of gaps and opportunities.

The objective of this work was to undertake a rapid and integrated review of agriculture risk management in Brazil, identifying gaps and opportunities for improving current public policies and programs at the federal level in the short and long term. Beyond potential improvements in specific agriculture risk management policies and programs, an improved coordination and integration of current tools can reduce the risk profile of the sector. In this context, the authors, within the context of a collaboration between the World Bank, the Brazilian Agriculture Research Company (Embrapa) and the Ministry of Agriculture of Brazil (MAPA), put forward a rapid an integrated review of agriculture risk management in Brazil. The rapid review had the following phases: (i) national electronic survey with sector stakeholders; (ii) analysis of the risk perception of agriculture sector experts; (iii) literature review; and (iv) validation of results from stakeholder institutions. This analysis intends to build a systemic vision of agriculture risk management and had the support of different institutions linked to the agriculture sector.

Extreme agriculture sector risks can be grouped by production, market and enabling environment risks. According to the typology set by the World Bank, extreme production risks refer to the actual production and includes weather, fire, animal and plant health, and any unexpected change in the management of the farm and/or its natural resources (such as

DOI: 10.2495/SAFE-V7-N3-419-430

changes in regulation and/or technical assistance for example). Extreme market risks include mainly significant changes in output and/or input prices, unexpected changes to access to credit, to markets, including external markets. Finally, extreme enabling environment risks include events that change the context of policies and institutions (changes in the legal/regulatory framework, sector institutions, interpretation of laws and regulations), including logistics and infrastructure (Table 1).

Agriculture production is one of the main pillars of the Brazilian economy. The participation of the primary agriculture sector to total GDP is close to 6% in Brazil and agribusiness represents 23% of GDP, being the backbone of an important value added chain that represents 32% of total employment. Beyond the multiplier effect in income and employment generation, historically, the agriculture sector has contributed significantly to the trade balance of Brazil. In the past 10 years, the volume of agroexports grew more than 200% and the trade balance 468%, resulting in agroexports reaching US$100 billion in 2013, representing almost 42% of total exports from Brazil. Brazil is the largest exporter of coffee, sugar, orange juice and meat (cow and chicken); the second largest in maize and soy (grains, oil and flower); and being one of the leaders in pork meat and cotton among other rural products.

Although the agriculture sector is one of the engines of the Brazilian economy, it is also more volatile in economic terms. This volatility has impacts in the country in terms of: (i) the

Table 1: Main extreme agriculture risks in Brazil.

EXTREME RISK GROUP	RISK	EXAMPLES OF HAZARDS
PRODUCTION RISK	Weather and Fires	Prolonged droughts, frosts, excess rain, strong winds and floods.
	Animal Health	Foot and mouth disease, BSE (mad cow disease), *Newcastle*, etc.
	Plant Health	Introduction of new pests and diseases (ex.: lagarta *Helicoverpa armigera*).
	Natural Resources and Production Management	Changes in water rights/allocation, in technical assistance, in control of regulations, in labor availability. Sudden mismanagement of natural resources (ex. drought in RS) like incorrect decisions on soil and water management.
MARKET RISK	Trade (price of inputs and outputs) and credit	Substantial changes in output or input prices, in exchange rates, in interest rates, in credit terms.
	International trade	Unexpected closing of export markets and changes in the access of imported inputs.
ENABLING ENVIRONMENT RISK	Infrastructure and logistics	Strikes at the port, closing of roads, railways, waterways, and changes in the incentives for storage.
	Regulatory, Policy or Institutional Framework	Changes in laws/regulations (environmental, labor, inputs, land), changes in the approach of public institutions or the change in interpretation of existing laws and regulations.

multiplier effect and its corresponding dependency in other agriculture subsectors; (ii) the importance of the sector in fiscal revenues; (iii) the impact in the increase of domestic prices, in particular food products; and (iv) the importance of the sector in rural incomes. Agriculture sector volatility is produced by different risk groups that can result in extreme economic losses.

Brazil loses annually, on average, 1% of the value of agriculture production due to extreme events. Following a simple methodology to calculate agriculture losses as a function of extreme events, we find that extreme events do not just cause losses in agriculture production, but also, in those given years of extreme events, agriculture GDP growth is negative. In the years that the largest losses have occurred, 2004–2006 and 2012–2013, they had been related to severe droughts. In 2004–2006 the impact was in the south and southeast region of the country, while in 2012–2013 the reduction in production was related to the extreme drought in the northeast region. It is important to consider that loss values from extreme losses could be higher when taking into account forgone producer income of those agriculture products and/or regions with limited or lack of access to risk mitigating mechanisms. Beyond the estimated losses due to extreme weather events, the agriculture sector of Brazil suffers other losses due to risks such as unexpected changes in the regulatory framework, in the management of production and natural resources and other risks that are difficult to quantify.

The consequences from extreme agriculture sector risks vary by region, by type of farmer and by value chain. Agriculture in the northeast region, saw losses of up to 90% of grain production in areas without irrigation in 2012, and had negative growth in the last three years, with average losses of three work-days/year for rural families due to low production and water shortages according to De Nys *et al.* [1]. Family farming, given they reduced access to financial risk transfer instruments, suffer disproportionally more from extreme risks that affect their revenues. This has contributed towards the increase in rural poverty in the State of Sao Paulo in precisely the poorest regions of the country, as well as towards the reduction of one standard deviation in the yields of permanent and temporary crops, reducing up to an estimated 42,000 jobs according to MBAgro [2]. Finally, certain risks can have severe consequences in entire value chains. For example, the foot and mouth disease outbreak in Mato Grosso do Sul and in Parana in 2005 closed the meat export market to Russia, which was restored only 28 months after according to Nogueira e Doliveira [3]. The reduction in meat exports during that period was one-third of the total exported volume, having a significant impact in farmer income and in also in the competitive position of the value chain in international markets.

For the federal government, as well as for the state governments, extreme agriculture risks imply important fiscal expenditures. From the public expenditure side, events that impact the agriculture sector income, such as natural disasters, pests and diseases, significant fluctuations in food, labor, input and energy prices, as well as agriculture production per sea, mean that agriculture emergency expenditures at all levels of government can have important impacts for the economy as a whole. The federal government estimates that, only for the northeast, the response to the extreme drought of the last few years has resulted in additional public expenditures of approximately R$17 billions according to the Senate of the Brazilian Congress [4]. The risks also have an impact in fiscal revenues. From the side of fiscal revenues, a reduction in 10% of the Brazilian production of soy can result in an annual reduction of R$1.6 billion in fiscal revenues, equivalent to 16% of the annual budget of MAPA (Fig. 1). If losses would occur in various agriculture products at the same time, the fiscal impact, in revenues and expenditures, would be significant.

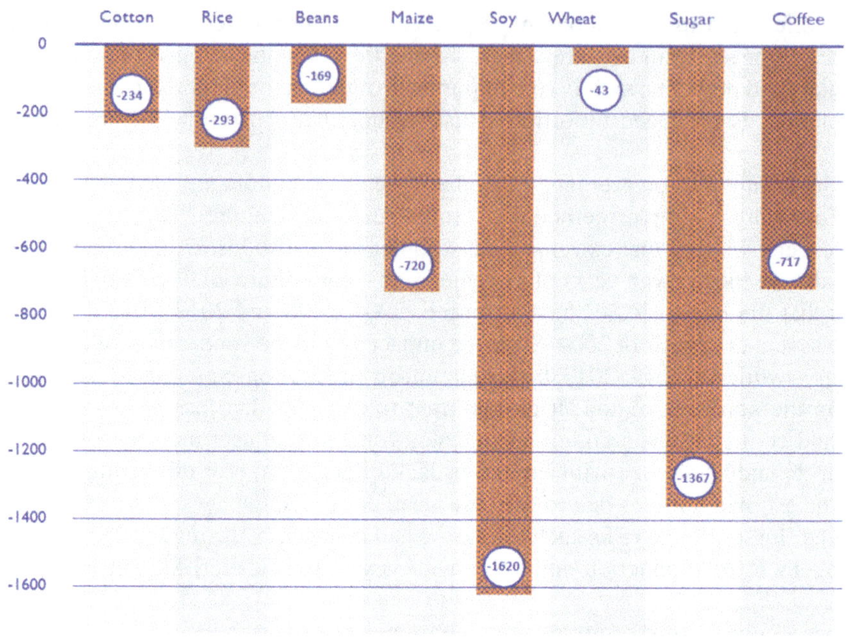

Figure 1: Loss in fiscal revenues due to a reduction in 10% in agriculture production (R$ millions) (MB Agro [2]).

2 METHODOLOGY

There are several methodologies and approaches to doing an integrated agriculture sector risk review. In the past few years, OECD proposed a holistic conceptual framework for agriculture risk management and the World Bank developed a methodology for evaluating in an integrated manner the risks of agriculture supply chains [5]. Recently, the World Bank has been using this latter methodology for an integrated assessment to support government in the prioritization of agriculture risk management public investments and actions.

A rapid review methodology was developed by the authors in order to be able to map the main public policies and programs and identify the challenges and opportunities to improve extreme agriculture risk management. Usually, the risk evaluation includes risk quantification, prioritization, analysis of capacities and the search of solutions to manage systemic risks. However, given the richness of previous information, studies and assessments before this study in Brazil, the decision was to undertake a rapid review based on consultations with more than 100 specialists and an electronic survey sent out to more than 5000 representatives of the sector, of which 10% responded. The work was backed up by an empirical evaluation of those risks in two states (Bahia and Paraiba) and in one Municipality (Piquet Carneiro, Ceara). The results of the consultations of the specialists and of the electronic survey were compared with the results of a literature review of the most recent studies in risk evaluation in the country and with the main public policies and programs, according to budget and coverage, dedicated to mitigating, transferring and responding to extreme agriculture risks. After they were categorized, the preliminary results were presented to various stakeholders related to the agriculture sector of Brazil and the implementation of public policies for validation and improvement of the results.

In order to reduce the complexity of the work, only events with farm-level impacts were evaluated due to limited resources, in particular time, only farm-level extreme risks were taken into account. This includes events that originate outside the farm-gate (for example, the unexpected closing of a road, changes in the regulations surrounding pesticides, etc.), as long as they have an impact in farmer income (through costs and/or sales).

Considering that the agriculture sector has cyclical variations, this work only considered the risk associated with extreme and systemic events that affect farmer incomes. The extreme and systemic events were defined as those that produce more than one standard deviation from the average or the historical trend. Income reductions less than one standard deviation (around 10%) were not considered. Thus, limitations of this methodology include: (i) not evaluating impacts across the supply chain (only on-farm); (ii) it is influenced by recent events (drought, road closures) that are more clear in the memory of the stakeholders, specialist and survey responders, versus events that have happened in the longer past; and (iii) not being able to ensure the correct interpretation by participants of the definition of risk, therefore referring to structural limitations (and not to unexpected adverse events).

The management of extreme agriculture risks requires different strategies, instruments, and has to acknowledge the differences in risks and stakeholders (Fig. 2). In order to reduce sector losses in an aggregate way, all extreme risks must be considered using different risk management strategies and instruments. The main management instruments available for the public sector can be summarized in investments, technical assistance and public policy actions. The different agriculture risk management actions can be grouped in the following broad strategic pillars.

Mitigation: actions that happen before the event (ex-ante) to prevent, reduce or eliminate the occurrence of extreme events or their negative economic impact on agriculture production, markets or enabling environment, such as infrastructure, animal and plant health, agroclimatic information systems, etc. Transfer: actions to transfer risks to a third party, by paying a cost (premium). Examples include financial instruments such as

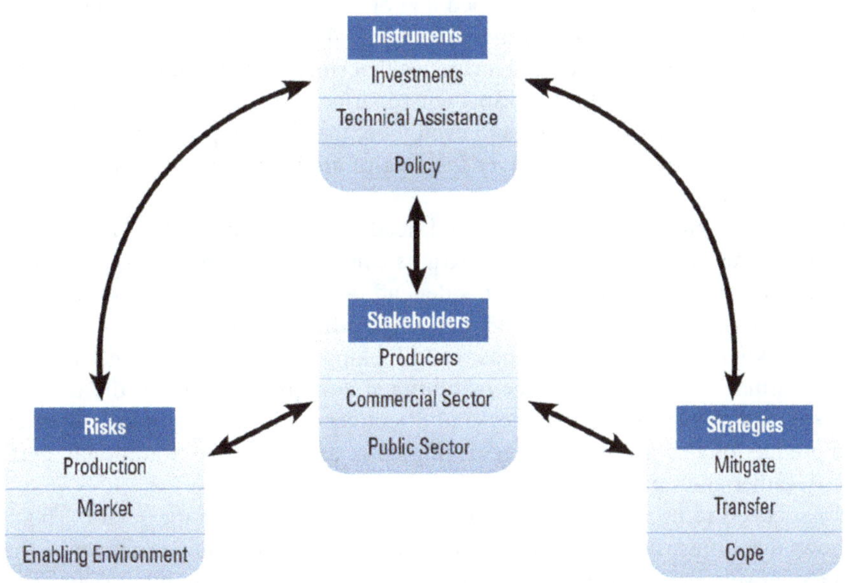

Figure 2: Integrated vision for Agriculture Risk Management (World Bank [5]).

insurance, reinsurance, derivatives, etc. Response: actions that happen during or after an event *(expost)*, geared towards rebuilding or compensating for losses caused by the event. Examples include emergency assistance to farmers, debt restructuring, infrastructure reconstruction, etc. An important response action that occurs ex-ante is the strategic planning or contingency planning, where response mechanisms are designed a priori in order to respond more effectively and efficiently after the event (ex: contingency plans for combating fires or for the control of disease outbreaks).

The rapid review of extreme agriculture risk management policies and programs was divided in five stages, which occurred over a period of 7 months. The Stages included: (i) the identification of the main risks and the specialists to be consulted, (ii) the literature review of recent studies on risks and the related federal-level public policies and programs; (iii) undertaking a national survey on agriculture risk perception from the sector; (iv) undertaking an event to consult with sector specialists (identified in Stage 1); and (v) systematizing the results.

3 RESULTS

The final results were a synthesis of the survey, of the workshop and the validation. After the validation with different institutions, the results were systematized in order to obtain a synthesis of the perceptions. The divergent results, when consensus did not happen, were eliminated. That synthesis of results, in particular during the workshop and validation, took into consideration the review of the 25 main public policies and programs.

Although there is a great number of public policies and programs in Brazil, there are gaps and opportunities. The mitigation strategy is the most representative in terms of number of public policies and programs. The predominance of mitigation is aligned with any risk management strategy. However, there are many important opportunities in the transfer and response strategies. In the case of response, specifically in cases of low frequency and high impact risks, such as sanitary problems not yet present in Brazil, the country has a relatively low number of contingency plans.

The public policies and programs mapped do not cover all losses from small and medium size farmers in an integrated manner. Larger commercial farmers have access to sophisticated financial instruments to cover price and insurance that small and medium size farmers do not. Family farmers depend on public programs where farmers receive income compensation for losses above 50% of expected yields, and even then, they do not get to cover all farm-level losses. Other public programs only cover the cost of credit, but do not compensate for the farmer income loss.

The operation of programs is complex and needs more coordination. The integration of price risk programs with the insurance programs can and should result in the protection of farmers' income. However, these programs depend on different institutions, and are implemented by different intermediaries, arriving through different vehicles to farmers.

In order to avoid duplication of actions – and looking to maximize services and coverage to improve farmers' income – an inter institutional coordination among the different programs is key to achieve more efficiency and effectiveness in the operation. Extreme risks linked to production management can be as important as other more 'traditional' agriculture risks, like weather and sanitary issues.

Risks associated to infrastructure and logistics are a priority for the sector. The sectoral vision shows that infrastructure and logistics risks have a relatively larger impact on the economy and a relatively lower level of public policy support in relation to other risks considered (Fig. 3 and Table 2). Nevertheless, the risks associated to extreme weather events were

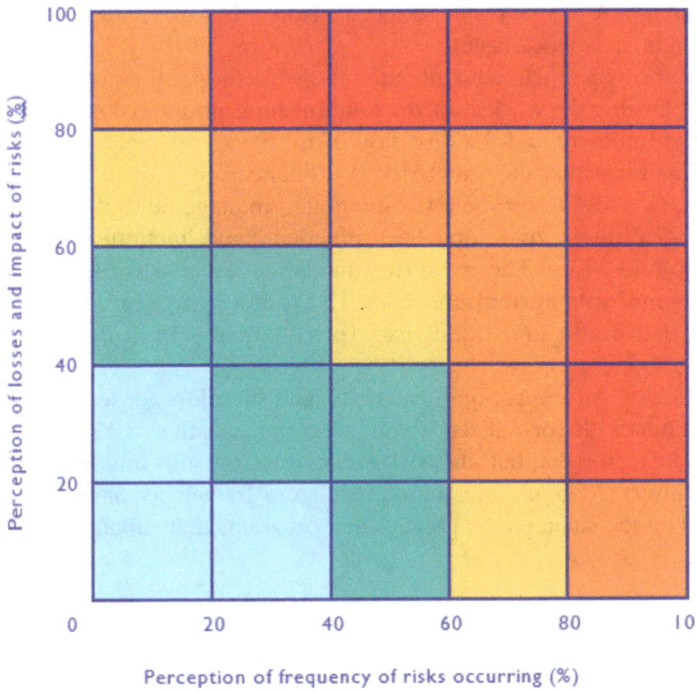

Figure 3: Perception of the impact and frequency (severity) of extreme agriculture risks.

Table 2: Perception of specialists regarding the extreme risk dimensions addressed in the study and their degree of support from current federal-level public policies in Brazil (Technical workshop, 2014).

PUBLIC POLICY SUPPORT	IMPACT OF RISKS		
	LOW	MODERATE	HIGH
HIGH		• International Trade • Credit	
MEDIUM		• Reg. framework • Nat. Res. and Prod. Mgmt • Market	• Animal Health • Plant Health
LOW		• Weather	• Logistics and infrastructure

considered as the second most important by specialists, receiving a classification of 'very high', as well as the credit and commercialization risks.

For all the extreme agriculture risk dimensions, opportunities were identified in order to improve current federal-level public policies and programs. Integration was the main subject of the discussions. An important conclusion, backed by specialists, was that, although gaps exist, Brazil currently has a series of good risk mitigation mechanisms, average risk transfer mechanisms, and a lack of risk response mechanisms – however, not being able to address each one of them in an isolated fashion.

According to the specialists who attended the workshop, all extreme agriculture risks are interrelated. During the workshop, the relationships among risks were identified (Fig. 3), including the intensity and the direction of these existing relationships (influences) among risks. It is clear that the credit risk is influenced by most other risks. Therefore, credit can serve as a tool to promote the integrated management of other risks, promoting/incentivizing farmers to adopt best practices and technologies to reduce the probability of future loses. The reduction in the credit and trade risks requires the management of multiple agriculture risks. The relationships among agriculture risks (Fig. 4) show that credit and trade risks (price) have a high dependency on other risks.

Weather risks have an impact on other risks, and therefore, deserve special attention. From consultations with specialists, it was clear that weather risks do not only impact agriculture production risks, but also risks related to logistics and infrastructure, prices and/or food quality). Also, changes in the regulatory framework have an impact on other risks, as well as on the same public policies and programs that support the management of those same risks.

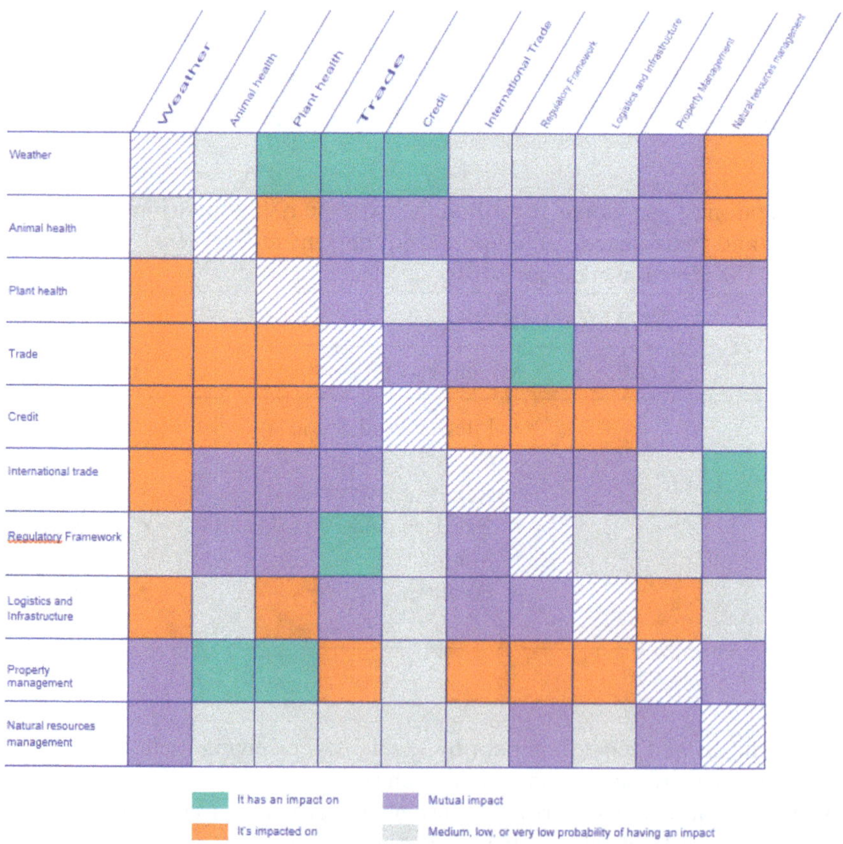

Figure 4: Relationships among the different agriculture risk dimensions in Brazil (source: author's calculations based on electronic survey, 2014).

It is important to note that beyond the direct impact established by the specialists, the relationships among these risks can create feedback loops in the system, resulting in several orders of impacts beyond the initial one. The work done did not identify such possible feedback loops in the risks interrelationships, but it was clear that they do exist.

4 CONCLUSIONS AND RECOMMENDATIONS

Brazil has a great diversity of policies and programs that support the agriculture sector in a counter-cyclical fashion, some of them being international success cases for both developing and developed countries. However, it is clear that there are several opportunities to improve the existing policies and programs, in particular in relation to their integration, which will not necessarily produce additional public spending, but can result in a better enabling environment for agriculture production, ensuring a more robust and sustainable sector growth.

According to the agriculture risk policies and program mapping, there are gaps in the management of extreme risks, in particular the lack of availability of contingency planning. In order to be able to manage risks in an integrated manner, the set of public policies and programs must consider mitigation, response and transfer strategies. Studies such conducted by the World Bank [6] show that although mitigation strategies, on average, have a higher economic return than transfer or response strategies (for each R$1 invested in mitigation, this saves up to R$7 in emergency response), there are cases in which mitigation and transfer are not optimal strategies. In Brazil, there are extreme risks and events (for example the sanitary events that occur in neighboring countries) that are not well known, but that have a potential for large economic losses for farmers. Although the mapping of those events and the adoption of mitigation actions could be recommended, it is always essential to adopt contingency plans. The actions of some of the northeastern states (Pernambuco and Ceará) in the preparation of state-level agriculture preparation plans for droughts are examples of such contingency planning in order to avoid future losses in the sector.

There is an insufficient level of support from public policies and programs for the management of risks such as: (i) logistics and infrastructure; and (ii) property and natural resources management. An agenda for the future needs to include the management of logistics and infrastructure risks and property and natural resources management risks (mitigation and adaptation actions) in order to reduce loses and the uncertainty of the sector. The creation of an environment and discussion forums among the various stakeholders will enable a qualitative leap towards an integrated management of those risks.

An integrated management of extreme risks by the government requires not only coordination among the different policies and programs, but also among the different levels of government. The current work focused on federal-level public policies and programs, but the coordination of federal actions with state and municipal agriculture risk initiatives is key, such as the coordination among Ministries and Agencies. In this work it was not possible to map subnational initiatives, but initiatives such as the state-level agriculture insurance premium subsidy that complements the federal-level subsidy and the preparation of municipal and state-level agriculture risk management plans (such as the experience of Pernambuco and Ceara) should be part of an agenda for the future.

A basic component for an appropriate management of extreme risks is to have policies and programs within a long-term strategic planning context. In the weather risk panel, although being mainly a technical group, the importance of multiannual agriculture plans for agriculture risk management strategic planning. Furthermore, in most validation meetings the recommendation extended beyond risk management, being clear the need of an agriculture law (as the Farm Bill of the United States), with a minimum horizon of five years, integrating issues related to the eight risk areas of this work, as well as other rural development and social mobility issues of the agriculture sector.

International experience and the number and diversity of existing public policies and programs in Brazil show that integration needs to happen gradually. Usually, in other countries, risks are managed in an integrated way by the private sector. As with most high-income countries, risk management policies and programs: (i) focus on farmers' income as a final objective; (ii) seek market solutions with public-private partnerships to adapt to producer demands and leverage private financing for risk management; and (iii) focus on coordination and information sharing among different participating institutions based on rural producers.

In order to promote and implement a long-term integrated management of risks, it is recommended to establish a risk management unit with the qualified staff and a national plan that would give sustainability for a permanent integrated vision. That agriculture risk management unit must be geared towards implementing a national multiannual plan, which can begin by the implementation of the six macro-objectives identified in this work and listed below.

4.1 An integrated agriculture risk management information system

Brazil already has several agriculture risk management information systems, some of them already integrating information on several risks. Embrapa, the National Meteorological Institute (INMET), and the Climate Forecasting and Studies Center (CPTEC/INPE) have information systems and probabilistic analysis about weather risks at the municipal scale, which are not being utilized to the fullest by farmers. These systems, beyond their main objective which is to support farmers production decisions in relation to the weather, they could be more integrated with other existing systems that are not yet geared towards risk management, such as market information, natural resources management, animal and plant health and logistics.

4.2 Integrated technology development and transfer system for risk management

The adoption of loss reduction technologies depend on the innovation and transfer capacity and the integration in the management of multiple risks. Today, in Brazil, the agriculture technology research and development system does not have tools for analyzing risks in an integrated way (see section 4.1 above), nor the capacity to interact with farmers for an integrated management of risks, in particular with family farmers.

It is necessary to seek the structuring of rural technical assistance ATERs, in particular in the north and northeast regions, promoting qualified staff, as well as the integrated approach towards managing risks in the sector. Although in Brazil there are some systems dedicated to technology transfer, such as the Information Technology Agency of Embrapa (Ageitec), it is not integrated and could be better utilized.

4.3 Integrated agriculture risk transfer system

There is an opportunity in Brazil to transfer a higher level of production risks through public-private partnerships. Agriculture insurance coverage provided by private insurances reaches 8% of cultivated area according to Buainain and Vieira [7]. However, the public insurance and income compensation programs represent more than that covered by private insurance firms. These income compensation programs do not transfer risks outside the public sector nor outside the country. The federal government absorbs 100% of the risks of catastrophic events, given that the premiums paid by beneficiaries are not enough to cover the payouts (the cost of the programs are paid in large part by the federal, state and municipal governments).

Furthermore, the rural insurance and price guarantee systems could be linked to offer an income insurance product to the farmer. The risks linked to price and pest/disease outbreaks are excluded from insurance programs and have particular policies and programs. The specialists who were consulted identified an opportunity to coordinate the insurance, price and animal and plant health programs to allow to have income insurance products for farmers.

4.4 Integrated agriculture logistics and rural infrastructure planning

The agrologistics and infrastructure system of Brazil suffers from extreme risks that could be reduced by the sharing of information and by the coordination with the management of other risks. These extreme risks cause on-farm losses, increasing the cost of transport and logistics, especially for farmers that do not have storage capacity. There are also risks along the supply chain given the uncertainty and seasonality of logistics costs, in particular transport. The bottlenecks in the logistics system (roads, ports and waterways) that generate uncertainties regarding time and transport and storage costs are well known. Therefore, with the sharing of information between the logistics and the planning of harvests and trades could reduce the uncertainty for the farmer.

4.5 Integrating weather risk management tools to natural resources management

There is an opportunity in the short term to reduce the natural resources management risk, in particular water resources and extreme drought risk, using existing weather risk management tools in the water allocation decisions. Challenges with water management in Brazil are long-standing, as the droughts in the northeast region, and increasing, as the recent uncommon drought in the southeast and the centerwest regions. Beyond the climatic risk, which depends on the management of an efficient and accessible information system, the risk of natural resources management persists given that the water allocation system do not use risk analysis tools such as the Agroclimatic Risk Zoning (ZARC) and the climate change models in decision making.

4.6 Integrating agriculture risk management into international trade promotion and monitoring

International agriculture trade policies in Brazil are not very well integrated to the agriculture risk management policies and programs. In particular, there are short term opportunities for more and better integration of animal and plant health policies with those of international trade, facilitating initiatives such as regional sanitary certification and their recognition in markets. There are also opportunities to reduce farmer losses due to the closing of international markets, coordinating international trade with trade risk management programs (price guarantees) and production (insurance). These agriculture risk management integration measures, along with international trade policies, could result in a more assertive participation of Brazil in international trade negotiation forums. In other words, Brazil could change its current position as a rule taker for rule maker in the international stage and a better management of agriculture risks can change that position.

REFERENCES

[1] De Nys, *et al., Drought Impacts and Cost Analysis for Northeast Brazil*, CRC Press, Boca Raton, Florida, 2016.

[2] MB Agro Consulting. Seguro Agrícola no Brasil. Presented at the *Agriculture Insurance Forum of Brasil*, 2012.

[3] Nogueira, JM. & Doliveira, CFD., Cost-benefit analysis of foot-and-mouth disease control strategies. Evaluation of a national operating program and assessment of a proposed alternative regional control program: Lessons from Brazilian experiences. *Presented at the Sociedade Brasileira de Economia, Administração e Sociologia Rural Conference*, July 2012.

[4] Website of the Senate of the Brazilian Congress. Available at http://www.senadorhumberto.com.br/governo-disponibiliza-r-27-bilhoes-para-combater-a-seca-no-nordeste/ (accessed 21 January 2017).

[5] World Bank. Agriculture Risk Management Framework. Available at http://www.agriskmanagementforum.org/content/ag-discussion-paper-10-agricultural-sector-risk-assessment-methodological-guidance-practitio (accessed 21 January 2017).

[6] World Bank. Natural Disasters: Counting the Costs. Available at http://siteresources.worldbank.org/INTEAPHALFYEARLYUPDATE/Resources/550192-1300567391916/EAP_Update_March2011_japan.pdf (accessed 21 January 2017).

[7] Buainain, AM. & Vieira Junior, PA., Seguro Agrícola no Brasil: desafios e potencialidades. *Revista Brasileira de Risco e Seguro, Rio de Janeiro*, **7**(13), pp. 39–68, 2011.

DISASTER NURSING: LOOKING TO THE FUTURE IN NORWAY

GUNN-MARI HOLDO[1,2], LEE MILES[2] & HEATHER HARTWELL[2]
[1]Faculty of Health, University of Tromsø, Norway.
[2]Faculty of Management, Bournemouth University, England.

ABSTRACT

Health resilience is an integral part of disaster management and lies at the nexus between this and public health considerations. As the largest group of professionals worldwide, nurses face continual challenges in further developing their competences in disaster response and recovery. This paper investigates the trajectory, role and future expectations of the Norwegian nursing profession in health emergency and disaster planning, and offers a future research agenda for those interested in investigating the complex inter-relationship between disaster management and nursing. The paper seeks to respond to observations made in a recent Norwegian report (2013) on behalf of the Norwegian nursing association that highlights the need to investigate nurses' knowledge in relation to emergency/disaster plans especially in the community health care system. The paper will explore the conceptual nuances between emergencies and disasters and then comments upon supplementary observations of the need for identifying disaster nursing practices and training programs to prepare nurses. By taking this approach, the paper also seeks to provide insights into how future demands of improving nurses´ capacities or implementation as leaders in disaster management – as highlighted by several International studies. In addition, by offering a future research agenda, the intention is that this may contribute to exploration of the role of nursing in handling long-term implications of disaster management.
Keywords: community nursing, disaster management plans, disaster nursing, education, nurses' involvement, resilient community.

1 INTRODUCTION

Disasters are defined as serious disruptions of the functioning of a community or society that exceeds its ability to cope using its own resources. Disasters have a huge impact on society and many of history´s great civilisations have been brought to their knees by the effect of natural disasters according to Coppola [1]. Due to development of new technology, such as air transport and internet, society has become more vulnerable [1, 2]. At present, there is an increase in terrorism, emerging natural threats and even threats of bio-nuclear terrorism [3]. The result of increased urbanization means a rise in the number of people affected when a disaster strikes [1, 4]. The unexpected may also happen such as the volcanic eruption in Iceland 2010, which paralyzed air transport in Europa for nearly two weeks. Because of these high risks it is important that communities and their local governments build up society's resilience. The present paper investigates Norway, which is a diverse country with both rural and urban regions where a large part of the rural regions lies in the vulnerable area north of the arctic circle. That gives Norway a challenge of building a good resilience in the community. Furthermore, Norway is an affluent country which is a member of UN (United Nations) and ICN (International Council of Nurses). Through these memberships Norway is bound by agreements such as the Sendai Framework for Disaster Risk Reduction [4].

To construct a resilient community there is a need to look at which resources the community have and take these resources into consideration and decide how they could be useful, [5, 6]. Since nurses are the largest healthcare workgroup in the health system their expertise should be taken in to consideration and they should be involved in the disaster planning work [7]. Nurses have, since Florence Nightingale´s work with soldiers in the Crimean War been

DOI: 10.2495/SAFE-V7-N3-431-442

linked to provision of care during crises [7]. At that time, Florence Nightingale saw the need for knowledge on how to treat and care for patients and how to prevent sickness. Also today nurses feel there is a lack of specific knowledge in disaster nursing competence [8–10], and where it is important that nurses have skills of disaster related nursing and are involved integrally in all phases of disasters. That also means nurses should be involved in effective preparedness such as planning for a disaster event.

Furthermore, there is a need to strengthen the nurse's skills for involvement in the aftermath of a disaster, the long-term care [7]. This was highlighted in the Norwegian report from nurses' work in the terror event of the 22nd July 2011 [11]. It is necessary to have more drills or exercises at the workplaces and further incorporate such work in the nurses' educational program in Norway and other countries.

The Norwegian government through parliament have decided that there should be disaster management plans and health preparedness plans for each county and for the local community [12, 13]. These plans are very important for the local government as they set minimum standards in terms of competence across levels of the health care system as well as a minimum level of risk assessment. This is despite the fact that Norway is a large, long and narrow country with only 5.1 million citizens. As shown in Figure 1 the northern part of Norway is located in the arctic and border to Sweden, Finland and northern part of Russia.

Most of the population lives in cities in the southern part of Norway. This means that there are large rural areas with a sparse population. In these regions, it is important that everybody who could be involved, knows what to do if a disaster strikes. For instance, Northern Norway constitutes a third of the Norwegian land mass, but less than 10% of the Norwegian population live in the Northern part of Norway. It is also a popular region for tourism. The community population in Northern Norway varies from 2000 to 20 000 inhabitants with the exception of two towns [15]. Despite such a small population, it is important to have good disaster plans because of the long distances between hospitals as well as the temporary or seasonal population increase that the current expansion in tourism brings. However, the level

Figure 1: Norway with major towns and its location is shown [14].

of services should be the same for the inhabitants anywhere in the country. Consequently, it is important that community nurses are well educated and know what to do when a disaster or a crisis strikes.

2 BACKGROUND

Since the time of Florence Nightingale, nurses have contributed to health care to help people in need and are at the forefront of the health care response to disasters [7]. The International Council of Nurses (ICN) defines nurses (short version) as follows:

'*Nursing encompasses and collaborative care of individuals of all ages, families, groups and communities, sick or well, and in all settings. Nursing includes the promotion of health, prevention of illness, and the care of ill, disabled and dying people. Advocacy, promotion of safe environment, research, participation in shaping health policy and in patient and health systems management, and education are also key nursing roles*' [16].

Nurses are the largest workforce in health care and they play an integral role in disaster response and preparedness according to Veenema *et al.* [17]. The goal of disaster nursing '*is ensuring that the highest achievable level of care is delivered through identifying, advocating, and care for all impacted populations throughout all phases of a disaster event, including active participation in all levels of disaster planning and preparedness*' [7]. With this in mind the role of nursing in handling long-term implications of disaster management, particularly in terms of nursing 'follow-up', is important especially in crisis aftermath. How this work is handled can have a large impact on the people living in the community and therefore also the resilience of the community [7]. In Norway, long-term care is often carried out by health care personnel in the local community. These are often nurses who meet affected people who need nursing care and treatment; there is no automatic follow-up for bereaved people. During a crisis/disaster, crises teams are often appointed based on who should take care of those affected by the event. Yet, when the crisis team is no longer involved, the follow-up process is dependent on the state of the victim/family member, but there are no specific follow-up plans available in the local community. Therefore, there is a need to identify disaster nursing practices and training programs to prepare nurses [18, 19]. To improve nurses as leaders in disaster management there is a need for more research [17]. This is supported by a Norwegian report by Andrews *et al.* [11], where findings highlighted the need for a better framework and outlining the length of the 'following-up' process, resourcing requirements and availability and access of competent staff, trained in leading this work in the aftermath of disaster [11, 20]. The aim of the present study is thus to investigate the Norwegian disaster health care plans for nursing expertise and involvement.

3 METHOD OF DATA COLLECTION

According to Polit and Beck [21] the investigation of documents and disaster preparedness plans is a method that provides useful insight and data collection. This data could give the researcher an insight to the extent that the nurses may be involved in developing disaster preparedness plans. Furthermore, this investigation will give an understanding of how nursing education in Norway reflects disaster/preparedness work in their curriculum. Data from documents and other non–peer-reviewed publication are also called Grey literature according to Ward–Smith [22], and it provides an alternative source of important information to the study.

This article focused on the Norwegian community nurses involvement in disaster management plans and how the training program prepares nurses for disaster participation and preparedness. 'Involvement' is here defined as the nurses' participation in writing the disaster plans, which means that nurses are mentioned, as part of the leadership team, or otherwise. Nursing education prepares nurses for working in the field therefore it should also reflect preparedness for disaster/crises events in their bachelor program. Involvement of nurses in local government disaster management plans can be at a strategic, tactical and operational level. The study of nurses' involvement in community disaster management plans and their training programmes was carried out by the following steps:

1. A review of all community plans in Norway that is available on the internet for the public. The focus has been on the two disaster/crises plans that the community has to have available by an Act of Government [13]. These are the overall disaster management plan and the health preparedness plan. The first is the overall plan where the management (leaders) are involved. The second is the community health preparedness plans [12] where the community nurses could be involved.
2. A review of Norwegian University and University college curriculum for BSc in nursing, as it represents the fundamental education for nurses in Norway.

The scope of disaster management plans in the community process involved reviewing all community plans available on the communities' home pages and cataloguing the involvement of nurses as leaders. This was an internet search on the communities' home pages. The search terms are given in Table 1 with both the Norwegian terms and the English translation.

The search was conducted in November/December 2016 to identify existing disaster and health preparedness plans in the Norwegian communities. Each community home page was studied to determine whether a disaster management plan and a health preparedness plan were available. Furthermore, if one or both plans were available, the next step was to determine if nurses had been involved and their involvement in the communities' crises team.

The internet search for the curriculum of BSc in Nursing was carried out in December 2016 to identify existing courses in disaster or crises management in the Norwegian bachelor program in nursing. Each curriculum was reviewed to determine whether there were courses directly pertaining to disasters.

4 RESULTS AND FINDINGS

These results are based on examination of the 428 Norwegian local government home pages in the 19 counties of Norway [23]. According to government regulations and a Government Act, every local government should have disaster management plans [13]. However, this

Table 1: Search terms used in Norwegian communities home pages.

Norwegian	English
Katastrofe	Disaster
Krise plan/krisehåndtering	Crises plan
Beredskap	Preparedness
Overordnet beredskaps plan/Samfunn og sikkerhetsplan	Disaster management plan
Helseberedskapsplan/helsemessig og sosial beredskapsplan/ helsemessig beredskapsplan	Health preparedness plan
Kriseteam/omsorgs team	Crises team

study shows that there are significant difference in how these plans are presented on the local government home pages and where they are presented (internet or intranet). For instance, some of the home pages have an icon representing preparedness and with one click one can obtain the information. Other local government home pages have no icon at all, and there was a need for carrying out further searches in order to obtain the information. The difficult part was that the necessary key words could also differ from one local government to another despite the words enshrined in the regulation from central government. An example is that whilst the word 'crises team' was the expression used by most of the local government web pages meaning the team that follow-up people who have emotional stress after a disaster or crises, but some web pages used the expression 'care team'. Unfortunately, on some of the local government home pages the information is not available at all. The same findings are also true for the local health preparedness plans.

A number of local governments choose to have their disaster management plans on their intranet rather than on the internet home pages; thus the information about who are involved in their teams are not readily available. Therefore, it has been difficult to be certain if nurses have been involved in the preparation and formulation of the plans. There are only a few cases where it is certain that nurses have been involved.

Another difficulty is that in many of the plans, the people who have been involved are named by their title and not by their profession. That could mean that the person who is the leader of the health department, and in most of the cases this was also a person who was a member of the crises leader team, and have been involved in preparing the plans could be an administrator or a nurse.

The findings from the examination of the local government/communities websites are summarised in Table 2.

The first column in Table 2 is the name of each county in Norway. The second column is the number of local governments within each county. The third column shows the number of disaster management plans available for the public on the home pages of local governments. The fourth column displays the number of health and social preparedness plans available on the home pages. The fifth column is the number of local governments that have involved nurses in the process of developing one or both preparedness plans. The fifth and final column gives the number of crises teams who have either or both psychiatric nurses or community nurses involved. In a few cases there are also ordinary nurses involved as team members.

In the northern part of Norway there are the three counties Nordland, Troms and Finnmark where the two last counties are north of the arctic circle, whilst Nordland county is divided with one part north and the other part south of the arctic circle. This region of Norway is a third in terms of the area of Norway, whilst the population is 481 994 and represents 9,2% of all Norwegian inhabitants [15], compared to Oslo, the capital, with 658 390 inhabitants or 12,57% of the total Norwegian population [15]. In another comparison, the area of Finnmark county is larger than the whole of Denmark, but the population is much smaller in Finnmark with 75 578 and Denmark with about 5 707 251 inhabitants [15]. Consequently, there are significant differences in the areas that are administrated by the local governments in Norway varying from the very rural in Finnmark to the more urban areas near the capital city, Oslo.

It is seen that not all local governments in the country have disaster management and health preparedness plans available on their home pages. Due to the level of transparency expected it is surprising that there is such a variation. Furthermore, there are no differences in numbers of plans on the home pages between the local governments in rural areas and the local government in more urban areas. There is a variation between counties from 60% to

Table 2: Disaster and health management plans for each county in Norway.

County	Local government	Disaster Management plan (%)	Health preparedness plan. (%)	Nurses involved in plans. (%)	Nurses involved in crises team. (%)
Østfold	18	2 (11.1)	2 (11.1)	2 (11.1)	1 (5.5)
Akershus/Oslo	23	14 (60.9)	5 (21.7)	5 (21.7)	0
Hedemark	22	9 (40.9)	2 (9.1)	1 (4.5)	4 (18.2)
Oppland	26	11 (42.3)	17 (65.4)	2 (7.7)	5 (19.2)
Buskerud	21	13 (61.9)	4 (19.0)	2 (9.5)	5 (23.8)
Vestfold	14	6 (42.8)	2 (14.2)	0	5 (35.7)
Telemark	18	6 (33.3)	1 (5.5)	0	5 (27.8)
Aust-Agder	15	3 (20.0)	3 (20.0)	0	6 (40.0)
Vest-Agder	15	8 (53.3)	4 (26.7)	0	5 (33.3)
Rogaland	26	11 (42.3)	5 (19.2)	3 (11.5)	12 (46.1)
Hordaland	33	8 (24.2)	2 (6.1)	2 (6.1)	16 (48.5)
Sogn og Fjordane	26	9 (34.6)	2 (7.7)	2 (7.7)	14 (53.8)
Møre og Romsdal	36	14 (38.9)	4 (11.1)	2 (5.5)	24 (66.7)
SørTrøndelag	25	9 (36.0)	6 (24.0)	0	11 (44.0)
NordTrøndelag	23	7 (30.4)	1 (4,3)	0	6 (26.1)
Nordland	44	14 (31.8)	4 (9.1)	3 (6.8)	17 (38.6)
Troms	24	5 (20.8)	3 (12.5)	2 (8.3)	6 (25.0)
Finnmark	19	3 (15.8)	3 (15.8)	3 (15.8)	6 (31.6)
Total	428	159 (37.1)	68 (15.9)	29 (6.8)	157 (36.7)

11% in terms of local government plans available on internet. Also, the nurses' involvement in the development of these plans is very small from 4% to 15% between counties. On the other hand, it is seen that nurses are more involved as members of the crises team. The variation is from 5% to 66% of local government that have involved nurses in this team. This could mean that nurses are involved mainly at the operational level.

A typical Norwegian BSc nursing education is shown in Table 3.

An investigation of BSc in Nursing at all Norwegian University and University Colleges that educate nurses and postgraduate/master in nurses' related program was carried out in order to detect any Disaster Nursing content. The results of the investigation are shown in Table 4.

The first stage was to investigate if the Norwegian nursing education program involved disaster/crises topic. All of the syllabus appeared to be based on the Norwegian framework for nurses' education [24]. On inspection, the framework does not include any specific reference to disaster nursing topics. None of the university or university college syllabus description appeared to have incorporated disaster nursing topics in the nursing studies. Exception was one MSc where Disaster topic is a part of the syllabus 'Pre-hospital Critical Care'. These syllabi have a module 'Major incident: Medical Management and preparedness'

Table 3: A Bachelor program in nursing.

1st Year	2nd Year Preparation, Skills and Reflection 2	3rd Year Preparation, Skills and Reflection 3
Universal Course For Health Faculty	General pathology, diseases and pharmacology 1&2	Integrated nursing, science and social science
Preparation, Skills and Reflection 1	Training in Medical or Surgical areas 1. Period.	Nursing, Health and Society
Anatomy, physiology and biochemistry	Training in Medical or Surgical areas 2. Period.	Training at nursing homes or psychiatric health care/ health protection 1. period 2. period
Foundation nursing with in ward training and social science Foundation pharmacology, drug calculation and handling	Common elements and social science	BSc Final Year Project

Table 4: Overview of Disaster Management related topics in nursing education in Norway.

Institutional Name	Bachelor	Postgraduate/MSc
University		
North University	0	0
NTNU	0	0
University in Agder	0	0
University in Stavanger	0	Part of MSc in Pre-hospital Critical Care
University in Tromsø	0	0
University College		
University College Bergen	0	0
University College Stord Haugesund	0	0
University College Molde	0	0
University College Buskerud og Vestfold	0	0
University College Østfold	0	0
University College Hedemark	0	0
University College Oslo og Akershus	0	0
Private University College		
University College Lovisenberg	0	0
University College Diakonova	0	0
University College Høgskulen i Sogn og Fjordane	0	0
University College VID vitenskapelig internasjonal diakonal høgskole AS	0	0
Total 16	0	1

where they highlight terror as an example, and that could indicate that they have disaster management in their education. This is, however, an MSc in the Health Faculty and not specifically for nurses.

5 DISCUSSION

Nurses are supposed to deliver an effectively and secured health care based on societies' needs. That should be recognised in the local government's disaster preparedness plans. This is highlighted by Powers *et al.*: '*An ineffective response can negatively impact on the community´s outcome from a disaster by failing to appropriately match nursing resources with need. Nurses must be well versed in strategies for overcoming disaster-induced adversities, and fully understand the different roles they may be required to embrace in order to deliver care effectively in disaster setting*'. [7]

This shows the reasons that it is important for nurses to take leading roles in disaster teams. These reasons are in summary:

- Matching nurses' resources with need, the Nursing leader know the nurses skills and can therefore order them to correct location.
- Nurses must be well educated and understand their roles in a disaster setting.

Norway, like a large number of countries, is a member of the International Council of Nurses (ICN), which promotes the concept of disaster nursing [25]. Furthermore, because Norway is largely rural with long distances between hospitals and in some regions, it is also far between medical practitioners it may take a very long time for them to reach a disaster area. Consequently, it is even more important in the remote regions that nurses have both medical and disaster management skills and be involved in strategic, tactical and operational roles. Examining the results with this in mind, several comments can be made.

Table 2 demonstrates that very few nurses are included in local government disaster planning teams. In most of the teams, the leader of health and social provision is a member of the team. It is not clear from the information available that these persons have background as nurses. Nevertheless, it is known that the leaders of the health and social provision may not be nurses.

The main aim of the disaster management plan is to prepare the inhabitants and reduce future risk in local government [26]. In Norway there is a Government requirement [13] that every local government should have a disaster preparedness plan. According to the Norwegian Directorate for Civil Protection and Emergency Planning (DBS), 94% of local governments have a disaster preparedness plan [23]. This agrees with the Norwegian NF rapport from 2012 [11] and could mean that the level has not improved. Only 59% of 428 local governments fulfil the requirements for updating plans once a year [23]. This should also be seen in the light of the population size in some of the local government regions. The smaller regions may only have a few thousand inhabitants. These rural regions are also located far from hospitals with an Accident and Emergency Department (A&E). Furthermore, in recent years many of the smaller hospitals have been stripped of their A&E department which has been transferred to larger hospitals further away from the smaller rural populations. It is still government policy in Norway to centralise A&E provision in order to make healthcare more efficient. In the Northern rural areas transport both by road and air is very challenging during the winter months. Weather conditions often make it impossible for helicopters to fly and transport by road could be slow or impossible because of snow and ice.

However most of the local governments have established psychosocial crises team and 37% nurses are involved in these teams. This is underlined by the Norwegian NF report [11], where 42% of the nurses were members of the psychosocial crises team and 25% were leaders of the crisis management team. Importantly, information extracted from the report shows that most of the informants come from urban areas. Consequently, it is presently even more important that nurses working in rural areas are well educated, trained and prepared [27]. The planning team of the preparedness plan should clearly contain nurses due to their knowledge of how to use the nursing resources in a crisis. This raises the question about why the nurses are not more visible in the planning of preparedness plans? Does it mean that nurses are seen as support staff only? Or does it mean that managers of local government do not have sufficient knowledge related to the skills that nurses have? Or finally does it mean that nurses need better education related to nursing disaster skills?

The result of the present study (Table 3) suggests that the Norwegian framework for nurses does not include any specific reference to disaster or disaster preparedness topics despite the fact that a large part of Norway is rural and disasters may happen anywhere. The Norwegian NF report [11] also shows that only 2,48% have sufficient knowledge about handling a crises or disaster based on BSc in nursing. The same findings were conclusions of a literature study by Rokkas and Steenkamp [28]. The Norwegian NF report [11] also shows that a little more, 3,09%, have competence in crises/disaster work from postgraduate education. Whilst there is an apparent need for disaster related topics in Norwegian nurses' education, this need for implementing disaster nursing topics is also proposed by several international nursing groups [10, 17, 19, 27, 29, 30]. A way forward in Norway could be to introduce Disaster Nursing topics within the Norwegian National Framework so that Nurses in general would have an understanding and background in Disaster Nursing. Involvement in disasters means strategic, tactical and operational roles. The future development of Nurses education should reflect this. To meet the future requirements in dealing with disaster and crises events, an MSc in Disaster Nursing is suggested in order to aid community resilience. This is also underlined by the NSF report [11]. However, it is possible that topics relevant to disaster nursing work may be part of other subjects such as those for A&E nursing. In order to investigate this, it is necessary to have available the details of each programme.

It is clear that disaster management will be strengthened by including nurses in the planning and management in the disaster management teams. This is also suggested by Miles [31]. This is especially important in rural and sparsely populated areas where the community nurse may be the only person representing the whole of the medical profession. In addition, there is a trend to centralise services such as A&E departments, this means that such services, as well as those pertaining to psychiatry, will be few and far between. Consequently, nurses in these communities will have a larger job description than what is presently the case. The need for a re-evaluation of their training to include disaster nursing topics as part of the BSc in nursing is certainly important in order to provide the same health care as in the cities, both for those living in rural areas and those visiting such regions. Developing standards and guidelines for incorporating disaster management and training in nursing education was in 2009 one of the highest priorities in ICN [25], and should certainly be the case for Norway.

6 CONCLUSION

The data available for the present study suggest that nurses have been less involved in disaster preparedness work than what would be expected since they represent the largest group of health workers in the community. It is suggested that they should be present to a much greater extent.

However, nurses with postgraduate education such as psychiatry and public health are found in crises teams.

Furthermore, the curriculum for the Norwegian BSc in Nursing does not include specific topics related to preparedness and disaster management. It is suggested that a revision to include such topics should be carried out.

Other conclusions are that the homepages for local government differ greatly between the communities. It is recommended that such homepages should be standardised in order to assist transparency and so that the public can easily find the information they need in a disaster/crises situation. Similarly, the terminology used by local government is also quite varied. Again, it is recommended that such terminology should be standardised so that the public can easily find the information they need in a disaster/crises situation.

7 FURTHER WORK

This study has highlighted the challenges that Norwegian community nurses have to meet in order to be prepared for disaster/crises. Furthermore, the study may indicate that the differing challenges from region to region or from rural area to urban area are not adapted to the community nurses work. To clarify this, the authors agree that the future research agenda of this work has to consider the following:

1. Further investigate if there are a minimum level of standards for all community nurses working in Norway.
2. Investigate the expectation regarding skills of nurses in risk assessment and disaster management at different levels for the various and diverse regions.
3. Further analyse how and where nurses are represented in local government disaster preparedness plans.
4. Investigate if nurses are represented in the strategic, tactical and operational level.

More broadly in terms of community resilience, exploring this agenda will help shape the need for nurses' special skills; what should be called disaster community nurses. If we can understand the need for nurses with special skills, then we can consider whether present day nursing education is adequate and take care of what local government can expect from them.

REFERENCES

[1] Coppola, D.P., *Introduction to International Disaster Management*, 3rd edn., Butterworth-Heinemann, Elsevier: Oxford, 2015.

[2] Comfort, L.K., Boin, A. & Demchak, C.C., Resilience revisited: an action agenda for managing extreme events. *Designing Resilience Preparing for Extreme Events,* eds L.L. Comfort, A. Boin & C.C. Demchak, University of Pittsburgh Press: Pittsburgh, pp. 272–284, 2010.

[3] Boin, A., Hart, P., Stern, E. & Sundelius, B., *The Politics of Crisis Management: Public Leadership Under Pressure*, Cambridge University Press: USA, 2005. https://doi.org/10.1017/CBO9780511490880

[4] United Nations Office for Disaster Risk Reduction (UNISDR) 2015. *Sendai Framework for Disaster Risk Reduction 2015-2030.* United Nations – Headquarters.

[5] Comfort, L.K., Oh, N., Ertan, G. & Scheinert, S., Designing adaptive systems for disaster mitigation and response: The role *of* structure. *Designing Resilience Preparing for*

Extreme Events, eds L.K. Comfort, A. Boin. and C.C. Demchak, University of Pittsburgh press: Pittsburgh, pp. 33–61, 2010.

[6] Hunter Revell, S.M. & McCurry, M.K., Postflood disaster management and the home health nurse: using theory to guide practice. *Journal of Community Health Nursing,* 27, pp. 126–136, 2010.
https://doi.org/10.1080/07370016.2010.494454

[7] Powers, R., Introduction to disaster nursing. *International Disaster Nursing,* eds R. Powers & E. Daily, Cambridge University Press: New York, pp. 1–11, 2010.
https://doi.org/10.1017/CBO9780511841415.004

[8] Yan, Y.E., Turale, S., Stone, T. & Petrini, M., Disaster nursing skills, knowledge and attitudes required in earthquake relief: Implications for nursing education. *International Nursing Review,* **62**, pp. 351–359, 2015.
https://doi.org/10.1111/inr.12175

[9] Chapman, K. & Arbon, P., Are nurses ready? disaster preparedness in the acute setting. *Australasian Emergency Nursing Journal*, **11**, pp. 135–144, 2008.
https://doi.org/10.1016/j.aenj.2008.04.002

[10] Nilsson, J., Johansson, E., Carlsson, M., Florin, J., Leksell, J., Lepp, M., Lindholm, C., Nordström, G., Theander, K., Wilde-Larsson, B. & Gardulf, A., Disaster nursing: Self-reported competence of nursing students and registered nurses, with focus on their readiness to manage violence, serious events and disasters. *Nurse Education in Practice,* **17**, pp. 102–108, 2016.
https://doi.org/10.1016/j.nepr.2015.09.012

[11] Andrews, T., Gustavsen, A., Waldahl, R.H. & Eide, A.K., *The role and action of nurses in a national catastrophy. About the work with victims of the event 22nd July 2011.* Bodø: NF- rapport. 2013. (in Norwegian) ISBN-nr.:978-82-7321-639-7

[12] *Health Preparedness Act. (Act No. 56 of 23 June 2000 relating to health and social preparedness.),* available at www.lovdata.no (accessed 4 November 2016) (In Norwegian)

[13] *Civil protection Act. (Act of 25 June 2010, No. 45 relating to municipal preparedness duty, civil protection and the Norwegian Civil Defense),* available at www.lovdata.no (accessed 4 November 2016) (In Norwegian)

[14] Statens Kartverk available at www.norgeskart.no (accessed 22 December 2016).

[15] Statistics Norway available at https://www.ssb.no (accessed 15 December 2016).

[16] International Council of Nurses, 2002. Definition of Nursing 2 (short version). available at http://www.icn.ch/who-we-are/icn-definition-of-nursing/(accessed 12th December 2016

[17] Veenema, T.G., Griffin, A., Gable, A.R., MacIntyre, L., Simons, R.A.D.M., Couig, M.P., Walsh, J.J., Lavin, R.P., Dobalian, A. & Larson, E., Nurses as leaders in disaster preparedness and response- a call to action. *Journal of Nursing Scholarship*, **48**(2), pp. 187–200, 2016.
https://doi.org/10.1111/jnu.12198

[18] Bahrami, M., Aliakbari, F. & Aein, F., Iranian nurses´ perception of essential competences in disaster response: A qualitative study. *Journal of Education and Health Promotion*, **3**, p. 81, 2014.

[19] Yamamoto, A., Development of disaster nursing in Japan, and trends of disaster nursing in the world. *Japan Journal of Nursing Science*, **10**, pp. 162–169, 2013.
https://doi.org/10.1111/jjns.12042

[20] Lereim, I., *et al.*, *Learning for improved emergency preparedness. Health intervention after the terror event 22 July 2011*, Directorate of Health, Oslo, 2012. (In Norwegian) ISBN-nr. 978-82-8081-256-8

[21] Polit, D.F. & Beck, C.T., *Nursing research Principles and Methods,* 7th edn., Philadelphia: Lippincott Williams & Wilkins, 2004.

[22] Ward-Smith, P., The fine print of literature reviews. *Urologic Nursing*, **36**(5), pp. 253–258. 2016.

[23] Norwegian Directorate for Civil Protection and Emergency Planning (DBS), Local government study 2016. (In Norwegian), available at https://www.dsb.no (accessed 5 November 2016)

[24] Norwegian government department of education and research. Framework for Nurses education, 25 January 2008. (In Norwegian) available at www.regjeringen.no

[25] World Health Organization and International Council of Nurses 2009. *ICN Framework of Disaster Nursing Competencies.* Available at: www.icn.ch/ (accessed 9 November 2016)

[26] Alexander, D., *Principles of Emergency Planning and Management,* Terra Publishing: United Kingdom.

[27] Baack, S. & Alfred, D., Nurses' preparedness and perceived competence in managing disasters. *Journal of Nursing Scholarship,* **45**(3), pp. 281–287, 2013. https://doi.org/10.1111/jnu.12029

[28] Rokkas, P. & Steenkamp, M., Disaster preparedness and response: Challenges for Australian public health nurses – A literature review. *Nursing and Health Sciences*, 16, pp. 60–66, 2014. https://doi.org/10.1111/nhs.12134

[29] Usher, K., Redman-MacLaren, M.L., Mills, J., West, C., Casella, E., Hapsari, E.D., Bonita, S., Rosaldo, R., Liswar, A.K. & Amy, Y.Z., Strengthen and preparing: Enhancing nursing research for disaster management. *Nurse Education in Practice*, **15**, pp. 68-74, 2015. https://doi.org/10.1016/j.nepr.2014.03.006

[30] Schmidt, C.K., Davis, J.M., Sanders, J.L., Chapman, L.A., Cisco, M.C. & Hady, A.R., Exploring nursing students' level of preparedness for disaster response. *Nursing Education Perspectives,* **32**(6), pp. 380–383, 2011. https://doi.org/10.5480/1536-5026-32.6.380

[31] Miles, L., The 'Five-C's': international disaster management and public health and place for entrepreneurial resilience. *Perspectives on Public Health,* **136**(6), pp. 323–325, 2016. https://doi.org/10.1177/1757913916672810

THE CAMPI FLEGREI DEEP DRILLING PROJECT (CFDDP): CALDERA STRUCTURE AND HAZARD

DE NATALE GIUSEPPE[1], TROISE CLAUDIA[1], SOMMA RENATO[1] & GIOVANNI PERILLO[2]
[1] Istituto Nazionale di Geofisica e Vulcanologia, Sezione Osservatorio Vesuviano. Naples, Italy.
[2] Department of Technology, Naples Parthenope University, Italy.

ABSTRACT

The recent investigation carried out on the west bound of the Naples metropolitan area and inside the Campi Flegrei caldera as part of the Campi Flegrei Deep Drilling Project provided new insight in order to reconstruct the volcano-tectonic evolution of this extremely populated area. Campi Flegrei represents the highest risk volcanic areas in the world, although its volcano-tectonic structure, eruptive history and eruptive style of the largest eruptions are intensely debatedby scientists since several decades. We present here a summary of stratigraphic and geochronological dating (40Ar/39Ar) allowing to define the age of intra-caldera deposits belonging to the two highest magnitude caldera-forming eruptions (i.e. Campania Ignimbrite, 39÷40 ka, and Neapolitan Yellow Tuff, 14.9 ka) and to evaluate the amount of collapse of the eastern sector of the caldera.

These results point out: (i) a reduction of the area affected by caldera collapse, which appears to not include the city of Naples; (ii) a small volume of the infilling caldera deposits, particularly for the CI; and (iii) the need for reassessment of the collapse amounts and mechanisms related to larger eruptions. Our results also imply a revaluation of volcanic risk for the eastern caldera area, including the city of Naples. The results of this study point out that large calderas are characterised by complex collapse mechanisms and dynamics, whose understanding needs more robust constraints, which can be obtained from scientific drilling.

Keywords: Ar dating, caldera dynamic, caldera-forming eruptions, Campi Flegrei caldera, CFDDP drilling, volcanic hazard

1 INTRODUCTION

Extreme volcanic eruptions from large calderas are the most likely hazard to generate global catastrophes [1, 2]. This topic has received considerable attention because of the millions of people living in high-risk volcanic areas [3].

Actually, conceptual models for many calderas are based on an incomplete reconstruction of their system [4]. A well-known case study is the caldera of Campi Flegrei in Southern Italy (Fig. 1), which was produced by large volume eruptions at c. 39 ka (Campanian Ignimbrite, CI, VEI=7) and c. 15 ka (Neapolitan Yellow Tuff, NYT, VEI=6) [5].

The mechanism of the above eruptions and the involved volume of magma (particularly for the CI eruption) are still uncertain. Recently, some authors proposed that the CI caldera limit enclosed the central sector of Naples [6, 7] while other authors [8, 9] do not incorporate the central part of the city. The latter models are mainly based on the hypothesis of partial reactivation of pre-existing regional faults along which the collapse occurred during the caldera formation.

Also, the magma volume involved in the CI eruption has been evaluated by various authors in a range of 80 to more than 300 km^3 of DRE. These different interpretations have important implications on the assessment of both the eruptive mechanism and the risk for people living in the surroundings. In this case, deep drilling provides the best mean to obtain direct information on processes occurring at depth and represents a key tool for reconstructing the caldera structure and understanding the volcano dynamic and volcanic risk assessment [10]. The highly urbanised Campi Flegrei Caldera have been experiencing decades of unrest, with uplift rates up to 1 m/y,

DOI: 10.2495/SAFE-V7-N3-443-448

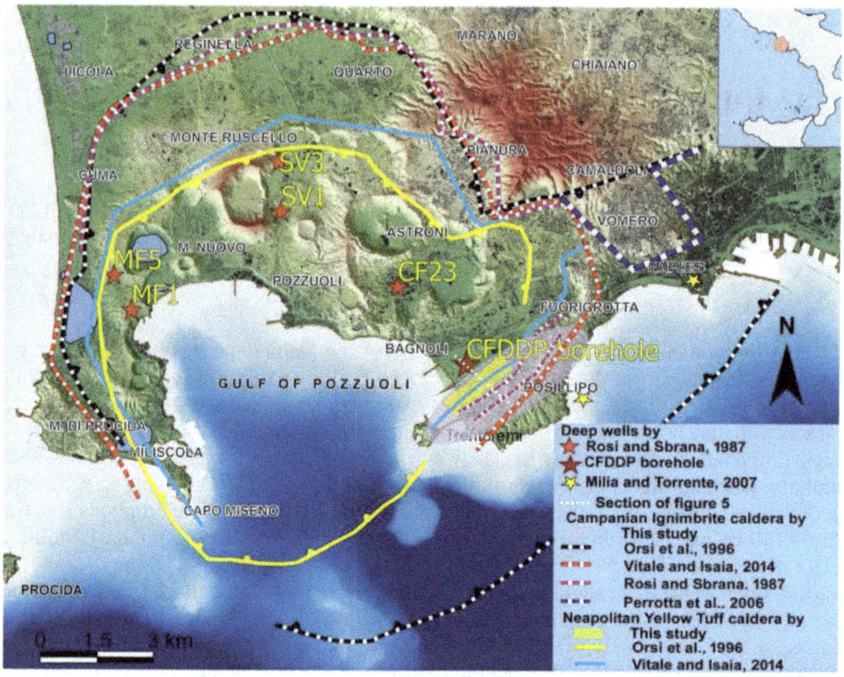

Figure 1: Morphological map of the Campi Flegrei caldera.The lines are the caldera limits as differently proposed. CFDDP drilling site (larger red star) is shown with the deepest boreholes realised by AGIP (SV1, SV3, MF1, MF5) andSAFEN (CF23) companies (smaller red stars).

thousands of microearthquakes per year [11] and strong geochemical anomalies [12], leading to people evacuation of Pozzuoli town in 1982–1983. At the present, the alert level for the caldera has been moved from 'base' (ordinary activity of volcano) to 'attention' status. However, any existing hazard planning requires a continuous implementation of geological data acquired with new researches, to attain a more reliable knowledge of the structure and dynamics of the volcano.

2 CALDERA DRILLING AND DATA RECOVERING

Among several scientific experiments in the caldera, the most recent and important one is the Campi Flegrei Deep Drilling Project (CFDDP), endorsed by ICDP [13, 14] and targeted to the eastern caldera sector with a borehole achieving a depth of 501 m below sea level (bsl). The CFDDP core/cuttings analyses were performed with the aim to provide new constraints to the volcanic setting in the eastern sector of the caldera, filling the 'information gaps' from previous deep exploration work in the western, northern and central sectors of the caldera.

Here, we report a summary of lithological, paleontological and geochronological studies on both cores and cuttings showing that deposits from the two caldera-forming eruptions (NYT and CI) are shallower and thinner than expected [14]. Our findings raise new questions about the structure and genesis of the caldera and the level of volcanic hazard for the Neapolitan area. The obtained results demonstrate the effectiveness of drillings in providing reliable constraints in the understanding of caldera dynamics.

The borehole was drilled in 2012 in the 2 km^2 old steel factory site in Bagnoli (western Naples) and nearby the Posillipo hill (Fig. 1), which represents a morphological limit of the caldera, possibly related to the collapse during the caldera-forming eruptions.

Drilling operations were carried out through technical and logistical mud services, thus allowing the collection of mud cuttings and cores. Two cores were extracted from 439 m and 501 m. In particular, 40Ar/39Ar geochronology was performed on feldspars from selected mud cuttings from the 191 m scoria and the 260 m pumice-rich levels, and the two cores.

3 ANALYTICAL RESULTS

Stratigraphy and volcanology of the drilled sequence as deduced by lithological, mineralogical and micropaleontological analyses are reported in Figure 2.

The sequence is fully composed of pyroclastic materials. The upper most part consists of subaerial pyroclastic deposits composed of variably vesicular and porphyritic fragments including Agnano Monte Spina tephra (c. 5 ka) at the base. Below, the sequence shows a c. 100m-thick (161–245 m depth interval) nearly monotonous deposit mainly composed by brown dense (obsidian-like) to vesicular glass fragments, mostly scoriae, and greenish dense clasts, beige pumices, and tuffs, with minor amount of sponge spicules. The remaining underlying sequence is composed of greenish tuffs overlying a gray-tuff at the bottom hole, both containing very low abundances of feldspars. The transition between the two tuffs is likely gradual and starts to occur at c. 450 m. Samples from the greenish tuffs are very homogeneous and strongly comparable to the core at 439 m representative of a fine- to coarse-grained heterogeneous fossil-free tuff, gray-green in colour. The core at 501 m is a chaotic tuff with coarse- to fine-grained ash matrix hardened by zeolitisation.

The 40Ar/39Ar age constraints for the four dated samples at the specific depth provide the absolute chronology of the sequence ranging between 16.9 ± 1.1/1.2 ka and 47.5 ± 0.8/0.9 ka (see Fig. 2).

Figure 2 shows a Synthetic stratigraphic reconstruction along the CFDDP drill hole as derived from lithological (description on the left) and micropaleontological (occurrence of fossils within the main column, i.e., sponge spicules that, only between 61 and 67 m of depth, associate with diatoms) analyses of samples (primary and neogenic mineral assemblage, and H2S spike are also indicated). The photos show the representative sampled materials (size between 2000 and 63 μm) and the cores (439 m and 501 m). Stars locate the Ar-dated feldspars. Shaded colours allow rapidly distinguish each stratigraphic unit; CI is related to the 439 core but we infer that the CI deposit can extend up to the 35.1 ka deposit.

4 RESULTS AND DISCUSSION

The stratigraphic sequence obtained and the 40Ar/39Ar dating constraints at the CFDDP borehole provide new insights on the eastern Campi Flegrei caldera deep structure and evolution. Based on the CFDDP data, we can point out the following main results in comparison with recent scientific literature. It is evident a reduction of the area affected by collapse and a surprisingly smaller volume of the infilling caldera deposits, particularly for the CI eruption, with respect to the previous hypothesis [7]. Therefore, it is possible to re-evaluate the collapse amounts and mechanisms related to the CI and NYT eruptions.

The obtained data do not support a collapse extending toward the Naples metropolitan areas and Gulf of Naples with a hypothesised drop of ~1500 m and show that the drilling site is very close to the caldera border (see Fig. 1). Furthermore, the inferred shallow depth of the CI layer, which in terms of hypothetical amount of collapses would indicate a ratio of 2.5 in

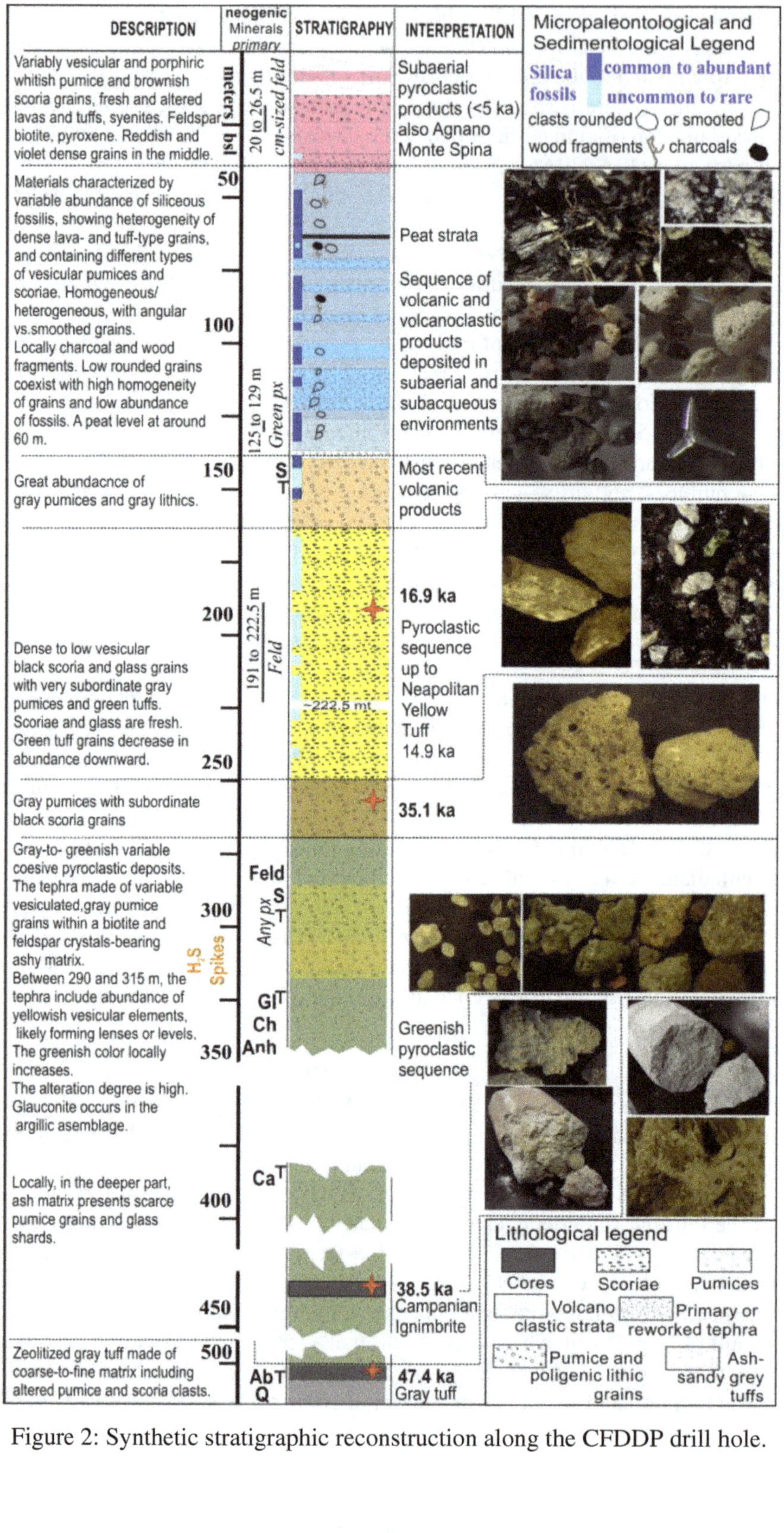

Figure 2: Synthetic stratigraphic reconstruction along the CFDDP drill hole.

favour of NYT with respect to CI, is largely unexpected because of the higher VEI of CI. This is a very crucial finding, which favour the hypothesis that CI was possibly erupted from vents also outside the Campi Flegrei caldera. An alternative model, which anyway should be reconciled with the low thickness of CI deposits, would be that a very large regional uplift occurred related to the CI, approximately several hundred meters.

The new findings are also helpful in orienting the future studies of the problem of risk posed to the city of Naples, indicating the Campi Flegrei caldera is delimited eastward by the Posillipo Hill, and do not extend in the central part of the city of Naples as stated by the most of recent literature.

The results of this study point out, in a more general way, that large calderas are characterized by complex collapse mechanisms, in which the collapse progressively deepens going from the borders towards the central part. Such a complex mechanism also emerges as the most likely one in recent theoretical studies on caldera formation. Finally, this study shows the importance of age-based correlation between intra-caldera and extra-caldera tephra in defining the eruptive mechanisms at the origin of collapse. This problem is common to several calderas worldwide [15]; for the above reasons, scientific drilling demonstrates as one of the most powerful tool to reconstruct the structure, the collapse mechanism and the volcanic history of calderas.

5 TABLE OF ABBREVIATIONS

ka = kilo annum;
bsl = below sea level;
40Ar = Argon 40 (Ar Isotope, used to compute, from relative radioactive decay, the absolute age of the sample);
39Ar = Argon 39 (Ar isotope, used to compute, from relative radioactive decay, the absolute age of the sample);
VEI = Volcanic Explosivity Index;
NYT = Neapolitan Yellow Tuff Eruption;
CI = Campanian Ignimbrite Eruption;
ICDP = International Continental Drilling Program (World Organization to promote and support scientific drilling);
DRE = Dense Rock Equivalent (Volume of pyroclastic erupted material compressed to a density equivalent to the corresponding rock);
CFDDP = Campi Flegrei Deep Drilling Project.

REFERENCES

[1] Hill, D. P., *Unrest in Long Valley Caldera, California, 1978–2004, in Mechanisms of activity and unrest at large calderas*, edited by Troise *et al.*, **269**, pp. 1–24, Geological Society, London, Special Publication, 2006.

[2] Lowenstern, J.B., Smith, R.B. & Hill, D.P., Monitoring super-volcanoes: geophysical and geochemical signals at Yellowstone and other large caldera systems. *Philosophical Transaction of the Royal Society A*, **364**, pp. 2055–2072, 2006. https://doi.org/10.1098/rsta.2006.1813

[3] Carlino, S. & Somma, R., Eruptive versus non-eruptive behaviour of large calderas: the example of CampiFlegrei caldera (southern Italy). *Bulletin of Volcanology*, **72**, pp. 871–886, 2010. https://doi.org/10.1007/s00445-010-0370-y

[4] Carlino, S., Somma, R., Troise, C. & De Natale, G., The geothermal exploration of Campanian volcanoes: historical review and future development. *Renewable and Sustainable Energy Reviews*, **16**, pp. 1004–1030, 2012. https://doi.org/10.1016/j.rser.2011.09.023

[5] Deino, A.L., Orsi, G., de Vita, S. & Piochi, M., The age of the Neapolitan Yellow Tuff caldera-forming eruption (CampiFlegrei caldera - Italy) assessed by 40Ar/39Ar dating method. *Journal of Volcanology and Geothermal Research*, **133**, pp. 157–170, 2004. https://doi.org/10.1016/S0377-0273(03)00396-2

[6] Rosi, M. & Sbrana, A., PhlegreanFields: Petrography. *Quaderni de La Ricerca Scientifica*, **114**, pp. 60–79, 1987.

[7] Perrotta, A., Scarpati, C., Luongo, G. & Morra, V., The CampiFlegrei caldera boundary in the city of Naples, in Volcanism in the Campania Plain: Vesuvius, CampiFlegrei and Ignimbrites, edited by De Vivo, B., pp. 85–96, Elsevier, Special Publication, 2006.

[8] Orsi, G., de Vita, S. & Di Vito, M., The restless, resurgent CampiFlegrei nested caldera (Italy): constraints on its evolution and configuration. *Journal of Volcanology and Geothermal Research*, **74**, pp. 179–214, 1996. https://doi.org/10.1016/S0377-0273(96)00063-7

[9] Vitale, S. & Isaia, R., Fractures and faults in volcanic rocks (CampiFlegrei, southern Italy): insight into volcano-tectonic processes. *International Journal of Earth Sciences*, **103**, pp. 801–819, 2014. https://doi.org/10.1007/s00531-013-0979-0

[10] Carlino, S., Kilburn, C.R.J., Tramelli, A., Troise, C., Somma, R. & De Natale, G., Tectonic stress and renewed uplift at CampiFlegrei caldera, southern Italy: new insights from caldera drilling. *Earth and Planetary Science Letters*, **420**, pp. 23–29, 2015. https://doi.org/10.1016/j.epsl.2015.03.035

[11] De Natale, G., Troise, C., Pingue, F., Mastrolorenzo, G., Pappalardo, L. & Boschi, E., *The CampiFlegrei Caldera: unrest mechanisms and hazards, in Mechanisms of activity and unrest at large calderas*, edited by Troise *et al.*, Geological Society, London, Special Publication, **269**, pp. 25–45, 2006.

[12] Moretti, R., De Natale, G. & Troise, C., A geochemical and geophysical reappraisal to the significance of the recent unrest at CampiFlegrei caldera (Southern Italy). *Geochemistry, Geophysics, Geosystems*, **18**, pp. 1244–1269, 2017. https://doi.org/10.1002/2016GC006569.

[13] De Natale, G. & Zollo, A., Statistical analysis and clustering features of the Phlegraean Fields earthquake sequence, May '83–May '84. *Bulletin of the Seismological Society of America*, **76**, pp. 801–814, 1986. https://doi.org/10.1785/01.20120036

[14] Lipman, P.W., Subsidence of ash-flow calderas: Relation to caldera size and magma-chamber geometry. *Bulletin of Volcanology*, **59**, pp. 198–218, 1997. https://doi.org/10.1007/s004450050186

Author index

www.ingramcontent.com/pod-product-compliance
Lightning Source LLC
LaVergne TN
LVHW081531300926
849161LV00006B/1384